Getting It Right

R&D Methods for Science and Engineering

Getting It Right

R&D Methods for
Science and Engineering

Peter Bock

Illustrations by Bettina Scheibe

ACADEMIC PRESS

A Harcourt Science and Technology Company

San Diego   San Francisco   New York   Boston   London   Sydney   Tokyo

This book is printed on acid-free paper. ∞

Academic Press
A Harcourt Science and Technology Company
525 B Street, Suite 1900, San Diego, California 92101-4495, USA
http://www.academicpress.com

Academic Press
Harcourt Place, 32 Jamestown Road, London NW1 7BY, UK
http://www.academicpress.com

Library of Congress Catalog Card Number: 00-2001089411

International Standard Book Number: 0-12-108852-9

PRINTED IN THE UNITED STATES OF AMERICA
01 02 03 04 05 06 HP 9 8 7 6 5 4 3 2 1

For Donna

CONTENTS

Part I   Introduction

Chapter 1   Research and Development   3

Chapter 2   Process and Preparation   13

Part II   Project Organization

Chapter 3   The Project Hierarchy   21

Chapter 4   The Project Task   35

Part III   Knowledge Representation

Chapter 5   An Epistemological Journey   71

Chapter 6   Categories and Types of Knowledge   77

Chapter 7   Roles of Knowledge Propositions   93

Chapter 8   Limits of Knowledge   109

Part IV   The Scientific Method

Chapter 9   Overview

Chapter 10   Analysis

Chapter 11   Hypothesis

Chapter 12   Synthesis

Chapter 13   Validation

Appendices   329

ACKNOWLEDGMENTS

First and foremost, I am eternally grateful to my loving wife Donna, who is my best friend and most treasured critic, and to my favorite daughter Ariana, for her lifelong support and encouragement. Second, to my dedicated doctoral students, especially Mark Happel, Kristin Heckman, and Terry Riopka, who spent countless hours with me thrashing out the logic and holding my feet to the pedagogical fire. And finally, to the many other people who made significant contributions to this book, especially Bettina Scheibe, whose mysterious and zany drawings breathe life into the subject matter; Fridolin Piwonka, the most talented R&D manager I have ever known, who wrote an inspiring Foreword and cajoled me into undertaking this enormous project in the first place; Murray Loew, my friend and colleague for more years than we care to count, who helped design the original seminar from which many of the ideas for this book emerged; and Joel Claypool and Julie Bolduc, my editors at Academic Press, who patiently waded through my crazy proposals to liven up the book's look and feel.

FOREWORD

I met Peter Bock in 1989 at a symposium he had organized on Adaptive Learning Systems at Schloss Reisensburg, a castle near Ulm in southern Germany, which had just been reconstructed to host seminars and conferences. I was then a department head in Corporate Research at Robert Bosch GmbH, Germany's largest automotive supplier. Peter and I got acquainted over drinks in the garden overlooking the Danube River, and this was the beginning of a warm friendship and a strong professional relationship.

At that time, Bosch Corporate Research was investigating pattern recognition and classification methods to process signals, images, and three-dimensional data representations like sonograms. Peter, who was on sabbatical leave from his home institution, The George Washington University in Washington DC, had been asked to set up a research project at the newly established Research Institute for Applied Knowledge Processing (FAW) in Ulm, Germany, not far from Schloss Reisensburg. The objective of the research project was to design and build a large-scale system for sophisticated image and signal processing, based on Collective Learning Systems Theory, a supervised adaptive learning paradigm that Peter had first proposed in the mid-1970s and has been developing and applying ever since.

The new image-processing engine that his research team was building at the FAW, called ALISA, was implemented on a Transputer-based parallel-processing computing engine. Each Transputer was programmed as a collective learning cell whose computational power could be compared to a few hundred neurons. In this way, a relatively powerful system could be achieved with existing hardware, overcoming the connectivity limits of a typical multilayer perceptron.

I was impressed by the rigor with which Peter had designed the ALISA engine and the project at the FAW. To implement his design, he brought together a group of graduate students from The George Washington University and scientists and engineers from Bosch. Strict adherence to the Scientific Method was the guiding methodological principle of the project. It was an enlightening experience for everyone involved in the project to witness the power of this approach in action. The project enjoyed quick success and was subsequently funded by Bosch.

Peter brought with him a rich background of knowledge and experience in the design and direction of R&D projects for many different kinds of organizations, including NASA, NIST, the US Navy, industrial firms, private research institutes, and academia. He started his professional career in 1965 with professional and industrial software development at IIT Research Institute (1965–1969), where he was responsible for the design of the experiment profile analysis (EPA) technique, a computer simulation used by NASA to determine the most efficient spatial and temporal sequences for remote sensing activities on the Apollo Earth and Lunar orbital missions.

When the Apollo program began to wind down, Peter joined the faculty of the Department of Electrical Engineering and Computer Science at The George Washington University (GWU) in 1970, designing and teaching computer science courses, including artificial intelligence, adaptive learning systems, cognitive science, robotics, and simulation. He was primarily responsible for the development of the graduate degree program in Machine Intelligence and Cognition. It was here that he pioneered the research in Collective Learning Systems Theory. Reflecting the neurophysiological and psychological processes of the mammalian brain, he designed a hierarchical network of learning automata to simulate the architecture and adaptive function of the human cerebral cortex: learning from nature how nature learns.

It was this research that eventually led to the joint American-German project ALANN (Adaptive Learning and Neural Networks) at the FAW in Ulm, Germany, which designed and built the first ALISA system (Adaptive Learning Image and Signal Analysis). Today ALISA is used in several learning and classification applications in industry and research, and is currently being expanded by Peter's research group at GWU into a complete hierarchical system for artificial cognition.

Over the years, Peter and I have sat together on many occasions and lamented the fact that so many of our young scientists and engineers leave the universities with a solid foundation of knowledge in their disciplines, but with little idea how to design and conduct real-life R&D projects successfully. Finally, Peter decided to try to do something about this problem. In 1996, he set up a seminar at GWU to teach the Scientific Method for R&D in science and engineering. We also brought this seminar to Bosch Corporate Research in 1997. His lecture notes quickly expanded into a complete set of basic principles and guidelines for designing and conducting effective research or development projects, in both industry and academia. In 1999, his department added a formal course in research methods to its graduate course curriculum, which is now required for all doctoral students.

At that point it seemed only natural that Peter should transform his lecture notes into a book to spread the word to the research and engineering community at large. By compiling his knowledge and experience into a reference book, his intention was to provide a complete and consistent methodology for scientists and engineers, project managers, and administrators to plan and run research and development projects. In 1999 Academic Press indicated its interest in just such a manuscript, and voila!, here it is. This book covers project organization, knowledge representation, and a modernized version of the classical Scientific Method, which has been adapted for the special requirements of research and development in the twenty-first century.

The question facing Peter, however, was how to present this complex material in a reasonably linear fashion, while avoiding the sterility of an IF-THEN-ELSE cookbook approach and a dry and overly academic presentation. It is a bit of a mystery to me exactly how he managed to do this, but in fact he did. He has succeeded in coming up with a smart, readable book that guides real-world practitioners toward the improvement of their daily R&D activities. It provides a clear perspective for people charged with the acquisition, organization, and application of knowledge, devices, and effects, which is the essence of research and development in both engineering and science. And it is also a well-structured textbook for an academic course.

A wealth of examples leads the readers to the underlying reasons for failures and successes in the performance of their projects. Carefully constructed summaries, lists of tips, tables, and figures give the reader brief answers about how to avoid common pitfalls and how to do things right. Amusing illustrations catch the reader's eye, add spice to the experience, and serve as convenient bookmarks and reminders for special problems and messages. Peter not only explains the subjects, but also puts the information in its historical and cultural context, making it part of something larger, an essential perspective in the modern multidisciplinary world.

Students will find well-defined and useful exercises to harden their theoretical knowledge, and professionals can use it as a handbook and a lexicon for answering questions that arise during their daily R&D activities. Project managers will learn how to design a project task tree and a milestone chart in evolving stages, using pilot tasks to explore new ideas before scarce resources are committed to formal investigations. Administrators will learn that every project needs a management reserve, an appropriate tool set, and the allocation of adequate time and money to acquire new tools and maintain the project environment. Scientists and engineers will learn about the categories and types of knowledge, arranged in a useful taxonomy.

From an industrial point of view, the most important guidelines in this book provide the means to conduct research and development optimally in terms of effectiveness and efficiency. The fundamental paradigm to accomplish this, the Scientific Method, proceeds from an analysis of the problem and precise statements of the governing propositions, to a clear formulation of the project objectives, to the creation of solutions with well-defined goals and hypotheses, to the design and conduct of rigorously controlled experiments, to the generation of reproducible conclusions or producible competitive products of the required quality. It is clearly acknowledged that research results must be validated by formal peer review within the scientific community, while industrial products must ultimately prove themselves in the marketplace.

Peter's book presents a complete and consistent methodology for research and development in science and engineering. Hopefully, it will encourage institutions of higher learning to include courses in methodology as part of their required curricula, with Peter's book at the heart of their efforts. Read, enjoy, and learn!

Fridolin Piwonka
Vice President, Corporate Research and Development
Robert Bosch GmbH
Stuttgart, Germany

BIOGRAPHIES

About the Author

Peter Bock is a professor of machine intelligence and cognition on the faculty of the Department of Computer Science at The George Washington University in Washington, DC. He is responsible for designing and teaching undergraduate and graduate computer science courses; serving on various academic and administrative committees; and directing independent research projects, master's theses, and doctoral dissertations. Professor Bock is credited with pioneering the research in Collective Learning Systems Theory, an adaptive learning paradigm for artificial cognition, and designing a hierarchical network of learning automata to simulate the structure and adaptive function of the human cerebral cortex. Bock has published many scientific papers and is the author of *The Emergence of Artificial Cognition: An Introduction to Collective Learning,* published by World Scientific Publishing Company in 1993. His research has been funded by several government agencies, including the National Bureau of Standards, the United States Navy, and NASA; several industrial firms, including Robert Bosch GmbH in Stuttgart, Germany, and Lockheed-Martin; and the Research Institute for Applied Knowledge Processing (FAW) in Ulm, Germany.

Peter Bock received his undergraduate degree in physics and mathematics from Ripon College. After completing his graduate studies at Purdue University in 1965, he joined the research faculty at IIT Research Institute (IITRI) in Chicago. As a member of the Astro Sciences Center, he developed high-speed computer simulations for the analysis of the proposed Earth and Lunar orbital remote-sensing missions of the NASA Apollo Applications Program. Following the first successful Lunar landing mission, Bock returned to academia and has been at The George Washington University since 1970.

Professor Bock's long-term research objective is the design and implementation of an artificially intelligent being whose cognitive capabilities are on a par with those of a human being. For the last 15 years, he has directed an international research project that has applied the principles of collective learning to the design and construction of ALISA (Adaptive Learning Image and Signal Analysis), a sophisticated adaptive parallel-processing engine that simulates the cognitive functions of the sensor cortex in the mammalian brain. A number of professional and industrial applications for the ALISA system have been developed by the project team, and it was during these R&D projects, which often involved complex human-machine interaction, that Professor Bock first became interested in organizing and codifying a rigorous methodology for conducting research and development in science and engineering. More information about Professor Bock and his research can be viewed on his web page at http://www.seas.gwu.edu/~pbock.

About the Illustrator

Bettina Scheibe, born in Mönchengladbach, Germany, has studied and worked extensively throughout Europe, specifically England and her native homeland. She experiments and produces work that spans several borders encompassing many practices, creating stage designs, installations, paintings, and film works, all of which root from her primary passion of placing pen against paper. Her textured approach allows her to create a unique signature throughout her craft. She consistently references the everyday, capturing the essence of humans in their daily existence. Within her work she literally allows her audience to "follow the line that draws the picture."

This is the first book that Bettina has illustrated, and she is looking forward to expanding her visual repertoire. If you would like to contact her, you can do so by e-mailing her at bettinascheibe@yahoo.co.uk.

Introduction

CHAPTER 1

Research and Development

1.1 Motivation

"When you get home from class tonight," the professor told his students at the end of the first class of his undergraduate course in research methods, "I would like you to conduct an experiment. Treat yourself to a glass of wine, or a beer, or whatever helps you relax. Put some rhythmic music on the stereo, classical or jazz or rock, whatever you enjoy. Plug in a set of earphones and turn off the loudspeakers. Light a candle and place it on a table about a meter in front of a comfortable chair. Sit down in the chair, put on the earphones, close your eyes, listen to the music, and relax for a few minutes. Then, open your eyes and see if the flickering of the candle flame keeps time to the music. Please be ready to report your observations in class next week."

The next week in class, the students were all abuzz. "Gee, Professor, it's crazy, but you were right! The candle flame keeps time to the music! Not all the time, but it happened fairly consistently. I saw it!" they chorused. "But how could that happen?" another quipped. "I was wearing earphones, and there was no sound coming from the speakers. There was no acoustical connection between the sound and the flame!" But of course, some said, "Nonsense. There was no correlation between the flickering of the candle and the music rhythms. That's impossible." But most of them saw it.

The professor let them rant for a few minutes. Then, when everyone finally quieted down, he spoke up. "How many of you observed the candle flame keep time to the music?" About 15 of the 20 students raised their hands. "And how many of you observed no apparent correlation?" The hands of the other 5 students shot up. "How many of you carefully followed the experiment protocol I gave you?" All 20 hands came up. "And how many of you are telling the truth about your observations to the best of your ability?" Again, 20 hands. "So, may we conclude," asked the professor, "that 15 of the candles kept time to the music, and 5 did not?"

There was a moment of silence while the students considered this possible conclusion. "I guess so," said one student, "but how did it happen? And why didn't it happen consistently?"

"Well," said the professor, "let's look at the conditions that governed this experiment. First, each of you used a different candle. Maybe some kinds of candles work, and others don't. Second, each of you was in a different room with different lighting conditions. Maybe the candle needs to be in a particular lighting condition to react to the music. (Some eyebrows went up on that one.) Third, each of you probably chose a different kind of music: Mozart, Al Jarreau, Hootie & The Blowfish, or whatever. Maybe different candles prefer different kinds of music. (Audible groans could now be heard in the classroom.) Fourth . . ."

"Maybe it wasn't the candle," interrupted Jack, one of the students. "I saw the candle flickering in time with the music, but maybe my eyes were flickering, not the candle. Or something like that." Some others in the class harrumphed. "Or maybe I just imagined it was flickering, but it really wasn't."

"No," said a classmate, Bill. "I didn't imagine it. I saw it."

"Well," Jack responded, "if you did just imagine it, you wouldn't know whether it was real or imagined. It would seem real either way."

"I don't know what you're talking about, dude. Maybe *you* imagined it. I know what *I* saw," countered Bill.

It was time for the professor to interrupt this exchange, which sounded like something straight out of the Middle Ages, and the means for settling intellectual disputes at that time were not pretty. "Let's get back to the original question: how come 15 of the candles kept time to the music, and 5 did not?"

"I think the question is wrong," ventured Jack.

"Oh," said the professor. "So you think your esteemed professor posed an ill-posed question, eh, Jack?" The students laughed nervously, and Bill swung around to shake his finger at Jack.

"Yes, sir . . . no offense," said Jack.

"None taken. So, how should the question be posed?"

"Maybe. . . . How come 15 of us saw the candle keep time to the music, and 5 did not?"

"That's the same as the question the professor asked, dude," said Bill.

"No, he asked how come the candle flickered. I asked how come we *saw* the candle flickering."

"Psycho-babble," said Bill. "If I saw it, it happened," he grumbled.

"I am absolutely convinced that you saw the candle flickering, Bill," said the professor. "I have no doubt that all of you have reported your observations faithfully. However, Jack may be right to reword the question. Understand that the brain is a massive computer that intercedes itself between your sensors, your eyes, and your conscious thoughts. A lot is going on beneath the surface that we are not aware of, and as the image of the candle makes its way from your eyes through your brain to your conscious thoughts, a lot can happen. A *lot* of processing takes place. All of us know firsthand how computers can screw up because their programs have bugs in them. Brains are no different."

"Are you telling us that we have bugs in our brains, professor?" asked Bill.

"Perhaps 'bats in your belfry' is a more appropriate expression, Bill," said the professor with a grin. The class, including Bill, laughed good-naturedly.

"So that's the answer?" asked another student. "It was all in our imagination?"

"No," said the professor. "We don't know that, Jane. It simply means that, as good researchers, we have to admit to that possibility. We have to understand that the task method that all of you applied is not objective and probably biased."

"Well, if we were biased during this experiment, where did the bias come from?" Jane asked. "Were we born with it? Did it come from our upbringing? Should we blame our parents?"

"No," the professor laughed. "It's much simpler in this case, Jane. You can blame me. I instilled this bias in you. Intentionally. *Mea culpa."*

"When?"

"When I first described the experiment protocol last week. If you recall, I told you what to expect. I told you to watch and see if the candle flickered in time to the music. And guess what? You saw it! Like good students, you dutifully saw what your professor told you to see. That's called a *self-fulfilling prophecy.*"

"Now, wait a minute, sir," said Bill. "I wasn't influenced by your remark. I know that objectivity is important for a good scientist."

"With all due respect, Bill, by definition you cannot know what goes on in your subconscious mind. I am pleased that you consider objectivity a critical criterion for the acquisition of knowledge. But like all the rest of us, you are influenced by unconscious motivations. This is not speculation on my part. There have been many, many carefully devised experiments that have confirmed the impact of unconscious biases on the behavior of humans and other intelligent animals."

"Then, how should you have described the experiment protocol to avoid biasing us with an expectation?" Jane asked.

"Why don't you give it a try, Jane," suggested the professor.

"Well, let's see. Put some music on the stereo, light the candle, sit down in the chair, and so forth, and then watch the candle . . . uh . . . to see if . . . uh. . . ." Jane's voice faltered as she searched for the right way to phrase it. "How can you word the protocol so that you don't give the goal away, but still make sure the subject knows what to do?"

"Tricky, isn't it?" said the professor, eyebrows arched.

"If you can't tell the subject to watch the candle, how is he supposed to know what to do?" Bill mused, finally realizing that the problem was more complicated than he had first thought.

"To begin with, the person watching the candle is *not* the subject," responded the professor. "The focus of the experiment, which is called the *task unit,* is the *candle,* not the person watching the candle. The person is simply the *sensor,* whose function is to acquire the data necessary to compute the correlation between the rhythm of the music and the flickering of the candle. One possible solution, then, is to use an electronic sensor and get the human out of the experiment entirely. The development of a vast array of mechanical and electronic sensors over the last two or three hundred years has allowed us to eliminate much of the bias resulting from the subjective interpretations of humans. That has been one of the most important contributions to research in science and engineering since the Scientific Revolution of the eighteenth century. "

"What kind of sensors would you use to replace the human eye and ear?" asked one of the class members.

"A video camera and a microphone might work. After acquiring the data, we would have to find a way to measure the periodicity of the flickering flame in the video images, and to measure the beat of the music on the audio track. If we could do that, then a simple correlation computation would tell us whether we can validate the hypothesis that the flame is at least partly synchronized with the music. In other words, that a nonzero correlation could not be attributed to chance."

"I've studied a little bit of electrical engineering, professor, and I think measuring the periodicity of the flame's flicker and the music's beat would be very difficult," said Jack. "Isolating the principal periodic components of the flicker and the beat of the music is not straightforward. That's pretty high-level information."

"You know, when it comes right down to it, this seems to be a very difficult experiment to design and conduct properly," said Jane.

The professor nodded. "Planning and conducting rigorous research and development projects in science and engineering is often very challenging."

That's what this book is all about.

1.2 Background

In the first couple of centuries of the second millennium of the Christian era, Europe began to break free of the stagnation of the Dark Ages. Commercial traders began to journey to the Orient in search of wealth, and they brought back tales of new and exotic lands and cultures. In the same time frame, at the behest of the Roman Catholic Church and the rulers of the Holy Roman Empire, crusaders gathered their armies and traveled to the Middle East to recapture Jerusalem from the Muslims, who had gained enormous strength and power over the centuries following the death of their prophet Mohammed in 632 AD, and had carried Islam into North Africa and Spain. Ship captains were ranging farther and farther from the coasts of Europe to transport the traders and crusaders to their distant destinations, and when the mariners returned to Europe, a growing number of them voiced their suspicion that the Earth was not flat, but round. It was the only way they could explain how their navigational beacons — the sun, the moon, and the stars — behaved as they did, with great regularity, but differently in the south than in the north. They perceived the heavenly bodies not as angels of their god, as suggested by Church doctrine, but as other worlds hanging in the sky far from the Earth. As early as 1200 AD, some brave European mariners began speculating about the possibility of circumnavigating the world.

Most of these travelers were smart enough to choose their friends carefully and keep their voices down. It was heresy to talk of such things, under penalty of death in one of several ghastly ways. Nonetheless, some young and daring monks in the Catholic Church, who enjoyed the rare privilege of being literate and had access to the few remnants of classical Greek and Roman literature, heard these stories and began to visualize the world in a new way. It was just possible, they thought, that not everything was under the *direct and immediate* control of their god, but functioned more-or-less automatically according to a set of laws, which was, of course, devised and even occasionally modified by their god. At great personal risk, they whispered among themselves about the mechanisms and phenomena of the universe.

But, of course, some conservative priests eventually got wind of this heresy and decided to crack down. To their surprise, however, the disease had already spread quite far; some of the high-ranking priests themselves were already infected with this new universal view. It was, therefore, not politically practical to employ the normal solution to such problems by simply burning the heretics at the stake (though many were at first). Besides, the old-timers understood full well the power of martyrdom. After all, Jesus had given his life as a martyr, and that had led inexorably to the vast and powerful empire run by their own organization.

The priests conferred and discussed and argued and finally decided that the church must be kept free of the poisonous ideas entertained by these monks and priests, not by destroying them, but by isolating them and keeping a close eye on them. So, the church stripped the conspirators of most of their clerical powers and expelled them from the monasteries, but allowed them to live and work in buildings near the church grounds, where they could be closely watched while they babbled on about the *universe*. Thus, as such things happen, these buildings soon became known as *universities,* and this *new search* for universal cause-effect relationships was eventually dubbed *research*. The oldest university in Europe is probably the University of Bologna in Italy, which was founded at the end of the 11th century.

Over the following centuries, as knowledge accumulated in the universities and their libraries, a trickle of practical applications began to emerge: the Gothic arch, the printing press, medicines and surgical procedures, the escapement, the cannon, handheld firearms, and so forth. By the beginning of the nineteenth century, emerging industries were investing enthusiastically in the *development* of practical industrial applications based on the results and conclusions of scientific research, and the Industrial Revolution moved into full swing. We are still in the midst of this revolution, although it seems to be quickly evolving into something new, which future historians may dub the Information Revolution.

1.3 R&D Problems

Today, at the beginning of the twenty-first century, the industries and governments of the industrialized nations spend an average of about 5% of their annual budgets on research and development in science and engineering, and the major universities expect their science and engineering faculty to spend about 25% of their time on research activities (funded, hopefully, by industry or government). Every large company has an R&D department, usually clearly separated from the production activities of the company with its own budget. R&D is big business.

R&D is the gestation phase of the overall industrial process, far removed in time and oversight from marketing and sales. For this reason, it is very difficult to trace business problems back to possible origins in R&D. The management and staff of R&D departments are seldom blamed directly for slumping sales, but are always the vaunted heroes of the success of the "killer app." In the protected and insulated R&D environment, scientists and engineers enjoy a remarkable immunity from business economics: when sales are up, they are congratulated for outstanding performance and given larger budgets; when sales are down they are exhorted to innovate and protected from budget cuts. The professional staff of R&D departments are often highly intelligent individuals, who have little knowledge of the exigencies of economic survival on the battlefield of business. Their arcane alchemies are little understood by managers and executives farther up the chain of command, and it is in the best interest of the R&D personnel to maintain and protect this technological mystique. Their intellectual prowess puts them in demand in the job market, and this implied threat insulates them from the strict oversight and control routinely imposed on other departments, such as production and sales. The evolution of modern-day scientists and engineers from the heretical priests and monks of the Middle Ages is often very evident.

For all of these reasons, R&D activities can be especially disorganized and free-wheeling. Top management often feels compelled to give R&D personnel their head, and like temperamental steeds, many of these high-strung intellectuals have a tendency to rear up and gallop off into the night. Table 1.1 presents a list of some of the common problems that result from this chaos, compiled by several high-level industrial executives. Some of them will probably strike you as hauntingly familiar.

The acronym *R&D* has become so pervasive in industry that many have lost sight of the important differences between the two component activities: **research** and **development**.[1] Therefore, before we go any further, we need to establish some working definitions for these two terms.

Definition 1.1 Research is a process that *acquires* new knowledge.

Definition 1.2 Development is a process that *applies* knowledge to create new devices or effects.

Research seeks truth, while development seeks utility. For this reason, research is often considered an effete activity by industrial executives, a luxury they feel they cannot afford. On the other hand, development is considered a necessary but uninteresting process by many academic researchers at universities and institutes, a menial activity unworthy of their intellects. Both views are myopic. The fact is, without research, nothing would get developed, and without development (and production and sales), no research could be funded. The two processes are as symbiotic as Tweedledum and Tweedledee. And often as chaotic.

Research, whose objective is to leach new knowledge out of the matrix of the universe, is often an insular process. A few individuals can work together on a research team to accomplish its objectives, but large centralized research teams are

Table 1.1 Sources of chaos and confusion in typical R&D groups

- Research results cannot be reproduced because of poor methodology and documentation.
- Speculations are not identified as such and are intermixed with supported conclusions.
- Knowledge is precarious, locked up in the heads of individuals.
- Data collection is haphazard and confounded with political issues.
- Experiment methods are chaotic, dominated by a "try this, try that" mentality.
- Experiment processes cannot be audited or reviewed due to a lack of logs and records.
- Reports are too long (or too short), poorly organized, incomplete, and confusing.
- Project documentation is sparse or nonexistent.
- Statistical analysis of results is missing or naive.
- Oral presentations are disorganized, confusing, and emphasize the wrong things.
- Data visualization techniques are poor and the presenters are hard to follow.
- Project activities are isolated and inbred.

---

[1]Throughout the book, **boldface** is used to highlight the first occurrence (or an important reoccurrence) of terms central to the principles set forth in this book. Many of these terms are included in the Glossary in Appendix B.

unusual and generally unworkable. The focus must be very tight, and progress is often painfully slow. Ambitious research topics, such as finding a cure for AIDS or uncovering the mystery of the disappearance of the pre-Columbian inhabitants of Mexico, require enormous investments of time and labor, but they are seldom conducted in one place. Instead, a number of small teams work by themselves in many different places, coming together now and then to share their results and conclusions through the well-established mechanisms of publication and peer review: professional journals and conferences. Sometimes a small group actually consists of a single person, working with little or no funding, occasionally joining his colleagues at professional gatherings to exchange knowledge. The early projects of Albert Einstein and Gregor Mendel are examples of such insular research activities.

Development, on the other hand, is a team activity that requires high efficiency and close coordination to be cost effective. The complete development staff, of course, may be distributed into small working groups, but constant communication and coordination among them is critical to ensure that the axle designed by one group fits the wheel designed by another group. The enormous development project for the invasion of Normandy in June of 1944, called Operation Overlord, required the coordination of thousands of teams and hundreds of thousands of people. More than 80,000 pages of planning documents were generated for this project; the Normandy Invasion was the largest military battle in history. The Apollo Project, which put the first humans on the Moon, was the largest peace-time development project in history at a cost of approximately $20 billion, three lives, and 100,000 person-years of professional effort expended over nine years. A good example of an extraordinary industrial development project was the design, construction, and testing of the Boeing 777 passenger aircraft over a three-year period in the early 1990s. At a cost of $4 billion, this development project set new standards in engineering design and management, which are now being applied to the design and construction of the International Space Station. An excellent film describing the management and technological challenges of the Boeing 777 development project can be obtained from the Boeing Corporation.

Research and development activities take place in many different fields, not just science and engineering. Table 1.2 gives some examples of the typical names given to the corresponding roles of research and development professionals in a number of fields.

Some may argue with the names suggested for the roles, but the point has been made: research and development is not limited to high-tech science and engineering. Many different fields of science, both hard and soft, engage in research and development, although they might not use these words. It should also be clear that individuals in these professions often play *both* roles. And, as will be discussed in later chapters, research projects almost always include a large number of essential development tasks, and *vice versa*.

What about art? Are there processes for artists that are similar to research and development in science and engineering? Perhaps so, but the methodology

Table 1.2 Corresponding professional roles in research and development

Research Role	Development Role
Historian	Politician
Scientist	Engineer
Psychologist	Therapist
Physiologist	Doctor
Linguist	Translator
Economist	Investor
Philosopher	Teacher
Cybernetician	Manager

described in this book is not intended to apply to the arts, which build on technique, just like science and engineering, but have a completely different objective: to express feelings. In the purest sense, the artist is unconcerned about the reaction of other people to the product; the process of validation takes place entirely in the mind of the artist. Art is a personal expression, and responding to demands for definitions and explanations, always obligatory in science and engineering, seems entirely irrelevant. On the other hand, many artists may genuinely want to (or need to) have their work reviewed and purchased. If so, for better or for worse, they must compromise and step into the arena of business; having explicitly sought external validation, they must deal with the reactions of their customers and peers.

What about religion? Most religions are based on faith and thus do not demand reproducible evidence for validating their conclusions. Religions are generally grounded in a value system that is dictated by an acknowledged and infallible deity or prophet. Deductive premises and new knowledge may only be the result of direct revelation. However, whenever challenges to religious precepts are no longer dismissed out of hand, but are accepted as valid counter-proposals to be considered on their own merits, then religion becomes philosophy, the seminal science from which all other sciences derive. This was precisely the motivation of the medieval monks who broke free of the church and established the first universities, and it is the closest point of epistemological connection between science and religion.

As will be discussed in Part III of this book, speculation and faith-based knowledge, *i.e.,* opinion and dogma, are not acceptable as a basis for achieving objectives in science and engineering. That battle has been hard fought for many centuries, and scientists and engineers must never give an inch on that fundamental principle. Clear and convincing evidence is necessary. Likewise, religious practitioners have the same right to be just as firmly dedicated to their conclusions based on faith. That's only fair. The two fields simply have different ways of doing things.

1.4 Primary Objective

We scientists and engineers have a lot of work to do to overcome our methodological and epistemological problems, as previously described. Without a rigorous and systematic approach to the planning and conduct of research and development projects, the increasing complexity in critical technological areas like genetic engineering, artificial intelligence, and human-machine interaction will preclude the cost-effective development of new products and degrade the reliability of the acquired knowledge. Given our ever-increasing vulnerability and dependence on highly sophisticated technology, sloppy methodology can only lead to very expensive mistakes if we don't clean up our act and get it right!

CHAPTER 2

Process and Preparation

2.1 The Methodology

During the Scientific Revolution in the early eighteenth century, Galileo Galilei, René Descartes, and Francis Bacon, working quite separately in three different countries, struggled to rationalize the structure and function of science. They came to almost the same conclusion at almost the same time: good science requires a rigorous and consistent methodology, from analysis to hypothesis to synthesis to validation, capped by Galileo's brilliant contribution, formal peer review, which was a brand new idea at the time. Over the next century, scientists in Europe and America integrated the ideas of these three methodological pioneers and dubbed it the Scientific Method. This method of scientific inquiry, somewhat updated to take into account the emerging needs of the twenty-first century, is the subject of Part IV of this book. It is preceded by a careful explication of a consistent framework for the representation of knowledge in Part III, and a set of organizational principles for R&D projects in Part II.

To understand the fundamental principles of the three elements of the proposed methodology is simply a matter of reading this book. To make it an integral part of your intellectual being as a scientist or an engineer, however, takes considerable practice, especially to overcome old habits. Although grasping the correspondence between notes in a musical score and the keys on the piano keyboard is a simple matter of learning the rules of music notation and the arrangement of the piano keyboard, learning to play music on this instrument requires years of diligent practice. Only after honing the hand-eye-brain coordination so that the process of translation from score to keyboard becomes effortless and automatic can the pianist hope to focus on musical expression and creativity. Only after several years of training and constant practice with the skills and discipline of the profession, can the doctor devote his full attention to the complex spectrum of needs of each patient, both physical and emotional. By the same token, only after sufficient exercise and practice with the knowledge representations and methodology set forth in this book, can the engineer or scientist undertake research and development projects with confidence that the processes will be conducted efficiently and reported objectively.

Some years ago the movie *The Karate Kid* illustrated the need for disciplined training and practice. As you may recall, the movie tells the story of a young teenager who wants to learn to defend himself with the martial art of *Karate*. He finds a gentleman of Japanese origin who agrees to be his trainer. For the young man's first lesson, the trainer asks the young man to wax and polish a score of classic automobiles parked on his property. The trainer demonstrates the required two-step technique. Moving the arm in a clockwise motion, he applies the wax thoroughly to a small area of the car; then, moving the arm in a counter-clockwise motion, he carefully removes the excess wax and polishes the surface: wax on, wax off.

The young man is disappointed that he is not going to learn to disable his opponents on the first day of training, but grudgingly accepts the task. For hours and hours he works on car after car: wax on, wax off; wax on, wax off. When he finishes all the cars, his trainer nods in terse acknowledgment and instructs him to do it all over again. The young man is outraged, claiming that the old man is wasting his time. The trainer tells him that he has a choice: wax all the cars again or abandon his training. When the young man asks for an explanation of the purpose of this boring and tiring task, the old man tells him that he would not understand now, but he will eventually — and just to have patience and follow orders.

By the end of the movie, the young man has distinguished himself as a talented athlete and has gained a deep understanding and appreciation of that initial training task. This film dramatically illustrates the need for diligent and extensive practice to acquire a complex discipline, sometimes without initially understanding the reason for the exercises. It also provides a brief and convenient expression for describing this training process: wax on, wax off.

Learning to apply the knowledge and methodology presented in this book must be accompanied by a similar dedication to discipline and practice. Even though some of the suggested practices may seem overly academic and unnecessarily rigorous at first, the serious reader is encouraged to make the effort. Read the book. Do the exercises. Follow the recommended procedures and apply the recommended techniques faithfully for several months in your professional work. Then, if you still regard some aspects as a waste of time, you can abandon (or modify) them with full confidence that you have made an informed decision. The danger is that impatience (either yours or your manager's) with constructing detailed plans, as well as business pressures to produce results as fast as possible, will compel you to abandon the effort before you have completed sufficient training to experience the primary benefit: becoming a much more effective engineer or scientist. As you sow, so shall you reap. But sometimes it's a long time from planting to harvest, with a lot of hard work in between. Try to stick it out.[2]

2.2 Tools and Resources

Table 2.1 lists a few tools and resources that will help to ease the pain of training and will continue to serve you well in your professional career. Some of them are easily acquired by simply purchasing them; others require formal study or training. All are essential for professional scientists and engineers working on sophisticated research or development projects.

Some managers may look at this list and decide that these resources and tools would be very nice to have in the best of all possible worlds, but in the real world, many are luxuries that are much too expensive to justify their acquisition. Be careful. Experience shows that being stingy about providing these resources and tools can be far *more* expensive by chipping away at the efficiency of the staff members in small ways that add up to considerable expense in the long run.

The first item on the list in Table 2.1, the research notebook, is easy and inexpensive to acquire, and worthy of special mention. A major cause of chaos in the R&D environment is the loss or inaccessibility of the bits of information created, acquired, and required by each staff member on a day-to-day basis. ("Where did I put that interface design I created yesterday over lunch?!" or "Has anybody seen a napkin with some math on it lying around here somewhere?" or "What was that guy's name who wanted to give us a half-million dollars in funding?") Writing or sketching stuff on the backs of envelopes or paper napkins, or trying to keep important bits of information in your head is a sure-fire way to sentence those good ideas to death.

---

[2]A dear friend of mine is a nun in the Russian Orthodox Church. She and I have had many conversations about requiring students (novitiates) to undergo exercises and training without first providing convincing explanations of the reasons and objectives of these exercises, which seem at first to be just rituals that are conducted for their own sake. Both of us agree that the expectation of blind obedience is morally and ethically dangerous, because abuse is so easy and, perhaps, so seductive. As a professor for many years, I am very aware of the power I hold over my students, and whenever I must apply this power, I make a conscious effort to examine and validate my motives. Do I have their best interests at heart? Am I simply rationalizing my actions? Am I abusing my authority?

I have no universal answer to this quandary. I do know that my doctoral students, with whom I generally have close relationships, have always thanked me for leading them down this methodological path. They readily acknowledge that they initially regarded the exercises and methods as somewhat "precious," but eventually came to understand and appreciate their value and necessity. It is, for example, unusual to see my students without their research notebooks clutched firmly in their hands, wherever they go, day and night.

There may be no convincing way to explain the purposes and benefits of the recommended methodological exercises ahead of time; a measure of trust is required.

Table 2.1 Tools and resources necessary for all professional scientists and engineers

- a research notebook that contains a complete record of *all* your activities, updated faithfully and regularly
- a thorough understanding of descriptive statistics to avoid blind reliance on statistical packages
- thorough knowledge of all functions and capabilities of a powerful spreadsheet program (*e.g., Excel* or *MatLab*)
- thorough knowledge of all functions and capabilities of a powerful slideshow design program (*e.g., PowerPoint*)
- access to a wide variety of specialized software packages for data analysis and visualization
- very fast and unconstrained Internet access
- a very fast and large-RAM-capacity desktop workstation at home and at work, replaced every three years
- convenient access to a well-stocked technical library, including books and professional journals
- a small collection of professional books, including handbooks and trade references
- regular attendance at and participation in professional conferences and workshops
- subscriptions to a few professional journals and magazines in your technical area
- the services of a professional technical editor for in-house technical reports and published papers

Each and every member of the R&D staff can make a substantial contribution to reducing the chaos and disorganization in the R&D environment by keeping a research notebook that is a complete record of the person's professional life. Put *everything* in your notebook, from the moment you wake up to the moment you go to sleep: ideas, creations, sketches, designs, graphs, reminders, reading and lecture notes, off-the-wall thoughts, lists, opinions, abstracts, copies of pages in documents and papers, spreadsheet printouts, program listings, pictures, photographs, scanned images, cartoons, doodles . . . *everything*. And don't hesitate to include personal stuff: grocery lists, phone messages, reminders, notes for the PTA meeting, *etc.* Keep your notebook by your bedside just in case you wake up in the middle of the night with some brilliant insight or critical reminder. You are the only one who will ever see the notebook, unless you decide otherwise. Table 2.2 lists some practical guidelines for the use of this record-keeping device.

A cautionary word to R&D managers and executives: some years ago, a large industrial firm in Europe decided to provide research notebooks to all their R&D employees, which was an excellent idea. However, they then proceeded to ruin this policy by requiring that all employees allow their managers to *inspect* their notebooks on demand. To the astonishment of management, they soon discovered from their inspections that no one was using the notebooks, so they bagged the whole idea. Please do not make the same mistake in your organization.

A word of advice to professional staff members: Do not wait for your company to buy a research notebook for you. First of all, you will be old and gray. Second, they will pick a format that you hate. Third, they may institute a ridicu-

Table 2.2 Guidelines for the use of the research notebook

- Put EVERYTHING in your notebook — use it like a diary and a scrapbook, both professional and personal.
- Keep your notebook with you at all times so that you are always ready to record your thoughts and ideas.
- Your notebook is your private record. No one else need ever see it, although you may choose to share items.
- Ask other people to sketch or write suggestions or information for you in your notebook, not on loose paper.
- Resist using loose sheets of paper for "casual" notes and ideas. Use your research notebook.
- Use paginated notebooks with high-quality quadrilled paper so that erasures will not wear out the paper.
- Put your telephone number and a request for return on the cover; loss of this record can be very serious.
- Number and date each notebook. You will fill one every few months. Archive the filled books in a safe place.
- Slate every new subject, and date every new entry. Specify the sources of nonoriginal information.
- Use right-hand pages for regular entries, left-hand pages for revisiting topics (or *vice versa,* if more convenient).
- Use preprinted (or manually entered) page numbers to cross-reference related entries.
- Paste or tape copies of important and useful documents into your notebook, including computer printouts.
- Do not erase major "mistakes." Cross them out and explain them. You may need to re-evaluate them later.
- When a notebook is full, put copies of important pages in a looseleaf binder using sheet protectors.

---

lous policy like the one cited in the previous paragraph. Buy your own research notebooks. Make them *your* property. Pick a format that appeals to you. Get used to using your notebook every day, all day, and after a couple of months, you will wonder how you ever got along without it. Guaranteed.

That's enough of an introduction. Open your research notebook, turn the page, and dig in.

PART II

Project Organization

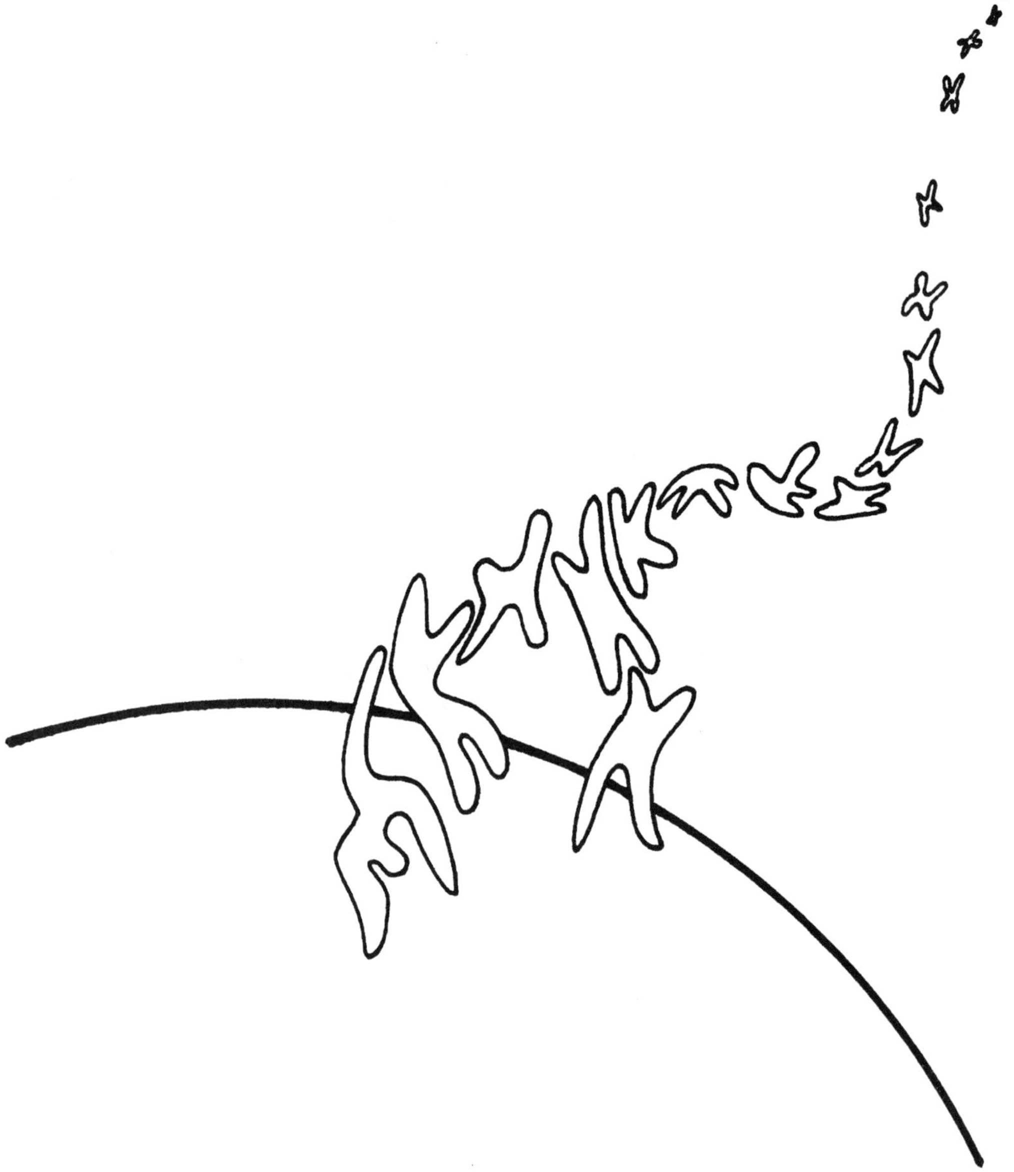

CHAPTER 3

The Project Hierarchy

3.1 Bottoms Up

As a strand of messenger RNA sweeps by a strand of DNA, it extracts a molecule of adenine from the DNA strand, which happily joins the project being carried out by the messenger RNA, the production of a new protein that is used in the manufacture of acetylcholine. Coursing down the interior tunnel of the axon of the neuron, the molecule of acetylcholine races to resupply the vesicle population in the synaptic endbulb, so that the project to excite the muscle fiber on the postsynaptic membrane can achieve its objective in time. The muscle fiber, in concert with thousands of other muscle fibers, initiates the project to curl the middle finger inward toward the palm. Joined by the other fingers, the hand grasps the doorknob, cooperating with the mind in the larger project to open the door. A bit frightened by the shouts and laughter of the boisterous first graders, the young girl steps hesitantly through the door into the classroom to begin the project of her primary school education.

Rapping sharply on his desk, the first grade teacher begins the challenging project of nurturing the new young minds with their first taste of formal knowledge. During recess in the playground, the pupils concoct a project to build a pile of plastic cubes as high as they can, laughing gaily as it finally tumbles to the ground. Watching from a window, the teacher is struck with a brilliant idea for his doctoral research project in educational psychology. Several years later, the established scholar accepts a position in the US Department of Education to direct a project for research in innovative education methods. Based on its success, the United Nations mounts an ambitious project to improve and extend educational resources for developing nations throughout the world.

As the educational level of the world population rises over the years, developing nations establish new projects to realize their own particular human, industrial, and economic potential. A team of young scientists in Brazil, empowered by a rich and diverse education, devises a project to counter the rape of the rain forests in the Amazon River basin with a sophisticated strategy that will support industrial land reform while still protecting and nurturing one of the world's major sources of oxygen and pharmaceutical drugs. Unable to resist the influences of worldwide instantaneous communication, the last withering projects of totalitarian oppression fall to the forces of freedom and fairness. Released from the chains of pragmatic compromise, artists across the world are supported to undertake creative projects whose sole objective is to celebrate the courage, the pain, and the triumph of the human spirit.

Its atmosphere repaired, its soil replenished, its forests rejuvenated, its oceans and lakes and streams revitalized, its people newly enlightened and empowered, the Earth breathes a sigh of relief and continues its never-ending project of exploring the universe.

3.2 Top-Down Project Planning

Every R&D **project** comprises a hierarchy of **tasks**, each of which has a single **objective**. Figure 3.1 represents a very simple task hierarchy as a tree, called a **project task tree**, or simply a **task tree**. Task 1 at the first (highest) level invokes six subtasks in the second level of the hierarchy (**descendant tasks**), whose **objectives** must be achieved to achieve the objective of the **parent task**, Task 1. Similarly, Task 1.2 at the second level invokes two descendant tasks at the third level, whose objectives must both be achieved to achieve the objective of the parent task.

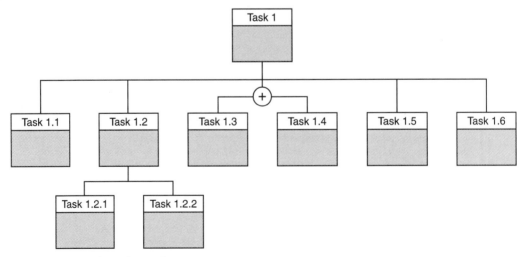

Figure 3.1 Sample project task tree

By definition, the objectives of *all* of the descendant tasks connected to the parent task by *uninterrupted* lines must be achieved successfully to achieve the objective of the parent task. For example, in Figure 3.1 the objectives of Tasks 1.2.1 and 1.2.2 must *both* be achieved successfully to achieve the objective of the parent Task 1.2. Sometimes, however, only one of several descendant tasks need be accomplished successfully to achieve the objective of the parent task. Such tasks are connected to their common parent task with *interrupted* lines that have a circled plus-sign ⊕ at their junction. This is called an **OR** operator, which means that the successful achievement of the objectives of *either* one task *or* the other is sufficient to achieve the objective of the parent task. For example, in Figure 3.1 Tasks 1.3 and 1.4 are connected to their parent task with an **OR** operator ⊕ at the junction, implying that Task 1 requires only the objectives of Task 1.3 *or* Task 1.4 to be achieved successfully.

The tasks at the lowest level, or **leaves**,[1] of a project task tree are called **terminal tasks**. This task tree provides no additional details of terminal tasks beyond their titles. This usually implies that the components of the terminal tasks are (or are expected to be) self-evident to those who must carry out these tasks, although these investigators may, of course, wish to break out each terminal task into a *more-detailed* task tree as a part of their own planning. On the other hand, a task tree may always be trimmed from the bottom up to any level to yield a *less-detailed*, high-level representation for those who are not concerned with low-level details, for example, for management briefings. Note that even a trimmed task tree is still a complete description of the project, lacking only in detail (often much to the relief of high-level management).

The highest level, or **root**, of a project task tree must consist of one and only one task, called the **primary task**. The reason for this restriction is that, by definition, each project must have *one and only one objective,* which is also the objective of the primary task; it is a high-level abstraction of the objectives of all the descendant tasks in the project. A common mistake in planning a project is to assign more than one objective to a project. When this happens, usually the project planner either: 1) has simply neglected to summarize the multiple objectives into a single high-level objective; or 2) has actually assigned several relatively unrelated objectives to the same project. The first case is easily corrected. The second case requires reorganizing the task hierarchy, so that each project has a single primary task with a single primary objective.

How many projects are represented in Figure 3.1? The most obvious (and incomplete) answer is one, namely the single project that comprises all the tasks depicted in the figure. However, any *complete* subtree of the hierarchy in Figure 3.1 is *also* a project, because a subtree always has exactly one task at its root. Therefore, there are a total of nine projects represented in Figure 3.1:

---

[1]Unlike its botanical counterpart, a task tree is drawn upside-down, with the *root* at the top and the *leaves* at the bottom. This is a standard convention for representing hierarchical information structures as trees.

Primary Task	Descendant Tasks
1) Task 1	all 8 subtasks
2) Task 1.1	none
3) Task 1.2	Tasks 1.2.1 and 1.2.2
4) Task 1.3	none
5) Task 1.4	none
6) Task 1.5	none
7) Task 1.6	none
8) Task 1.2.1	none
9) Task 1.2.2	none

It is important for managers to understand that even the smallest project will probably have many subprojects, and that a corresponding hierarchical management structure of some kind may be needed to coordinate the efforts of the interrelated projects.

Referring to the root task as the primary task is a relative assignation; the root task may actually be a low-level task buried in a much larger task tree for a much larger project that has a very large and complex task hierarchy. In fact, this is almost always the case in large organizations with many departments and layers of management. Although an R&D engineer quite properly refers to his root task as the primary task of his project, in the larger scheme of things his project is probably just one of many being carried out as part of the overall plans of the R&D department, other departments, and the company as a whole.

To assume that a small task tree like that shown in Figure 3.1 must represent a rather small and uncomplicated project is wrong. At one extreme, Figure 3.1 could represent a high-level summary of the entire NASA Apollo Project, whose objective was to land the first humans on the moon at a cost of no more than $20 billion over a period of no more than eight years. At the other extreme, Figure 3.1 could represent a project to peel a potato. The number of levels and tasks in a project task tree does not reveal the complexity or size of the project, simply the level of detail being shown.

A simple example will help put all of this in perspective. For several months, a homeowner had been nagged with an irritating problem: the hinges on his guest bedroom door squeaked. Analysis of the problem revealed that simply lubricating the hinges would not solve the problem; the hinges had gotten bent somehow, and they had to be straightened or replaced. Procrastination successfully delayed the project for several months, until finally, with house guests arriving that evening, the homeowner set aside a Sunday morning for the project and laid out a plan. Figure 3.2, which is an instantiation of Figure 3.1, was the task tree for this project.

The primary task was, of course, *Repair Squeaky Hinges.* The homeowner expected the six subtasks at the second level, *Unmount Door, Remove Hinges, Straighten Hinges OR Buy New Hinges, Reinstall Hinges,* and *Remount Door,* to be sufficient to accomplish the primary task. Likewise, the descendants of these tasks specified a set of more-detailed subtasks, such as *Remove Screws* and *Pry Off Hinges.* The terminal tasks, such as *Unmount Door* and *Remove Screws,* the

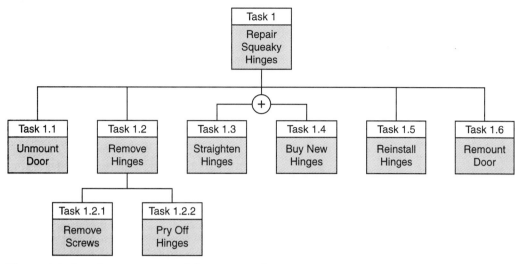

Figure 3.2 Sample instantiation of the project task tree in Figure 3.1

homeowner thought, provided sufficient detail to allow him to carry them out without explicitly specifying additional task levels. As it turned out, Murphy's Law proved him wrong.

After finishing his morning coffee, the homeowner fetched his largest screwdriver from the basement and trudged up to the guest bedroom to begin the project. Addressing the first task (*Unmount Door*), he slipped the pins from the hinges of the door and laid it against the wall. Next, he proceeded to remove the two halves of the top hinge, one on the door and the other on the doorframe, each half held in place by three large screws. Even though the screwdriver was not really large enough for the job, all six screws for the top hinge came out without too much trouble. Unfortunately, however, as he strained to loosen the first screw on one of the halves of the bottom hinge, which refused to budge, a big chunk broke off the tip of the screwdriver.

The task *Remove Screws* had failed. The homeowner looked at his project plan, but there was nothing there to provide any guidance for what to do if the task failed. He didn't have another screwdriver that was large enough for the task, and there were no hardware stores open yet on this Sunday morning. Rising to the challenge, the homeowner spawned a new project: *Repair The Screwdriver*. The objective of this project was to repair the screwdriver by grinding down the chipped tip to form a new tip. He went down to the basement, dragged out his electric grinder, plugged it in, and turned it on. Nothing. It just sat there. Once again, a task had failed. Once again, he was forced to set aside the current project and initiate a new project: *Repair the Grinder*.

He proceeded to analyze the problem with the grinder and quickly discovered that an electrical connection had come unsoldered. No problem. He whipped out his soldering gun, plugged it in, and . . . you guessed it . . . it did not work. The heating element on the tip of the soldering gun was broken. Another task failure.

Many of us would have surrendered at this point, chucked the whole process, and retired to the living room and Sunday afternoon football. But the homeowner's blood was up; he was *determined* to succeed before his guests arrived. By now, it was already early afternoon. So he climbed into his car, drove to the nearest hardware store, and purchased a new heating element for his soldering gun. After returning home, he fixed the soldering gun, re-soldered the broken connection in the grinder, ground down the tip of the screwdriver until it was again serviceable, returned to the guest bedroom door, and removed the last few screws on the bottom hinge halves. Now he was back on track with the original project. Things were looking up!

Until, of course, it happened again. As it turned out, the homeowner could not straighten the hinges sufficiently; the steel was just too strong to bend back into shape with the tools he had available. In all fairness, he had anticipated *this* possibility in the project plan (Task 1.4 in Figure 3.2), but what he hadn't anticipated was that the project would derail earlier for other reasons, delaying Task 1.4 by so much that the hardware stores would be closed by the time this task failed. He could no longer successfully execute the alternative subtask (**Buy New Hinges**) and thus could not accomplish the primary task (**Repair Squeaky Hinges**) before his guests arrived.

To make a long story short, the unmounted door lay against the wall for several more days until the homeowner could find the time to go back to the hardware store to buy a new hinge and finish the repair. In the meantime, his guests had to sleep with the door to their room wide open all night. They assured him it was just a minor inconvenience, but were really rather annoyed.

What are the important lessons to be learned from this tale (the likes of which each of us has most certainly experienced at one time or another)? First, it is impossible to specify alternative plans for every possible type of failure; there are just too many ways things can go wrong. No project plan can be expected to include contingency plans for all kinds of all-too-common failures. Imagine the extraordinary foresight necessary to predict all the problems with the screwdriver, grinder, and soldering gun! When he was planning the project, the homeowner would have had to imagine: "Okay, I know that when I try to remove the screws I might break the tip of the screwdriver, which I will have to repair with the grinder, because I don't have a another large screwdriver. But just in case the grinder doesn't work because of a broken electrical connection, I will have to repair it with the soldering gun. And just in case the soldering gun does not work because the element just happens to be burned out, I better go to the store and buy a new one before I begin the project." Now that would have been truly prescient!

And that is just for the problems that *did* happen. What about all the problems that didn't happen, but *could have*? Clearly, anticipating all possible problems ahead of time and making comprehensive sequential contingency plans to avoid or sidestep them is just not possible. The only way to handle such problems is to be ready to *suspend* the current task and invoke a remedial task *recursively*, just as the homeowner did. **Recursive task invocation** means suspending the current task to invoke a new task that must be accomplished before the suspended task can be resumed. To process recursive tasks successfully requires that the **invocation point** of the suspended task be remembered (in a knowledge structure called a push-down stack). For example, when the homeowner finished buying the replacement soldering-gun element, if he had not remembered the invocation point of the suspended task (***Repair The Grinder***), he would have just stood there in the hardware store with a confused look on his face and said to himself: "Now, why did I just purchase this soldering-gun element? What am I supposed to do with it?" We humans take this powerful recursive task-suspension capability for granted, but it is an indispensable part of our almost unique ability to plan on-the-fly and not get lost in the process.[2]

On the other hand, if the homeowner had been in the profession of solving problems like the squeaky door, one could appropriately criticize him on several counts: not having the right screwdriver for the job, not having a spare screwdriver, and not properly maintaining his grinder and soldering gun. A professional mechanic who does jobs like this every day clearly understands the importance of having a complete set of high-quality and well-maintained tools. Therefore, because the subject of this book is *professional* R&D methods, the following tip is offered:

> **TIP 1**
> Provide every professional with a comprehensive set of high-quality state-of-the-art tools for planning, designing, and implementation, as well as the time and resources to update and maintain them.

Quite obviously this piece of advice is aimed primarily at bureaucrats and managers, not the researchers, engineers, and technicians who do the actual project work; they already *know* that this tip is right on the money, but often are powerless

---

[2]The part of the mammalian brain that contains the cognitive machinery for recursive task planning is the frontal lobes of the cerebral cortex. Most mammals have very small frontal lobes or none at all and thus have very limited capabilities for recursive task planning. In humans, however, the frontal lobes occupy one-fourth of the cortex.

to do anything about it. Chapters 2 and 12 provide several specific recommendations for providing R&D personnel with appropriate tools for their professional activities.

A second important lesson derived from this anecdote is that it is often advisable to execute "dry runs" to explore ways in which the project can be conducted more efficiently and to identify possible pitfalls in the project plan. Such trial passes through the task space are called **pilot tasks**, or simply **pilots**. Conducting pilots can often save a lot of time, money, and frustration.

Consider the project to fix the squeaky hinges. The original plan had been to spend the morning trying to straighten the bent hinges. If that failed, then the homeowner figured he would still have time in the afternoon to go to the hardware store and buy one or two new hinges, as needed. But his time estimate was wrong; by the time he had dealt recursively with the screwdriver, grinder, and soldering-gun failures, the entire day had slipped by, and the hardware store was closed.

On the other hand, if the homeowner had realized that his plan could easily go awry (like so many plans of mice and men), then he would have reordered the sequence of the tasks, beginning with a pilot task. Instead of removing both hinges and then repairing or replacing both of them, he would have first removed both halves of one hinge and immediately tried to straighten them. (This is the pilot.) When that failed, he would have learned much earlier in the day that straightening the hinges was not possible (given the tools he had at his disposal; see Tip 1), and that he had to buy new ones. Then, when he broke the screwdriver removing the second hinge and recursed through the unexpected sequence of repair projects, he still could have gone out much earlier in the day while the hardware store was still open and bought *both* the new soldering gun element *and* the new hinges. He would have finished the project on schedule, and his guests would have had their privacy and gladly come to visit him again.

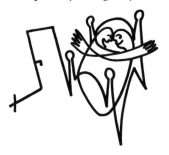

TIP 2
Use pilot tasks throughout a project to identify unexpected consequences, forestall failures, avoid brittle procedures, determine operating ranges, and estimate the feasibility of proposed solutions.

The third important lesson to be derived from the hinge-repair project has to do with the timing and sequence of project tasks. Although the arrangement of a project task tree indicates the *hierarchy* of tasks, it does not explicitly specify the *order* in which the tasks are to be executed. The project task trees given in Figures 3.1 and 3.2 do not provide any information about the sequence of the tasks. Although the semantic names of the tasks in Figure 3.2 seem to imply that the tasks should be executed in order from left to right across the task tree, there is no reason to assume that this will be true in general. In many projects, for example, sets of tasks can and should be executed in parallel. Moreover, if we ignore the semantic information in the names of the tasks in Figure 3.2, there is nothing in this task tree that rules out buying new hinges *before* attempting to straighten the existing hinges, or obviates the absurdities of remounting the door *before* unmounting it, or reinstalling the hinges *before* they are removed.

Clearly, the project plan must also provide explicit information about the task sequence, both estimated initially during project planning, and modified periodically based on results of pilots and other task activities. A project task tree, however, does not lend itself well to the encoding of this sequencing and scheduling information. For example, if we used arrows to indicate the sequence of the tasks, it still would not provide a clear picture of their relative start and end times or any overlap that would permit parallel execution.

Fortunately, there is a very straightforward way to transform the project task tree into a structure that is particularly well suited for scheduling and sequencing the project tasks, as well as for visualizing those sensitive and precarious paths through the task space that signal important candidates for pilot tasks. This structure is really nothing more than a graphical *outline* annotated with timing information, which is called a **milestone chart**.[3] Figure 3.3 is the *proposed* milestone chart that the homeowner might have prepared before beginning the squeaky-hinge project. The first gray bar indicates the expected length of the entire project, including a break for lunch, and ending at 4:00 pm. The question mark (?) in the bar for Task 1.3 indicates where the homeowner expected to decide whether straightening the bent hinges was feasible, *i.e.,* after trying this method for no more than an hour. After a break for lunch, he would have either finished straightening the hinges or gone to the hardware store to buy a set of new hinges. Either way, at 2:00 pm he expected to reinstall the hinges and remount the door, finishing the project successfully at about 4:00 pm.

The proposed milestone chart shown in Figure 3.3, however, does not allow much time for flat tires, given that the hardware store closes at 4:00 pm. As a result, when the homeowner tried to execute this precarious plan, the *Remove Hinges* task failed unexpectedly, and the plan failed along with it. As shown in Figure 3.4, by the time the homeowner had completed the repairs on the screwdriver and *then* discovered that he could not straighten the hinges, the hardware store had closed. Unforeseen delays had defeated him, and his guests would suffer the consequences.

---

[3]There are several project-planning software packages available that can automatically transform tree structures into outlines, and *vice versa,* such as Microsoft Project. However, like many automated tools, they often do everything except what you want. I used Microsoft Excel to make the milestone charts for this book (and many other projects) and have found it quite satisfactory. *Chacun son goût.*

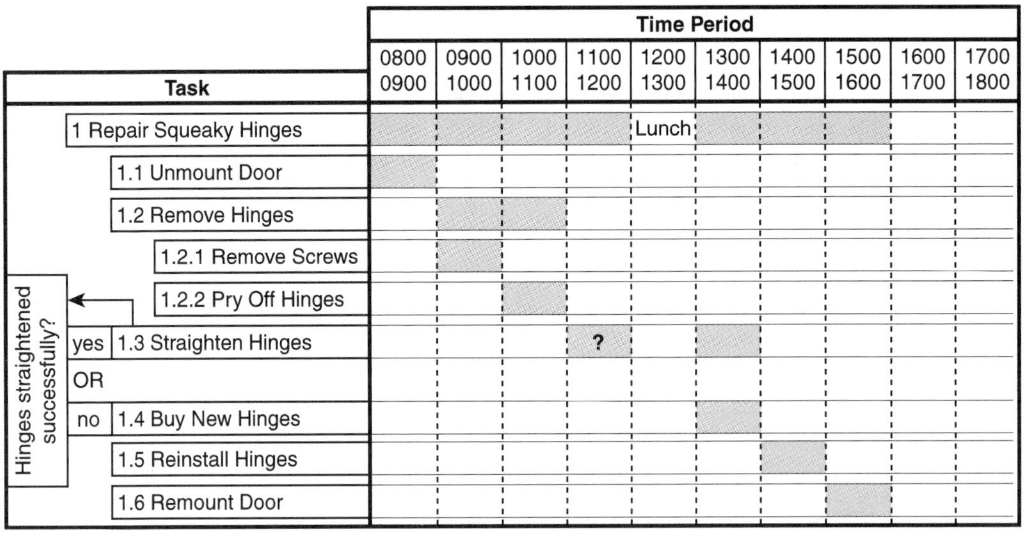

Task	Time Period									
	0800 0900	0900 1000	1000 1100	1100 1200	1200 1300	1300 1400	1400 1500	1500 1600	1600 1700	1700 1800
1 Repair Squeaky Hinges					Lunch					
1.1 Unmount Door										
1.2 Remove Hinges										
1.2.1 Remove Screws										
1.2.2 Pry Off Hinges										
yes 1.3 Straighten Hinges					?					
OR										
no 1.4 Buy New Hinges										
1.5 Reinstall Hinges										
1.6 Remount Door										

(Left margin label: Hinges straightened successfully?)

Figure 3.3 The initial milestone chart for the squeaky-hinge project without a pilot

Task	Time Period									
	0800 0900	0900 1000	1000 1100	1100 1200	1200 1300	1300 1400	1400 1500	1500 1600	1600 1700	1700 1800
1 Repair Squeaky Hinges					Lunch					
1.1 Unmount Door										
1.2 Remove Hinges										
1.2.1 Remove Screws		**Recursive Repairs**								
1.2.2 Pry Off Hinges										
yes 1.3 Straighten Hinges								FAIL		
OR										
no 1.4 Buy New Hinges									FAIL	
1.5 Reinstall Hinges										
1.6 Remount Door										

(Left margin label: Hinges straightened successfully?)

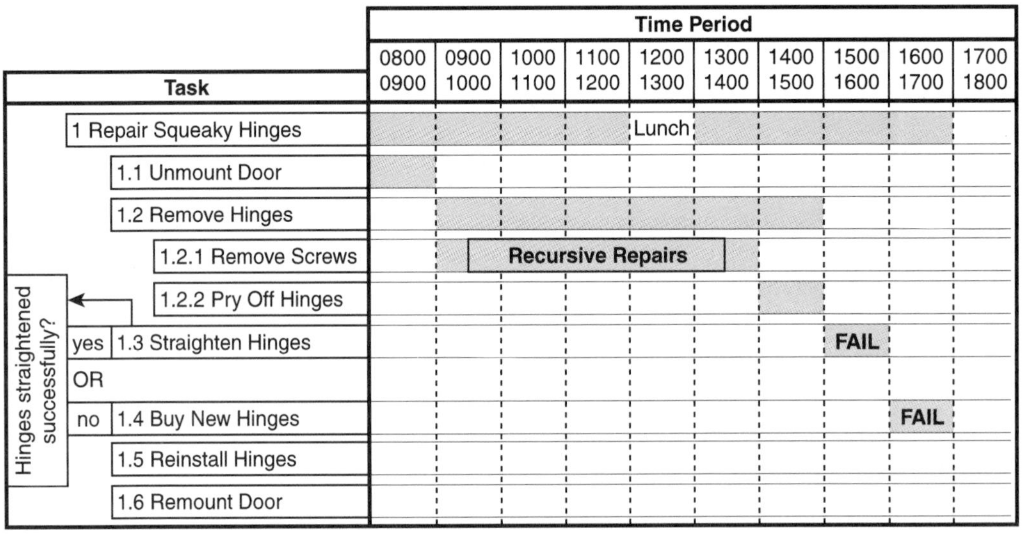

Figure 3.4 The final milestone chart for the squeaky-hinge project without a pilot

As described previously, the homeowner could have avoided this failure by conducting a pilot early in the day to find out if the hinges could be straightened. As shown in Figure 3.5, during this brief pilot he would have removed the screws from one of the hinges (before, it is assumed, the screwdriver broke), pried off the hinge, and tried to straighten it to determine whether that solution was feasible. Either way, he would have still completed the project by 4:00 pm.

Of course, in fact, the screwdriver would still have broken during the second instantiation of Task 1.2, *Remove Hinge 2*. However, unlike the plan without the pilot, the plan with the pilot still allows the homeowner to complete his

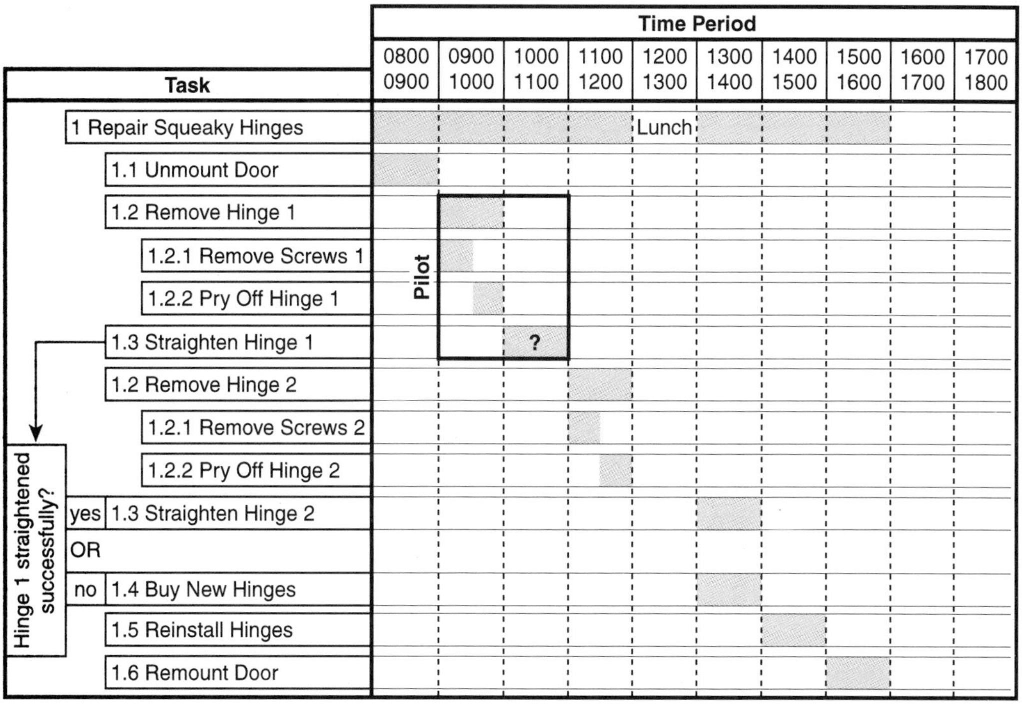

Figure 3.5 The proposed milestone chart for the squeaky-hinge project with a pilot

project before his guests arrive, even though he must still spend the same four hours repairing the screwdriver. As shown in Figure 3.6, at the conclusion of the pilot at about 11:00 am, the homeowner concludes that straightening the hinges will not work. However, instead of going to the store immediately to buy new hinges, he starts to remove the second hinge and, of course, breaks the screwdriver. He then starts his recursive screwdriver-repair project, eventually discovering at about 1:00 pm that he must buy a new heating element for his soldering gun. He goes to the store and buys the heating element *and, of course, the new hinges, because he already knows he needs them.* He returns home with plenty of time to complete the screwdriver repairs and go back to the squeaky-hinge project. It takes a couple of hours longer than expected, but by 6:00 pm he has successfully completed project, well before his houseguests arrive.

Compared with straight shots through a project task, pilots usually increase the project time and cost a bit, but in the long run, they can spell the difference between costly nonrecoverable failures and slow-but-sure success. By using pilots to test and refine the procedures and logistics of critical tasks, the enormous costs of catastrophic failures can often be avoided. For instance, if an experiment using human subjects fails because the processes and consequences of the task were not well understood to begin with, the original cohort is often rendered useless for re-running the experiment, because the subjects are no longer **naive**; they know too much about the experiment and will respond quite differently and probably inappropriately. Obtaining and preparing human cohorts for experiments is usu-

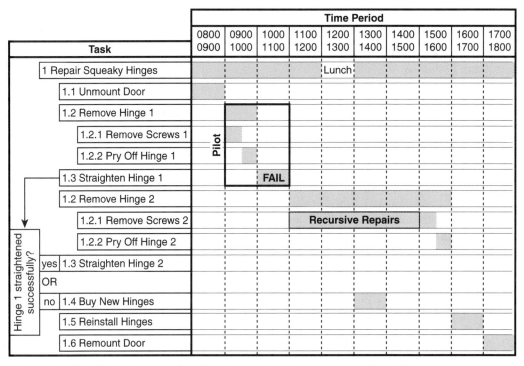

Figure 3.6 The final milestone chart for the squeaky-hinge project with a pilot

ally enormously expensive and time-consuming. Tainting a cohort without obtaining useful results often spells the premature end of a study. Pilots conducted with small representative cohorts, whose members are not used again for the actual task, can help avoid this problem. Much more will be said about this in Chapter 12 (synthesis).

The milestone chart is an ideal mechanism for inserting and planning the pilots into a project plan. Even Gantt-chart-type information can be superimposed on the milestone chart as line plots to present estimated and actual project costs.

> **TIP 3**
> Use task trees for the top-down design of projects; use milestone charts for planning the logistics, schedules, and costs of these projects as they evolve.

The task tree and the milestone chart are useful tools for designing, planning, and documenting an R&D project. Unfortunately, given the informal and whimsical ways in which too many R&D activities are planned and conducted, usually only two task trees (if any) are prepared: one when the task is planned and proposed, and the other for documentation purposes when the project is over. Typically, these two task trees bear little resemblance to each other, because the project goes through enormous changes during its lifetime. Ironically, this fact is often used to support the notion that detailed planning is a waste of time, because the project will most likely play out in a completely different manner.

That prophecy is both myopic and self-fulfilling. The main purpose of planning a project is not just to predict the likely course of events at the outset, but to *estimate* the organization, time, and resource requirements of a project *on a continuing basis,* so that some obvious and expensive pitfalls can be foreseen and avoided, scarce resources can be allocated in a timely and efficient manner, and management can be kept appraised of unexpected roadblocks and delays. But this will only work if planning is, indeed, a regular, ongoing activity that is undertaken throughout the lifetime of the project, accompanied by task trees and milestone charts that continually evolve, reflecting the best current estimates from all cognizant participants. And then, as a natural consequence, if these planning documents are frequently updated, the final version will not be very different from the one that immediately preceded it. In retrospect, much can be learned about the design and conduct of research and development tasks in specialized fields from the archived sequence of planning documents that were in force at various times throughout the lifetime of the project.

Exercise 3.1

Based on a simple and short household project that you perform routinely and successfully, design a task tree and a milestone chart for this project so that it can be performed by someone who has never performed it before. Insert contingency tasks to forestall failures and avoid brittleness in the plan. Use your repeated experiences with this project, including common pitfalls that can and do happen, to insert pilots in the initial phases of the project to orient and educate a person who must perform the task successfully as quickly as possible with no prior experience.

If possible, perform this task using your plan and record your observations.

CHAPTER 4

The Project Task

The discussions in the previous chapter relied on an intuitive understanding of the word **task**. This concept is so important and central to the conduct of R&D projects, however, that some formal definitions are required before we go much further. A task is analogous to a function in mathematics, having inputs, outputs, and a process that transforms the inputs into the outputs. There are special words in mathematics for these three parts of a function. The inputs are called the **domain** of the function; the outputs are called the **range** of the function, and the process is called the **method** of the function. If the domain includes quantities that may be intentionally varied, these quantities are called the **independent variables**. The range comprises the quantities that are expected to vary as a direct result of applying the method to the domain, called **dependent variables**. These names imply that the outputs (range variables) are dependent on the inputs (domain variables), but that the inputs are not dependent on anything in the function and may be varied freely (albeit subject to the rules of mathematics and perhaps limited by appropriate bounds). Consider the following mathematical function:

$$z = f(x, y)$$

The independent variables x and y comprise the domain of this function, the dependent variable z is the range of the function, and the actual mathematical operation represented by the letter f is the method of the function. For example, in the familiar quadratic equation $y = ax^2 + bx + c$, the constant exponent 2 and the variables *a, b, c,* and *x* comprise the domain of the function, and the variable *x* is (usually) the independent variable. The method is expressed by the algebraic operators (addition, multiplication, and exponentiation), which define how the domain variables are to be combined. The dependent variable *y* is the sole member of the range of the function.

In an effort not to invent too many new words, it is both practical and convenient to use the same terminology, with analogous meanings, to define an R&D task:

> **Definition 4.1** An R&D **task** applies a specified **method** to the **domain** of the task with the objective of obtaining a satisfactory result in the **range** of the task.

It is important to note that the method, domain, and range, when applied to an R&D task, are usually much more complex than they are for a typical mathematical function. For instance, the task domain may consist of many quantities required as inputs, including both constants and independent variables; the task range may consist of several different quantities (dependent variables) generated as outputs; and the method may consist of several complex interrelated mechanisms and processes that transform the domain into the range of the task. Each of these three elements is so important that a separate section is devoted to explaining each of them.

4.1 Task Domain

The task domain includes everything that is brought to the task:

> **Definition 4.2** The **task domain** comprises the **task unit** and the **resources** necessary to achieve the **task objective**.

Once again, let us break down this definition into its three components, the task objective, the task unit, and the task resources, and consider each one in detail.

4.1.1 TASK OBJECTIVE

> **Definition 4.3** The **task objective** is a statement of what the task is intended to achieve, expressed as an infinitive phrase.

As was pointed out in Section 1.1, research concerns itself with the *acquisition of knowledge,* represented by its intellectual results and conclusions, while development concerns itself with the *application of knowledge,* represented by the practical devices or effects it creates. Thus, it might be tempting to assume that research projects do nothing but acquire knowledge and pursue intellectual truth, and development projects do nothing but apply knowledge and make things. Nothing could be further from the truth. The fact is that research projects must often devote a great deal of time to building devices and generating effects to achieve the research objective, and development projects almost always include a number of tasks to acquire some fundamental knowledge that is essential to achieve the development objective. There is no contradiction here. It is quite appropriate and usually absolutely essential to embed some development tasks in every research project, and *vice versa.* This important juxtaposition is one of the primary reasons why the terms *research, development, project,* and *task* have been so carefully defined, and why the relationship between projects and their component tasks is given so much emphasis.

It is often convenient to name a task with a very brief summary of its singular objective. A simple convention is to omit the word "To" from an abbreviation of the infinitive phrase that specifies the objective, so that the name becomes a short imperative or command. For example, the *objective* of the primary household task described in the example of project organization in the previous Chapter (see Task 1 in Figure 3.2) might be: "To repair the squeaky hinges on the guest bedroom door before the guests arrive at 8:00 pm." The *name* assigned to this task in Figure 3.2 is a three-word summary of this objective: ***Repair Squeaky Hinges***. The names of the subtasks in the task tree in Figure 3.2 are also very brief summaries of their complete objective statements. This constraint on the length of the name of a task may also help to ensure that a task has a single objective; it is difficult to compose a very brief summary of task that has mistakenly been assigned multiple objectives. Thus, if one encounters difficulty expressing the name of a task in very few words, this may be a warning that the objective is not singular.

The critical process of formulating a meaningful and realistic objective for a task is presented at the end of Chapter 10 (Analysis).

4.1.2 TASK UNIT

> **Definition 4.4** The **task unit** is the set of objects or concepts that undergoes some required alteration before or during the task and measurement during the task.

The task unit is the central focus of the task. It is the object or concept to be investigated in an effort to achieve the objective of the task. First of all, it may be a physical object or set of physical objects, such as:

- a car seat (whose comfort is being evaluated);
- a rock layer on an exposed cliff face (whose geological age is being estimated);
- a highway with badly congested traffic (whose speed is being measured);
- a cohort taking a new pharmaceutical drug (whose effectiveness is being measured); or
- a sample of fruit (whose freshness is being evaluated).

In each of these examples, you can reach out and touch the task unit. It has a physical existence. But the task unit may also be an abstract concept or theory, such as:

- the algebraic expression of a mathematical conjecture (that is to be proved);
- the content of journal articles (that are being summarized into a set of short abstracts);
- a software graphical user interface (GUI) (whose effectiveness is being evaluated);
- a linguistic principle (that explains structural differences among related languages); or
- a new mechanism for government-citizen interaction (whose efficacy is being measured).

Each of these examples is an abstraction that can exist only in the mind, such as an idea or a theory, or in an information structure, such as an algorithm that is implemented as a program in computer memory. Just because a task unit is abstract, however, does not detract from its reality or importance, and the methodology described in this book is just as valid for abstract concepts and ideas as it is for concrete gadgets and potions.

Where does the task unit come from? As will be explained in Chapter 11 (Hypothesis) and Chapter 12 (Synthesis), it is fully disclosed in the Hypothesis Phase of the Scientific Method and then acquired and processed in the Synthesis Phase of the Scientific Method. Sometimes the task unit already exists when the task is undertaken. Often it is acquired or purchased by the task team. Sometimes it is the product of the creativity of the R&D scientists and/or engineers who design a solution for a task and then construct a prototype for testing and validation. It is important to emphasize that the task unit is *not* the implementation of the solution for the task, but rather the object or concept that *undergoes some required alteration and measurement* necessary to achieve the task objective.

The task unit can take the form of a cohort, a device, an algorithm, a mathematical equation, a sociological prediction, *etc.* For example, Albert Einstein wanted to develop a General Theory of Relativity, and the four-dimensional space-time continuum became his task unit. NASA needed to launch the Apollo spacecraft into earth orbit, and the giant Saturn V rocket became its task unit. Clyde Tombaugh searched for a new member of the family of planets in the solar system, and the region in space at his predicted position became his task unit. Andrew Wiles set out to prove Fermat's Last Theorem, and the age-old conjecture became his task unit. Jonas Salk created a polio vaccine, and a cohort of trial subjects who had been injected with the new vaccine became his task unit. Victor Frankenstein was obsessed with creating life, and a collection of dead body parts became his task unit.

As stated in Definition 4.4, the task unit must undergo some required alteration before or during the task. **Required alterations** must be either *artificially* or *naturally* applied to the task unit *before* or *during* the task to achieve the objective of the task. **Required artificial alterations** are applied to the task unit by humans (often the task personnel themselves) or by devices that humans operate, such as the injection of a drug, the application of a precise load to a mechanical structure, the algebraic manipulation of an equation, or the programmed display of an input request on a computer screen. **Required natural alterations** are usually applied, sometimes unavoidably, by some fundamental force of nature that is essential for the task, such as the force of gravity acting on a rocket, the emission of radiation from a radioactive sample, the gradual accretion of sedimentary layers in the earth's crust, or the growth of a tumor.

Past alterations occur *before* the task is performed. Geologists look for sedimentation layers in rock strata that were deposited millions of years ago, which is a natural past alteration. Insurance adjusters inspect your car to estimate the damage incurred during a recent traffic accident, which is an artificial past alteration. **Concurrent alterations** occur *during* the performance of the task. The cooling of a liquid in a vessel during a task is a natural concurrent alteration. During the test of a rocket, the force of the exhaust is a required artificial concurrent alteration, while the force of gravity is a required natural concurrent alteration.

The task unit is also often exposed to alterations that are not required for the task. In fact, it is important to realize that a task unit *always* undergoes some alteration during the task, because whenever the task unit is measured (observed), the laws of physics dictate that it must undergo some alteration. And because it is absurd to imagine a task during which nothing about the task unit is measured (nothing would be accomplished by the task), it is impossible not to alter the task unit in some way during the task. Even if only photographs are to be taken (a so-called "passive" measurement), the task unit must be illuminated to do so, and the photons from the illumination source will unavoidably alter the task unit as they bombard it. Of course, for mechanical engineers who are photographing the deflection of a long steel beam under load, the mechanical effects of the photons bombarding the steel beam are negligible and may be safely ignored. Such insignificant effects are called **irrelevant alterations** and are not included in the resources of the task. On the other hand, when nuclear physicists conduct experiments at the atomic or subatomic level, they must often worry about and compensate for effects of alterations that other tasks would consider irrelevant; energetic photons can disrupt the delicate balance of events at the subatomic level. Such deleterious effects are called **unwanted alterations**.

Unwanted alterations of the task unit are also common in experiments with animal and human subjects, who are extraordinarily sensitive, both consciously and unconsciously, to a wide variety of visual, chemical, haptic and aural stimuli. Consider, for example, a human subject who is taking an intensive intellectual test, and at the same time, unbeknownst to the subject or the researcher, is being exposed to other spurious internal or external stimuli. Perhaps the whining noise of an air conditioner in the room subconsciously irritates the subject. Or perhaps the subject's stomach is digesting a recalcitrant bit of food, causing physical discomfort. Both of these unwanted alterations of the task unit (the subject) might well affect the subject's cognitive processes, causing significant distortion of the results of the experiment.

The source of an alteration can be **external** and/or **internal** to the task unit. The injection of a drug into a test subject, the gradual deposition of layers of minerals in the earth's crust over millions of years, and the application of a test load to a steel beam are examples of external alterations. The organic decay of a piece of fruit during a test of the fruit's freshness, the effects of a good night's sleep before a stressful day, and the natural emission of alpha particles from a radioactive sample during an experiment in nuclear physics are examples of internal alterations.

Figure 4.1 summarizes the types of alterations that may be applied to a task unit. A great deal of care must be taken to identify *all* of the alterations that a task unit may undergo and to minimize the effects of unwanted alterations. Often the source and effects of unwanted alterations are very difficult to identify and isolate, and the only way to do so is to observe their effects in a series of especially designed tasks and then try to deduce their origins. In addition, deciding that an alteration is irrelevant must be done very cautiously, lest a significant alteration end up being inappropriately ignored. More will be said about identifying irrelevant and unwanted alterations and potential sources of error in Chapter 12 (Synthesis).

> **TIP 4**
> Inventory the external and internal alterations that the task unit undergoes before or during the task, carefully deciding whether each is required, unwanted, or irrelevant, and whether it is artificial or natural.

4.1.3 TASK RESOURCES

Now that we have a clearer notion of the task unit, we consider the next component of the task domain specified in Definition 4.2: the resources used for the task. There are five categories of task resources:

- inducers
- sensors

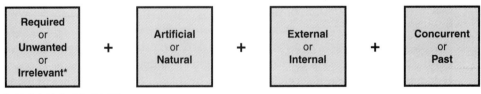

*not included in the task resources

Figure 4.1 Types of alterations and inducers that can affect the task unit

- supervisor
- channels
- domain knowledge

We will examine each of these resources, one at a time.

4.1.3.1 Inducers

> **Definition 4.5** An inducer is a device or mechanism that alters the
> task unit during or before the task.

Inducers are the devices that cause the types of alterations just defined, and the same words may be used to describe the corresponding inducers as well. A *required alteration* of the task unit, for example, is caused by a *required inducer*. Examples of required *external* inducers might include an x-ray machine scanning a suitcase, an infrared light source illuminating a target on the battlefield, a mechanical stress tester applying a load to a steel beam, a centrifuge separating two chemical compounds, a new pharmaceutical drug given to a test subject, and so forth. Examples of required *internal* inducers might include the atoms of a radioactive test sample that emit gamma radiation, the algebraic operators in a proof being constructed by a mathematician, the neurotransmitters in the brain of a rat running a maze, and so forth.

Unwanted inducers (*i.e.,* those that cause *unwanted* alterations of the task unit) are often very difficult to identify and isolate. Examples of *external* unwanted inducers might include the noise that a fan makes in the room while a human subject is performing an intellectual task, winds that buffet a rocket during a flight test, the heat that penetrates the insulation surrounding a vessel containing a liquid to be held at a constant temperature, or the multitasking overhead that steals computing cycles from a computer program designed to measure the speed of an algorithm. Examples of *internal* unwanted inducers might include prejudices in the minds of human subjects that influence their decisions during the task, the organic decay in an apple during an evaluation of its freshness, impurities in a drug being evaluated for effectiveness, or unnecessary code in a computer program that must run as fast as possible.

Once the inducers that cause unwanted alterations of the task unit have been identified (which is not always easy!), they must be dealt with. Often their effects can be minimized or eliminated. For instance, unnecessary code in a program can usually be identified and eliminated, and perhaps the concentration of impurities in a drug can be reduced. But sometimes ignoring or minimizing the effects of unwanted inducers is not possible or extremely expensive. For instance, it is generally not feasible to eliminate a subject's prejudices. Instead, such unwanted inducers must be handled proactively. This is usually done by conducting a series of tasks to measure performance of *the task unit in the absence of the required inducers, the required inducers in the absence of the task unit, and the sensors in the absence of the required inducer and the task unit,* so that the impact of the unwanted inducers, the task unit, and the sensors by themselves can be isolated and measured. These are called **control tasks**. Prejudices of the test subjects in a study, for example, can often be measured using a special control task called a **pretest**. Based on the results of this task, either the material that triggers the prejudicial reaction can be eliminated, or the test subjects who exhibit this prejudice may simply be excused from further participation in the study. When the effects of biased tasks units, noisy sensors, and unwanted inducers have been identified, isolated, and then either minimized, eliminated, or measured and taken into account, then the task is said to be **in control**. More will be said about this important topic in Chapter 12 (Synthesis).

4.1.3.2 Sensors

Just as the inducers alter the task unit, the **sensors** measure the task unit to decide whether or not the objective of the task has been achieved.

> **Definition 4.6** A **sensor** is a device that acquires the required data from the task unit.

Every task employs at least one sensor. Something must be measured or received, or there is no way to obtain results for the task. The sensor may be rather imprecise, like the meat thermometer a chef sticks into a lamb roast, or it may be highly precise, like the micrometer a mechanic uses to measure the thickness of a titanium plate. Regardless of the precision of the sensor, something must always be sensed.

Sensors, of course, need not be mechanical or electronic devices. Very often, the sensors are the eyes, ears, taste buds, or the other senses with which humans or animals are endowed. Psychotherapists observe the body language of their patients under various conditions. Gourmets taste different brands of ice cream to decide which they prefer. Pet food specialists gauge the reactions of a cohort of dogs as they are offered different brands of dog food. Physicians listen to their patients describe their symptoms. Clothing designers feel the texture of a new fabric.

The opinions of humans or other animals are usually referred to as **qualitative** data, while counts or physical quantities, such as votes or lengths or temperatures, are usually referred to as **quantitative** data. There is nothing improper or unscientific about tasks that acquire qualitative data, such as opinion surveys.

They are often extremely important data in an investigation. It simply must be understood that opinions obtained from humans or other animals are *extremely susceptible to unwanted alterations,* which is another way of saying that their opinions may well change with changing conditions, consciously or unconsciously, depending, for example, on what they had for breakfast or the weather that day. Although electronic and mechanical sensors are generally very consistent and reliable because they are usually free from such unwanted influences, human and other animal subjects are not, and thus are often *very difficult* to deal with. To make matters worse, even electronic measuring instruments are often read directly by humans who can easily misread them or even "fudge" the readings, albeit unconsciously. The task chain is only as strong as its weakest link. Much more will be said about this in Chapter 12 (Synthesis).

Finally, sometimes the sensors for a task also act as unwanted *inducers* whose alterations of the task unit cannot be ignored. This is especially true with human and animal subjects, who are often keenly aware of the sensors and change their behavior based on this knowledge. For example, human subjects often exhibit uncharacteristic behavior when they are aware that they are being videotaped as part of an experiment. Some become narcissistic; others become shy and unresponsive. Most other animals that are fairly high on the cognitive ladder (primates, dolphins, dogs, rats, *etc.*) exhibit similar psychological behavior patterns as pet owners and zookeepers know very well. Such behavior can taint the results of the task, rendering the conclusions questionable or even meaningless.

Electronic and mechanical sensors can also affect inducers or other sensors. Electronic sensors often emit radio frequency interference (RFI), which can disturb other electronic instruments. Mechanical sensors sometimes emit acoustic noise and/or heat, which can affect other sensors, the inducers, and the task unit itself. Fortunately, assuming that such "leaky" sensors can be identified (which is by no means guaranteed!), their emissions can usually be measured through control tasks and minimized or eliminated, or failing that, their effects can be taken into account.

> **TIP 5**
> Be cautious with tasks using humans or other animals. They are usually very difficult and expensive to manage and control, and often introduce complex ethical and legal issues that must not be overlooked.

Before we continue with more definitions and discussion of the remaining task resources, perhaps some illustrative examples will help to consolidate the concepts presented so far. Each example represents a different kind of task in a different discipline with its own special issues. We will use these tasks as running examples periodically throughout the book to illustrate the methods and concepts. For each of these sample tasks, the task unit and principal task resources (except knowledge) are summarized in Table 4.1. Note that each inducer is followed by a four-character code, which defines the inducer as (R) required or (U) unwanted, (A) artificial or (N) natural, (E) external or (I) internal, and (C) concurrent or (P) past. In addition, a code in the name field designates each task as either a research task (r) or a development task (d).

Remove Screws. Consider the subtask *Remove Screws* in the project *Repair Squeaky Hinges* (see Figure 3.2). The **task unit** for this task was the set of screws, because their alteration (their removal) was required to accomplish the task. The required **inducer** for the task was the screwdriver, because it caused the required task-unit alteration (artificial, external, concurrent). In addition, the screw threads functioned as an **unwanted inducer** (natural, internal, concurrent), inhibiting the removal of the screws through friction. In fact, because the effect of this unwanted inducer was so strong for one of the screws, it caused a chip on the tip of the screwdriver to break off, necessitating the temporary suspension of the current task, *Repair Squeaky Hinges*, so that a new project, *Repair Screwdriver*, could be launched.

What was the sensor for this task? What device gathered data from the task unit (the screws)? The answer is, of course, that the homeowner himself was the **sensor**. First, as he unscrewed each screw, he watched it and felt it until it was loose enough to pull out of the hole (qualitative data). Second, when he counted that all six screws on the first hinge had been removed (quantitative data), he moved to the second hinge and repeated the process. Finally, when his count told him that all the screws from the second hinge had been removed (quantitative data), the task *Remove Screws* had come to a successful conclusion.

Launch Spacecraft into Orbit. As part of Project Mercury in 1963, NASA launched the first manned flight in Earth orbit. The capsule, *Friendship 7,* piloted by astronaut John Glenn, was launched by an Atlas rocket from Cape Canaveral. The capsule orbited the Earth three times before reentering the atmosphere and splashing down safely in the Atlantic Ocean. Consider just one of the many tasks undertaken for this enormous project: *Launch Spacecraft into Orbit*. The **task unit** was the spacecraft (including John Glenn), which was altered by two **required inducers**: the Atlas rocket (artificial, external, concurrent), and gravity (natural, external, concurrent).

The most powerful **unwanted inducer** during the rocket launch was the atmosphere that caused friction and buffeted the rocket (natural, external, concurrent). However, the emotions of John Glenn were also an **unwanted inducer** (natural, internal, concurrent) that could not be dismissed as irrelevant. In case of a catastrophic failure, NASA realized that this unwanted inducer might well play an unpredictable and pivotal role, even though John Glenn was a highly experienced test pilot. Over the years, the psychological effects of this unwanted inducer were very carefully studied by NASA, and all the astronauts were thoroughly trained to handle emergencies with a steady hand. During the emergency aboard Apollo 13, these studies and the training paid off!

The **sensors** for the task *Launch Spacecraft into Orbit* included the radar systems that tracked the position of the launch vehicle (quantitative data), the

Table 4.1 Task units and major resources for several sample tasks

Task Name		Task Unit	Inducers		Sensors		Supervisor	Channels
	CODE			CODE		CODE		
Remove Screws	d	screws	screwdriver thread friction	RAEC UNEC	fingers counter	L N	homeowner	fingers
Launch Spacecraft into Orbit	d	spacecraft	rocket gravity atmosphere emotions	RAEC RNEC UNEC UNIC	radar attitude fuel quantity John Glenn	N N N L	ground personnel ground computers John Glenn onboard computers	telemetry cables fingers
Identify Faster Traffic	r	speed samples	average standard deviation	RAEC RAEC	standard error confidence	N N	traffic engineer computer	keyboard screen
Prove Fermat's Last Conjecture	r	conjecture	math tools emotion fatigue	RAIC UNIC UNIC	eyes	N	Andrew Wiles	mind pencil & paper
Find New Planet	r	region of predicted position	gravity	RNEC	telescope camera	N	astronomer	space atmosphere telescope
Identify Collision Damage	r	car	reported accident other accidents normal wear & tear	RAEP UAEP UNEP	eyes	L	insurance adjuster	car frame
Measure Fossil Age	r	cliff face	sedimentation erosion	RNEP UNEP	eyes	N	geologist	water air
Measure Drug Effectiveness	r	cohorts	drug unrelated sickness stress awareness	RAEC UNIC UNIC UNIC	blood test	N	computer medical staff	hypodermic sample vial
Evaluate GUI Effectiveness	r	GUI	user actions user physical state user mental state lab conditions display conditions	RAEC UNEC UNEC UAEC UAEC	computer questionnaire	N L	computer technician	mouse keyboard screen language
Compile Summaries	d	database	articles distractions	RAIC UNIC	student	L	student computer	mind internet
Measure Liquid Temperature Variation	r	liquid	refrigerator external heat thermocouple	RAEC RNEC UAEC	thermocouple	N	computer	cooling coils vessel walls cables
Measure Age of Organic Sample	r	organic sample	carbon-14 atoms radiation sources	RNIC UNEC	radiation detector	N	computer	vacuum air

Task Codes:
d = Development Task
r = Research Task

Inducer Codes:
R = Required or U = Unwanted
E = External or I = Internal
A = Artificial or N = Natural
C = Concurrent or P = Past

Sensor Codes:
N = Quantitative or L = Qualitative

sensors that continuously monitored the state of the rocket (quantitative data), the onboard spacecraft sensors that monitored John Glenn's vital signs (quantitative data), and John Glenn himself, who observed and recorded the experience of space flight from the human perspective (qualitative data). In this case, note that an element of the task unit, John Glenn, is also a sensor. The assumption here was that the effects of the inducers (both required and unwanted) did not affect the accuracy of his sensors. Although this may be reasonable with a highly experienced and trained test pilot, such assumptions can be very precarious in experiments with typical human subjects!

Note that if both the spacecraft *and* the rocket had been included in the task unit, the *only* required inducer would have been the force of gravity, and the objective of the task would presumably have included measuring the performance of both the rocket and the spacecraft. However, the Atlas rocket used for Project Mercury had already been developed for other purposes by the United States Air Force, and therefore its development was not a part of the current task; the Friendship 7 spacecraft was simply plopped on top of an off-the-shelf launch vehicle, so to speak. Later, during the Apollo Program, NASA developed its own specially designed rocket, the Saturn V. For this subsequent reenactment of the task *Launch Spacecraft into Orbit*, the task unit included the rocket and the spacecraft, both of which required extensive research and development.

Identify Faster Traffic. As mentioned previously, the task unit need not be a device or even a physical object. Often it is an abstraction, such as data, or an algorithm, or a philosophical concept. To illustrate this common circumstance, consider a research project whose objective is to decide which of two different highways should receive additional funding for widening the road. One criterion for making this decision is to know which highway has the higher average traffic speed. To achieve this objective, a previous task for this project measured and recorded random samples of the speeds of 100 vehicles for each highway. Following this measurement task, a traffic engineer is then asked by her boss to decide which highway has the higher traffic speed.

To accomplish this computational task *Identify Faster Traffic*, the traffic engineer calculates the average and standard deviation of each sample of speeds. These results allow her to estimate the statistical confidence that the traffic on one highway is traveling at a significantly higher average speed than the traffic on the other highway. The **task unit** is the set of two samples of 100 vehicle speed measurements, one for each highway. The traffic engineer alters the task unit with two **required inducers** (artificial, external, concurrent): 1) the transformation of each of the samples of 100 speeds into its average; and 2) the transformation of each of the samples of 100 speeds into its standard deviation.

Could there be any **unwanted inducers** during this task? Probably the only candidate is some diversion that causes the engineer to make an arithmetic error during the computations, although with the excellent and easy-to-use computer-based computational tools that are available these days, this seems unlikely. However, before the advent of computers and spreadsheets, such errors were extremely common, and if the task was critical, often much effort and expense was expended to check the computations by asking several people to repeat them independently. The subject of acquiring and using appropriate and efficient tools for R&D projects was covered in Section 2.2.

Sensors are not always mechanical or electronic devices. Sometimes they are quantitative or qualitative observations of abstract information made by humans or computer software. The **sensor** for the task *Identify Faster Traffic* is a computer program (perhaps an Excel spreadsheet) that computes two averages and two standard deviations, which are the **results** of the task. From these results, a **performance metric** is computed, namely the standard difference between the two averages, expressed as a z-value. Then, based on a couple of assumptions about the distributions of the data samples (called **governing propositions**, which will be introduced in Chapter 7), the engineer may convert the z-value into a statistical confidence.

This statistical confidence is *not* the conclusion of the task (although it could have been); the conclusion requested by the traffic engineer's boss is a *decision* about which highway has faster traffic. The process of arriving at a **conclusion** for a task or a project is carried out in the Validation Phase of the Scientific Method, which will be covered in detail in Chapter 13 (Validation). In this case, the traffic engineer will presumably use the statistical confidence to *decide* whether the difference in the average traffic speeds on the two highways is too large to be attributed to chance. For instance, if the confidence turns out to be 99%, the decision may be straightforward. However, if the confidence is only 85% (which is quite low by conventional standards), this may suggest that the difference between the two average speeds is not significant. In general, someone must decide what level of significance is sufficient to achieve the objectives of the task without running an unacceptable risk of committing a Type I error — someone who has the knowledge, authority, and responsibility to make such decisions.

Prove Fermat's Last Conjecture. This is an appropriate place to reemphasize that the principles and methods presented in this book are *not* limited to projects that have concrete task units. They can be applied equally well to the most abstract intellectual investigations that take place entirely in the mind. Consider, for example, the mathematical conjecture known as "Fermat's Last Theorem" proposed by the French mathematician Pierre de Fermat in 1665, which states:

> For $n > 2$, there are no non-zero integer values of x, y, and z
> that satisfy the equation $x^n + y^n = z^n$

The proof of this conjecture[4] eluded mathematicians for over 300 years until 1993, when Andrew Wiles of Princeton University, published an elegant

---

[4]As will be discussed in Chapter 6, a **theorem** is a mathematical statement whose validity has been rigorously proved. Until it is proved, the *proposed* theorem is called a **conjecture**. Although Fermat claimed to have proved his conjecture, he did not record this proof (which he stated could have been jotted down in the margin of a single page of his text!). Therefore, it seems more appropriate to call it "Fermat's Last Conjecture."

proof after a decade of exhausting work. The **task unit** for the task *Prove Fermat's Last Conjecture* was the equation $x^n + y^n = z^n$ subject to the conditions stated in the preceding text. The **required inducers** were the tools of modern mathematics available to Wiles (artificial, internal, concurrent), which he used to alter the task unit by transforming it into equivalent forms that finally led to a solution. The **sensor** was, of course, the conscious logical recognition (quantitative) by Wiles himself during the project that his proof was on-track or not on-track, so that finally he could confidently pronounce those victorious words in mathematics: *Quod Erat Demonstratum* (which was to be demonstrated).

But even though Andrew Wiles "sensed" that the proof was complete and correct, that was not sufficient. Only after it had been rigorously reviewed and eventually accepted by the international mathematics community was the proof **validated**. The process of **peer review** is perhaps the most critical (and most modern!) component of the Scientific Method. It is presented in detail in Chapter 13 (Validation).

We know from Andrew Wiles himself that several **unwanted inducers** affected his work, including exhaustion, frustration, and discouragement, as well as a modicum of jubilation now and then (natural, internal, concurrent), while he wound his way through the labyrinth of mathematical options. Although he was faced with many dead ends and roadblocks during his journey, the effort paid off in the long run. If there were a Nobel Prize for Mathematics (which Alfred Nobel intentionally proscribed, maintaining that the study of mathematics was largely a waste of time), Andrew Wiles would have surely won this coveted award for his elegant work.

Find New Planet. At the beginning of the twentieth century, astronomer Percival Lowell (1855–1916) calculated the angular momentum of the known bodies of the Solar System and concluded that it was not conserved. Because the laws of physics will not allow that, the implication was that the data was incomplete, *i.e.,* there was a large amount of undiscovered mass in the Solar System. Assuming that this mass was concentrated in a single planet, Lowell computed the most reasonable orbit for this planet. In 1929 Clyde Tombaugh, a young amateur astronomer from Kansas, was hired as an assistant at the Lowell Observatory in Flagstaff, Arizona. His task was to search for evidence of a planet in the predicted orbit, which was his **task unit**. Using the camera of the 13-inch refracting telescope (**sensor**), Tombaugh photographed the sky night after night, for months on end, while gravity, which was the required **inducer** (natural, external, concurrent), played its music of the spheres. After analyzing hundreds of photographs using a new device called a blink comparator, on 18 February 1930 Tombaugh found indisputable evidence of a ninth planet. His discovery was soon corroborated by other astronomers, and the new planet was named Pluto.

Identify Collision Damage. Following a traffic accident, an insurance adjuster uses his eyes (**sensor**) to inspect a client's car (**task unit**) to decide how much his company will pay to have the damage repaired. The **required inducers** for this task are the object(s) that altered the task unit, *i.e.,* caused the damage to the client's car (artificial, external, past). For many years, the insurance companies were very concerned about the source and cause of the damage to the car. Was it caused by another car running into the client's car, or *vice versa*? Was the damage caused by *this* accident or a *previous* accident? Today most companies neatly sidestep the problem of sorting this out by writing "no-fault" insurance policies, making claims generally much quicker and easier to service, and therefore less expensive. In terms of the definitions presented here, the insurance companies simply decided it would cost less in the long run to classify these inducers as *irrelevant,* instead of *unwanted.* (Anthropologists and archeologists cannot be so efficiency-minded. As they sift patiently through physical evidence bit-by-bit, they are constantly concerned about what caused what, and what came before what. Cause-and-effect is everything to them.)

Measure Fossil Age. A geologist wants to measure the age of a fossil she finds inside a rock on a cliff face. She does this by climbing down the cliff face (**task unit**), using her eyes (**sensor**) to determine the height, kind, and number of layers of sedimentation from the top of the cliff to the level at which she found the fossil. The **required inducer** is the geological process by which the ocean deposited layer upon layer of sedimentation over millions of years (natural, external, past). The geologist knows the results will be distorted by an **unwanted inducer**, the erosion of the cliff face by air and water for hundreds of millions of years (natural, external, past).

Measure Drug Effectiveness. A large pharmaceutical company has asked its medical staff to conduct a study to measure the effectiveness of a new cancer drug. The **task unit** is a cohort of human subjects. Half are treated with the new drug, and half are given a placebo as a control. The study is double-blinded, which means neither the subjects nor the medical staff treating them know which subjects have received the drug or the placebo. The **required inducer** is the drug under test (artificial, external, concurrent). The **sensor** is the device that tests the samples of blood taken from each subject. The **unwanted inducers** include unrelated illnesses that afflict the subjects (known or unknown to the staff or the subject), high levels of emotional stress in the subjects (who are all very ill), and the *suspicion and guesses by each subject and each staff member about who is probably receiving the drug and who is not.* This last unwanted inducer, which severely jeopardizes a double-blinded protocol, is of enormous methodological importance, because there is almost no way to eliminate or minimize it. Based on their improving or worsening health during the study, humans quickly guess (rightly or wrongly) whether they are receiving the drug or the placebo. The distortions introduced by this stress often invalidate the conclusions of the study, especially when the potential benefits of the drug are modest and not very consistent. This difficult problem is discussed in Chapter 12 (Synthesis).

Evaluate GUI Effectiveness. The manufacturer of an Internet browser wants to evaluate how well a new graphical user interface helps its users surf the Internet. At first glance, this seems very much like the previous task *Measure Drug Effectiveness*, but there is an important role reversal. For the task *Measure Drug Effectiveness*, the task unit is the cohort, in which each subject is altered by the required inducer, the new drug. On the other hand, for the task *Evaluate GUI Effectiveness*, the **task unit** is the GUI, not the cohort. The subjects are asked to configure (alter) the GUI to make it easier for them to accomplish a specified task, *e.g.,* to search the Internet to compare the features and prices of several brands of VCRs. Each subject may create buttons to access sites, open new windows, resize them, set up lists, submit questions to the sites, and so forth. Thus, each subject in the cohort, which is selected at random from the general population, is the **required inducer** (artificial, external, concurrent). The subject (**sensor**) is connected to the GUI via the keyboard, the mouse, and the display screen. Anything the subject might do during the test, even if highly inappropriate (such as banging on the keyboard in frustration), is fair game for evaluation of the interface. The **unwanted inducers** include the ambient conditions of the laboratory and the CRT display (artificial, external, concurrent) and the variation in the subjects' physical and mental states (natural, external, concurrent). Every keystroke and mouse click made by each subject is recorded, and a video recording is made of the subject's interaction with the computer and the GUI. At the beginning of each trial, demographic information about the subject is recorded, and at the end of each trial, the subject's opinions about the GUI's effectiveness are solicited using carefully prepared questionnaires.

Compile Summaries. As part of the Analysis Phase of her research project, a doctoral student in psychology is constructing a database (**task unit**) of summaries of scholarly articles, books, and reports on Multiple-Personality Disorder published since 1850. The student (**sensor**) reviews each relevant article (**required inducer**) (artificial, external, concurrent) and then summarizes it into a few qualitative categories that she has defined for the database. She continually reviews each summary to make sure it is complete and accurate, occasionally revising her summarization categories to capture the significant information in a consistent manner. Distractions, emotions, and fatigue are powerful **unwanted inducers** (natural, internal, concurrent).

Measure Liquid Temperature Variation. Based on complaints of temperature instability, an engineer wants to measure the variation in the temperature of a liquid (**task unit**) stored inside a refrigerated vessel. The refrigeration engine (artificial, external, concurrent) and the heat source outside the vessel (natural, external, concurrent) are the **required inducers**, which, when properly balanced, should maintain a steady-state temperature of the liquid. The thermocouple, infrared (IR) detector, and thermometer that the engineer uses to measure the temperature of the liquid in the vessel are the **sensors**. But the sensors also act as **unwanted inducers** (artificial, external, concurrent), because they transfer heat to the liquid from the cables that *must* penetrate the vessel wall to connect them to the computer in the laboratory.

Measure Age of Organic Sample. An archeologist wants to measure the age of an organic sample (**task unit**) that has been unearthed from an archeological dig. The intensity of the gamma radiation of the radioactive isotope carbon-14, which is present in all dead organic materials, varies inversely with the age of the organic sample. The **sensor** for this task is a radiation detector, that is used to measure the intensity of the radiation from the carbon-14 atoms in the sample, which are the **required inducers** (natural, internal, concurrent). However, the radiation detector will also measure the intensity of the radiation from all the gamma sources in its field of view, which includes the entire universe. These external and spurious sources of gamma radiation are **unwanted inducers** (natural, external, past) and cannot be eliminated entirely, because no amount of shielding can prevent some radiation, albeit very little, from entering the detector. The older the sample, the less accurate the age measurement will be, because the less intense the gamma emissions of its carbon-14 atoms, the more likely these emissions will be buried in the background radiation. Because the half-life of carbon-14 is about 5700 years, this technique is limited to the measurement of ages less than several tens of thousands of years.

> **Exercise 4.1**
> According to the legend for Table 4.1, there are 16 possible codes for the types of inducers, but only 11 of them are used in the table. For each of the five codes not used in this table (RAIP, RNIP, UNIP, UAIP, and UAIC), describe a task that might include an inducer of this type.

> **Exercise 4.2**
> Describe a typical task for each of the following occupations, specifying the task objective, task unit, inducers, and sensors for each: carpenter, teacher, student, doctor, parent, banker, philosopher, writer, politician, soldier, sculptor.

Having defined the task unit, sensors, and inducers, we can now define a very important concept:

Definition 4.7 The **system** for a task is the intersection of the regions altered by the required inducers and the regions sensed by the sensors, as long as this intersection contains the task unit.

The word "intersection" has a precise mathematical definition; it specifies a subregion that two or more regions share in common. For physical regions, the noun "overlap" is synonymous with "intersection." However, the application of the word "intersection" is not limited to physical regions; it can be applied equally well to abstract spaces. For instance, the intersection between the set {1,2,3,4,5} and the set {1,3,4,7,8,9} is the set {1,3,4}, which are the symbols that the two sets share in common. Once the intersection has been found, the task unit must be present in the region of intersection, or the system is not completely defined. Figure 4.2 illustrates a simple system formed by the intersection (overlap) of the region altered by a single inducer and the region sensed by a single sensor.

The main purpose behind this rigorous definition of the system for a task is to *establish an unambiguous boundary that separates the region containing the task unit from the rest of the universe,* which is called the task **environment**. This tessellation of the task domain allows us to identify (and possibly minimize or eliminate) any unintended and unwanted influences that could cross the system boundary, carrying spurious information and energy in and out of the system and thus possibly interfering with the task. It also allows us to identify (and possibly minimize or eliminate) elements inside the system *other than the task unit* that will be unintentionally induced and sensed, becoming possible sources of error in the results of

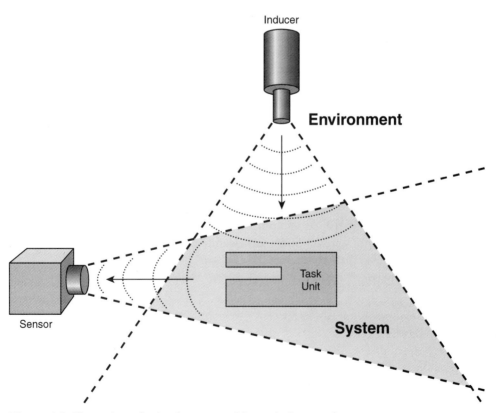

Figure 4.2 Illustration of a simple system with one inducer and one sensor

the task. Often the existence of these spurious influences and unwanted elements is unsuspected until a comprehensive search is undertaken during a control task.

In the simple example in Figure 4.2 it is very easy to determine the required intersection and to ascertain that the task unit is contained within this intersection, thus defining the task system. For complex tasks, however, it is often much more difficult to determine the intersection and the relative location of the task unit. To illustrate this, let us consider a few of our running examples.

Remove Screws. The system for the homeowner's task was the region defined by the tip of the screwdriver (required inducer) and the homeowner's field of vision (sensor), as long as it contained the screw (task unit). As the homeowner visually guided the tip of the screwdriver toward the task unit before the screw actually came into sight, the system for *this* task did not exist, because the required intersection did not yet contain the task unit. A system did, however, exist for another task *Find the Screw*, which was really the homeowner's task at the moment. Then, as soon as the tip of the screwdriver slipped into the screw slot under the critical eye of the homeowner, the system for the task *Remove Screws* sprang into existence, and this task officially began. Note that the advice "keep your eye on the ball" is just a more literary way of saying "maintain the integrity of the system."

Launch Spacecraft into Orbit. For this task, the system is defined by the intersection of the points of application of the forces of the Atlas rocket and gravity (required inducers) and the fields of view of the radar tracking equipment, the onboard sensors, and John Glenn's senses (sensors). The requirement that the task unit (the spacecraft) be contained within the system is easily satisfied in this case, because the spacecraft was rigidly attached to one of the required inducers (the rocket) and could not escape the influence of the other required inducer (gravity). The determination of the boundary of this system is somewhat simplified if the significant influences of the required inducers (the rocket thrust and the force of gravity) are assumed to be limited to the structure contained within the physical envelope of the spacecraft and the launch vehicle. This

assumption is probably fairly accurate, because neither inducer could have had much effect on anything external to and surrounding the launch vehicle and its payload, *i.e.,* the atmosphere during liftoff or the vacuum of space shortly thereafter. Under this assumption, the rigid shape of the launch vehicle and spacecraft completely determine the boundary of the system. Such an assumption permits the R&D personnel to mount a well-organized and methodical search of a well-defined region for potential sources of error and distortion.

It was stated previously that a good way to make sure a task is in control is to mount a control task to observe the behavior of the task unit, required inducers, and sensors in isolation. The task *Launch the Spacecraft into Orbit* is a perfect example of when such advice is *not* very useful. Although it is easy to remove the influence of the rocket on the task unit (just don't light it off), it is very difficult to avoid the influence of gravity. In fact, it is for exactly this reason that NASA spent so much money on swimming pools to train their astronauts under simulated conditions of weightlessness. But clearly, such techniques could not be applied to the rocket or the spacecraft. There just was no easy way to use control tasks to measure the effects of the unwanted inducer, the atmosphere, on the task unit in the absence of both thrust and gravity. Despite the enormous amount of time and money spent on ground-based stress tests, giant wind tunnels, and atmospheric sounding experiments, it is no wonder that NASA had so many catastrophic failures during their early rocket tests (take a look at the films sometime!).

Identify Faster Traffic. The system for the traffic engineer's task is the intersection of the average and standard deviation transforms (required inducers) and her statistical confidence metrics (sensor), as long as this intersection contains the two samples of vehicle speeds (task unit). Note that the system in this case is not a physical region, *per se;* it includes an external set of data, a computer program that calculates the averages and standard deviations, and the traffic engineer's conclusions based on the statistical confidence metric. Try to visualize this system boundary! This situation highlights one of the great difficulties in many research tasks: isolating the system so that unwanted influences of the task resources can be identified and avoided, and the task can be kept in control. For instance, it is clearly infeasible to erect any barriers between the relevant intellectual process and the other emotional and cognitive activities that are going on in the mind, such as preoccupation, proprioceptive distraction, faulty knowledge, and emotional stress, to name just a few. Nevertheless, at the time the project was planned, a thorough analysis of the engineer's mental state might have allowed the task planners to eliminate some known extraneous influences, correct some subtle lack of knowledge, or, if necessary, assign the task to someone else who was less encumbered by such disturbances or flaws. But clearly, such an approach raises a number of thorny ethical issues!

Find New Planet. The system for Clyde Tombaugh's task was defined by the intersection of the field of view of the Lowell Observatory telescope (sensor) and the gravitational fields (inducer) acting on the task unit (the predicted position). Because gravitational fields exert their influence throughout the universe, the system included everything the telescope could see, a cone spreading out from the focal point of the telescope's lens to the limit of the resolving power of the telescope (many light-years). Within this vast volume of space, Tombaugh had to search for a tiny point of light that inched slowly, from night to night, across the brilliant profusion of billions of fixed stars. No one works with larger systems than astronomers.

Evaluate GUI Effectiveness. The most significant parts of the system boundary, *i.e.,* the intersection of the subject's actions (the inducer) and the subject's observations recorded on the questionnaire (the sensor), are the CRT display screen, the mouse, and the keyboard. The rest of the boundary of the system is the physical enclosure of the workstation, which contains the task unit, *i.e.,* the GUI software and the browser application.

Measure Age of Organic Sample. Unlike the astronomer Clyde Tombaugh, the archeologist can collimate his sensor (a radiation detector) and bring it arbitrarily close to his task unit (the organic sample). Because the required inducer is internal to the task unit, the resulting system volume is tiny. Therefore, by enveloping his tiny system with very effective shielding so that the background radiation is greatly attenuated, the researcher can *almost* eliminate the effect of this unwanted inducer entirely. Unfortunately, a collimated detector placed very close to the sample will exhibit a tiny detection cross section, which decreases the total flux of the required inducers, the carbon-14 atoms. There are always design tradeoffs.

4.1.3.3 Supervisor

> **Definition 4.8** The **supervisor** for a task is the set of human and automated agents that operates and monitors the task unit, the concurrent inducers, the sensors, and itself.

Clearly, the task unit, sensors, and the required inducers must be supervised during a task. A task supervisor has two major responsibilities: 1) sequencing and monitoring the **parameters of the system**, including the task unit and the system boundaries; and 2) sequencing and monitoring the **conditions of the environment**, including the concurrent inducers, the sensors, and itself. The agents that perform these supervisor functions can include humans, computer programs, electronic devices, and mechanical devices. Note that the supervisor functions are applied only to the *concurrent* inducers for the task, *i.e.,* those inducers in Table 4.1 whose descriptive code strings end with the letter C. Past inducers, although just as important, obviously cannot be monitored by the task supervisor at the time the task is being performed.

For simple tasks, very often all supervisor functions are performed by the task personnel, because limited funds proscribe more complex and expensive mechanisms. There is a danger, however, in allowing the task personnel to supervise their own tasks; their knowledge of the desired outcome of a task can unconsciously influence the way in which the task is carried out and, thereby, possibly distort the results of the task. This is a very common source of error in research and development tasks. It is far better for the investigators to design an unbiased and fair procedure for the task ahead of time, called the task **protocol**, and then let the protocol be applied either automatically by a mindless machine or manually by technicians whose knowledge about the objective of the task has been purposely limited. Such a strategy, called **blinding** the task, helps to ensure that the task is in control. Much more will be said about controlling sources of error in Chapter 12 (Synthesis). The next to last column in Table 4.1 lists some of the principal supervisor agents for each task. Several are worth brief discussion.

Launch Spacecraft into Orbit. The supervisor for this task was enormously complex, comprising a massive network of computers and technical personnel at NASA Mission Control in Texas, Cape Kennedy in Florida, and onboard the spacecraft. The astronaut John Glenn supervised some of the functions of the Mercury capsule and the Atlas rocket, but not very many. He was primarily just along for the ride to achieve an objective that was as much political as technological.

Find New Planet. Clyde Tombaugh exerted manual supervision over this task, which took place in 1930 long before the advent of computers (which, he later admitted, would have made his task much easier!). His supervisor functions were limited to the operation of the task unit (the equations that predicted the hypothesized position of the new planet) and the sensor (the telescope at the Lowell Observatory). Although he could not exert any influence over the major required inducer of the task (gravity), the conclusions of the extensive research that had preceded his task assured him that this inducer was predictable, well-behaved, and in control.

Identify Collision Damage. The supervisor functions of the insurance adjuster are limited to operating the sensor (his eyes); the required inducer (the traffic accident that caused the damage) is out of the realm of his supervision, because it happened in the past. On the basis of his sensor observations, he must make sure that the damage was caused by the alleged traffic accident, and not simply as a result of normal wear-and-tear, which the insurance policy does not cover. Having done that, he must then accurately estimate the cost of repairing the damage. It is safe to assume that he has been well trained to achieve both objectives.

Measure Fossil Age. Like the insurance adjuster, the geologist as the supervisor of this task has no influence whatsoever over the required inducer, which was the gradual process of sedimentation over countless millennia. Unlike the astronomer Clyde Tombaugh, however, the geologist has only very tentative **knowledge** about the behavior of this natural inducer, based on studies of similar sedimentation processes at geological sites where some other independent method of dating was available. These studies allowed the geological community to **postulate** a sedimentation model that is, hopefully, reasonably accurate and generally applicable.

Evaluate GUI Effectiveness. The supervisor for this task includes two agents: the computer that runs and monitors the experiments with the cohort of test users, and the technician who must work with the subjects before the formal experiment begins, introducing them to the browsing exercise, familiarizing them with the equipment, and so forth. It is essential that the ergonomicist who designed the user interface and the technician who supervises the experiment are *not* the same person, because the ergonomicist could and would bias the results of the experiment by unconsciously alerting the subjects to difficult traps and time-saving options intentionally built into the scenario of the exercise. In fact, to ensure that the experiment is in control, very strict isolation protocols must be established and enforced during the conduct of the experiment to protect the subjects from unwanted inducers. In general, it is important to remember that a task may have only *one* supervisor, and even though this supervisor may include several agents, arrangements must be made so that these agents can and will act independently or in concert, as appropriate, to achieve the objective of the task. In large projects where the supervisor function includes several humans, such coordination is often compromised by politics, rivalries, hidden agendas, and personality conflicts. This important subject is discussed in detail in Chapter 12 (Synthesis).

Measure Liquid Temperature Variation and Measure Age of Organic Sample. These are examples of tasks that can be supervised entirely by a computer, which goes a long way toward ensuring unbiased operation and measurement. As an added benefit in its role as supervisor, a computer has the unique and enviable ability to check its own efficacy by running periodic system diagnostics. On the other hand, humans are notoriously poor at accurate cognitive self-diagnosis, and worse yet, are often firmly convinced of their objectivity and lack of bias. To expect reliable and unbiased self-checking protocols from human supervisors is unrealistic. However, a technique called a **randomized block design** can help overcome some of the difficulties in tasks that are vulnerable to human biases. This topic is covered in detail in Chapter 12 (Synthesis).

4.1.3.4 Channels

> **Definition 4.9** The **channels** in a task domain interconnect the resources of the task to provide the means for exchanging energy and information among them.

It is clear that the various components of the task domain must be interconnected via a set of **channels** to transmit and receive information and to supply the energy necessary for some devices. Clearly, the inducers and the task unit must be connected with channels to transmit the forces and energy that alter the task unit. Similarly, the task unit must be connected to the sensors with channels to transmit the responses of the task unit caused by the actions of the inducers. In addition, the supervisor must be connected to the inducers, the sensors, and the task unit to transmit operational and sequencing commands to these devices and to receive status information from the devices. Electrical power must also often be provided to various devices. Finally, although often unwanted, extraneous sources of energy, such as heat and acoustical noise, are transmitted and received via channels that connect the various devices in the task domain.

It may seem unnecessary to place such special emphasis on these channels; after all, how could the devices communicate if they were not connected together, and how could the devices operate without power? However, one glance at Figure 4.3, which is a duplication of Figure 4.2 with several examples of typical information and energy channels added, should highlight their importance. Many of these channels must cross the system boundary, which can threaten the necessary isolation

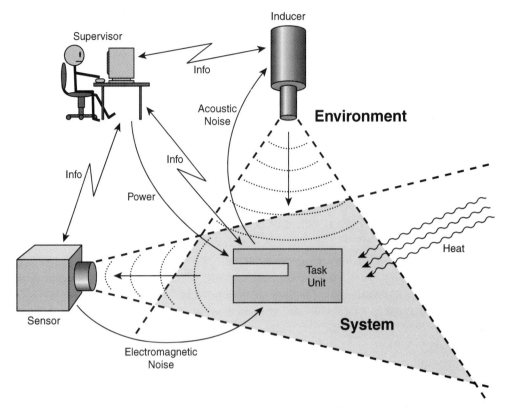

Figure 4.3 A simple system connected to its environment by information and energy channels

of the system from the environment. Such channels can introduce spurious signals that interfere with the operation of the system elements and distort the results of the task. Suppose, for example, that the task unit in Figure 4.3 makes a loud sound during normal operation. If this acoustical noise crosses the system boundary, it can cause the inducer to vibrate sympathetically and fail to cause the correct alteration of the task unit. A common example of this kind of interference occurs when the microphone of a public address system is inadvertently placed in front of the loudspeakers, and unintended feedback causes an electronic shriek that drowns out the speaker. Placing the microphone in front of the loudspeakers is equivalent to enlarging the field of view of the inducer to include the sensor, which is not a good idea.

No system can be completely isolated from its environment. At the very least, something about the task unit must be measured during the task, or nothing is accomplished by the task. And when something about the task unit is measured, information must pass across the system boundary to a sensor in the environment. In addition, the unwanted transmission of energy to and from the system cannot be eliminated entirely. There is no such thing, for instance, as perfect heat insulation, so the heat energy will unavoidably flow down the temperature gradient as nature strives to maintain thermodynamic equilibrium. The best we can hope to do is to inhibit this unavoidable heat transfer sufficiently so that the effects of the energy losses or gains during the task are less than the required precision of the temperature sensor measurements.

Once again, let us consider a few of our running example tasks in Table 4.1 that employ unusual or interesting channels.

Remove Screws. The principal channel that connects the supervisor and the sensor for this task (the homeowner) with the task unit are the fingers of the homeowner. Anyone who has ever removed screws with a screwdriver knows that the eyes are not a very important sensor for this activity; it is done primarily by feel. Even as the homeowner counts the number of screws he has removed, the major channels for gathering this data are his fingers.

Identify Faster Traffic. Like the homeowner, the traffic engineer is connected to her task unit (car speed samples) and sensors (average and standard deviation functions), both of which reside in the computer, via her fingers, which operate the computer keyboard. All of us know full well how easily characters may be reversed or mistyped while entering information on a computer keyboard. Often such mistakes are detected by the computer as statements whose syntax is illegal, but almost as often such typing mistakes just happen to result in perfectly legal commands that are duly, and inappropriately, executed by the computer. If the results are unreasonable, the supervisor may recognize this fact and reenter the correct commands. Now and then, however, the incorrect entries are *not* detected by either the computer or the supervisor, and the errors are accepted as *bona fide* results of the task. From this source of error was born the familiar programmer's lament: "Computer, please do what I mean, not what I say!"

A similar kind of error was committed in 1999 when the Mars Climate Orbiter was sent a series of navigation commands expressed in English units instead of metric units, causing the $125 million spacecraft to enter the Martian atmosphere at the wrong angle and either crash on the surface or burn up in the atmosphere.[5]

Prove Fermat's Last Conjecture. As pointed out previously, the construction of Andrew Wiles' proof of Fermat's Last Conjecture took place primarily in his mind. However, as all of us who have toyed with mathematics problems know, the use of pencil and paper to support this intellectual process is essential. We also know that a momentary lack of focus and concentration can easily cause an inadvertent transposition of a sign or a similar syntactic error, which, when propagated downstream, can taint many hours and pages of laborious derivations. On the other hand, using other people to periodically check our derivations is invaluable! Thus, the channels that connect the various parts of our minds to each other and to the outside world are both useful and distracting. The trick is to know when and how to turn these channels on and off, *i.e.,* to keep our cognitive tasks in control.

---

[5]Apparently, for purposes of expediency, no provision had been made to transmit the *units* along with the *values* of the navigation data from the contractor to NASA for this important course correction. If I had a dollar for every time I have had to remind one of my students to include the *units* of a measurement along with its *value,* I'd be a rich man!

Find New Planet. The channel that connected Clyde Tombaugh's task unit (the predicted position of the missing planet) with the telescope camera included most of the Earth's atmosphere. The distortions caused by our atmosphere are enormously frustrating for earth-bound astronomers. The Hubble Space Telescope has successfully eliminated this short, but ruinous section of the channel between astronomers and their task units.

Measure Drug Effectiveness. The channels that connect the sensors and inducers with the task unit for this task include the hypodermic needles for injecting the test drug and collecting blood samples, and the sample vials for storing the blood samples. The enormous impact of the vulnerability of these channels was driven home in recent decades when it was learned that the deadly disease AIDS had often been vectored from one person to another through reuse of these devices without proper sterilization. In fact, as a direct result of this worldwide tragedy, hypodermic needles in most clinics and hospitals are no longer reused for any reason.

Evaluate GUI Effectiveness. The most obvious channels for this task are the mouse, the keyboard, and the display screen, which connect the subject (inducer) to the GUI (task unit). Another set of critical channels communicates heat, noise, light, and other disturbances in the laboratory to the subjects. Excessively high or low temperatures, loud or unusual noises or sounds, changing lighting conditions, and other distractions in the environment can easily disturb the cognitive processes of the subject, which, of course, could directly affect the subject's interaction with the GUI. Either some attempt must be made to eliminate excessive variation and extremes in these conditions, or a series of control tasks must be conducted to measure their effects and, hopefully, to counteract or minimize them.

Another important channel is the language and layout of the questionnaires (sensor) that acquire the demographic data and the qualitative evaluations of each subject. If the wording of the questions resonates with irrelevant prejudices or emotional biases of the subjects, the acquired data can easily be rendered meaningless. To address this problem, pretests should be used to detect cognitive or emotional triggers in the composition of the questionnaires. Because these potential sources of channel bias can have such an overwhelming effect on experiments using human subjects, they are given special attention in Chapter 12 (Synthesis).

Compile Summaries. This task uses two channels, one input and one output. The input channel, which connects each article (inducer) to the doctoral student's computer, is the internet. This channel also connects the computer screen to the student's mind via her eyes (sensor), allowing her to read each article and compose the entry for the database (task unit). The output channel connects the student's mind to her hands, with which she manipulates the mouse and the keyboard to navigate on the internet and enter information in each category of the summary.

Both of these channels pass through the brain, interfacing with the student's summarization and evaluation functions (**solution**). The images of the text are converted by the brain into *symbolic language,* and it is this information that the student manipulates and records. (This innate ability to *record* thoughts symbolically through the written word, diagrams, or pictures is uniquely human.) Also present simultaneously in this powerful cognitive machine, however, are countless other cognitive processes, including emotions, which can easily and unconsciously influence the intellectual task at hand, often inappropriately. Minimizing these sources of error is extremely difficult, but acknowledging their existence and power is an important step toward objectivity.

Measure Liquid Temperature Variation. This task clearly illustrates a problem common to all measurement tasks: as soon as you try to measure the task unit, you disturb it, rendering your measurement inaccurate to some degree. If the sensor that measures the temperature of the cold liquid in the vessel is connected by a cable to a recording instrument outside the vessel in the laboratory, this cable will conduct some heat across the system boundary into the vessel and raise the temperature of the thermocouple, causing the measured temperature to be too high.

What about eliminating the cable altogether? Suppose we attach a small radio transmitter to the thermocouple and broadcast the measured temperature through the vessel wall to a receiver connected to the recording device in the laboratory. This would avoid a breach in the vessel wall and a conduit for external heat to flow into the vessel. This is true, but irrelevant. The radio transmitter must be supplied with power, and even if we avoid using a power cable by installing a battery in the transmitter, the heat dissipated by the battery and transmitter will (must!) heat up the liquid, once again resulting in an incorrect measurement of the temperature of the liquid.

This is a fruitless quest. The laws of physics tell us that there is absolutely no way to completely eliminate the effects of this unwanted inducer. Measuring something changes it. However, a careful design of the sensor, the channels, the vessel, and the laboratory can *minimize* the heat loss through these devices, perhaps sufficiently to make the temperature measurement error smaller than required by the **performance criterion** for the task.

4.1.3.5 Domain Knowledge

The last (but certainly not least) important resource for the task domain is knowledge. All R&D tasks are built on a solid foundation of **domain knowledge**, which the task presumably proposes to supplement with **range knowledge**. We spend much of our formal and informal education acquiring knowledge that we can bring to the various tasks that we undertake in our professional and personal lives. In fact, one of the major purposes of the Analysis Phase of the Scientific Method is to gather and organize the existing knowledge that is necessary to design and execute an R&D project. This process is fully described in Chapter 10 (Analysis).

Because of the importance and extent of this topic, the fundamental concepts that specify the kinds of knowledge representations and their specific roles in R&D projects are presented in Part III (Knowledge Representation).

The preceding presentation of the resources of the task domain leads us to a definition that is a companion to the previous definitions of the system and the environment of a task:

Definition 4.10 The **laboratory** for a task is the set of physical regions containing the task unit and all the task resources required during the experiments.

Most of the resources that have been specified for our running example tasks are included in their laboratories. There are, however, a few noteworthy exceptions. Usually much of the domain knowledge is not required during the experiments, even though it may have been essential to assemble and equip the laboratory. In fact, often some items of domain knowledge are *intentionally* excluded from the laboratory to ensure the experiment is in control; *e.g.,* isolating the GUI designer from the cohort during the experiments for the task *Evaluate GUI Effectiveness.* On the other hand, any data files to be used during the experiments must be in the possession of the supervisor, who is by definition *always* present during the experiment. Finally, it is obvious that past inducers will not be present in the laboratory. For example, the inducers for the tasks *Measure Fossil Age* (sedimentation and erosion) and *Identify Collision Damage* (accidents and normal wear-and-tear) altered the task units long before the experiments were conducted.

> **TIP 6**
> For a given task, carefully define the system, the environment, and the laboratory, and then examine them thoroughly and methodically to identify:
>   a) unwanted influences generated by and acting on the task resources;
>   b) detrimental ways in which the system boundary can be violated; and
>   c) extraneous elements within the system that influence the task unit.

The best way to implement Tip 6 has already been briefly explained: conduct a complete series of control tasks to determine whether the task is in control. Any unexpected results or effects observed during these control tasks may indicate the presence of one of the problems listed in Tip 6, implying that the task is not in control. Armed with this information, much can often be done to rectify the task design. More about this subject is presented in Chapter 12 (Synthesis).

Let us take a moment to review where we are. We have completed the description of the **task domain**, which comprises everything that is brought to the task: the **task unit**, which is the set of objects or concepts that undergoes some required alteration and measurement before or during the task; and the **task resources**, which are devices and knowledge that are brought to the task, so that the objective of the task can be achieved. Specifically, the task resources include the **inducers** that are required to alter the task unit, the **sensors** that intentionally measure the task unit in some way, the **supervisor** that monitors and operates the components of the task domain, the **channels** that transmit information and energy to and from the components of the task domain, and the **domain knowledge** that is necessary to perform the task.

> **Exercise 4.3**
> For each of the tasks you described in your answer for Exercise 4.2, identify the system, the supervisor, and the important or unusual channels.

> **Exercise 4.4**
> 1) Select an interesting research or development task of some historical importance about which a great deal is known. Describe the major components of the task domain.
> 2) Define and describe the components of the task domains of the two tasks undertaken in part 1 of this exercise:
>   Select Interesting Task
>   Describe Major Task Components

4.2 Task Method

Now we are ready to define the process by which an R&D task is carried out, known as the **task method**. The task method transforms the task domain into the **task range**, which comprises the products of the task in the form of knowledge, devices, and effects. If the products of the task meet the requirements specified in the task objective, this objective has been achieved.

> **Definition 4.11** The **task method** comprises the **solution** specified to achieve the task objective and the **experiments** designed to determine the effectiveness of the solution, *i.e.,* everything required to transform the task domain into the task range.

The task method has two parts: 1) the **solution**, which includes the mechanisms and procedures necessary to achieve the objective of the task; and 2) the **experiments**, which include the resources and protocols necessary to measure the performance of the task unit when the solution is applied to it. The solution is specified in the Hypothesis Phase of the Scientific Method, as described in Chapter 11, and is implemented in the Synthesis Phase, as described in Chapter 12. When the experiments are conducted, the task unit is exposed to the concurrent inducers (if any) as specified by a formal protocol, while the sensors gather the data that become the basis for the results and conclusions of the task.

The design of the experiments is one of the most critical and complex components of many tasks, but it can be codified to a large extent, which is a major topic in Chapter 12 (Synthesis). Codifying the process of inventing solutions, on the other hand, is very difficult. Fortunately, for most tasks, the solution is a well-understood, tried-and-true approach that is well suited for the task, and the project team is clearly much more interested in the products of the task than in the solution. For example, an architect may use reinforced concrete to construct a very tall building, but his primary focus is the architectural success of the building, not an evaluation of the superiority of reinforced concrete as a building material. Less often, the solution is invented by the project team, and the primary objective of the task is to determine its effectiveness. Between these two extremes are the tasks that require new solutions, but the focus is still on a practical objective, and the invention of the solution is regarded simply as a hurdle that must be overcome.

New and clever inventions are the offspring of the marriage of creativity and expertise; they exemplify the ingenuity of the scientist or engineer. It is very difficult to distill the elixir of invention into a universal potion for increasing efficiency or productivity in R&D environments; the process of creativity is far too elusive and individualized to yield to reductionism. This is one of the reasons that our institutions of higher education focus primarily on teaching classical solutions, rather than inventing new solutions, claiming that creativity cannot be taught. Acknowledging the seemingly ineffable mystery of intellectual invention and innovation, this book assumes that the investigator already knows or will somehow learn from life itself how to ignite and nurture the flame of inspiration. Instead, the focus here is on organizing the knowledge and processes required to realize the solutions. Some practical suggestions for the all-important processes of disclosure and documentation of the solution, whether mundane or revolutionary, are offered in Chapter 11 (Hypothesis).

The mechanisms and procedures of a solution are usually very strongly interdependent. Anyone who knows how to make pancakes knows this. It is not enough to have the ingredients for making pancakes. You have to know how much to stir the batter, how to test the temperature of the skillet to make sure it is just right, how to pour the batter quickly onto the skillet in puddles of just the right size, how far apart to put the puddles, and when to flip the pancakes. If the procedure is incorrect or sloppy, the pancakes may be an embarrassment. Making pancakes is a task with a simple mechanism and a tricky procedure.

In 215 BC, in his work *Pappus of Alexandria,* the famous Greek mathematician Archimedes said: "Give me where to stand (and a lever), and I will move the Earth." This task, which Archimedes conducted entirely in his mind,[6] is told to school children as an example of one of the basic principles of mechanics, known as *mechanical advantage.* Included in this statement are the task objective (to move the Earth), the inducers (a lever, a fulcrum, and a place to stand), and the sensor (presumably some instrument that measures the position of the Earth). The solution *mechanism* is the set of inducers. What is not mentioned is the *procedure* that Archimedes proposed. Of course, as children we assumed (or, rather, our teachers assumed) that Archimedes intended to plant the tip of a long lever securely on the Firmament (?) slightly below the Earth, raise the lever until the shaft contacted the Earth a short distance from the tip of the lever, and then,

---

[6]An experiment conducted solely in the mind is known as a "Gedankenexperiment," a German word meaning "thought experiment."

grasping the far end of the lever, shove it forward so that the Earth was pried out of its position.

But maybe instead, Archimedes meant to swing the lever like a baseball bat, and knock the Earth out of the park, so to speak. Or perhaps, he was planning to cut the lever into two pieces, place them side-by-side, parallel to each other, and roll the Earth down this makeshift track. Or perhaps Archimedes was going to bend the lever back like a big spring, and then release it so that it would thwack the Earth hard enough to move it a bit. Or maybe, Archimedes was privy to the mysteries of nuclear physics, planned to extract a critical mass of some fissionable element from the huge lever, and detonate a small atomic bomb to move the Earth. Certainly none of these procedures is any more fanciful than the normally accepted one.

We can shed some additional light on the important relationship between the solution *mechanism* and solution *procedure* by speculating about their roles in several of our running example tasks.

Remove Screws. The *mechanism* the homeowner used to remove the screws from the hinges is a screwdriver. The *procedure* was to grip the screwdriver, insert its tip into the screw slot, rotate the screw counterclockwise, and so forth. Clearly, if the homeowner had wanted to invest sufficient planning time, he could have documented this procedure in the form of highly detailed instructions, such as flow charts. For a task as simple as removing screws, however, he probably felt he could keep this all in his head. Note that other solutions for the task were possible; instead of using a screwdriver, the homeowner could have drilled the screws out or pried them loose with a crowbar. He could even have considered using a small charge of plastic explosive.

Launch Spacecraft into Orbit. The solution for this project was, of course, orders of magnitude more complicated than removing a few screws. The enormously complex set of *procedures* for this task included launching the rocket with its payload (the task unit), controlling its trajectory in powered flight, accurately inserting the capsule into low Earth orbit, and on and on. It also included carefully designed contingency plans to deal with every possible in-flight system failure and emergency that could be anticipated, as well as a team of skilled and versatile professionals to handle unexpected failures, not unlike the ones that the homeowner experienced when he discovered his screwdriver was broken. On the other hand, the cost of the *mechanisms,* which included designing and building the Mercury Capsule and purchasing the launch vehicle from the US Air Force, was less than 10% of the total Mercury mission cost; the rest of the budget was spent on the design and execution of the other part of the task method, the experiments.

Identify Faster Traffic. The use of the averages and standard deviations of the vehicle speed samples to compute a statistical confidence under the assumption of normality is the *mechanism* that the traffic engineer selects as the solution for this task. The *procedure* for this solution is the detailed, step-by-step process that the traffic engineer uses to calculate the statistical confidence that the two speed distributions have different averages. For instance, her procedure might specify converting the speed samples to tab-delimited text files, reading these files into a Microsoft Excel spreadsheet to perform the calculations, printing out the results, and generating various plots of the speed sample distributions and results.

Find New Planet. The *procedure* that Clyde Tombaugh used to search the heavens for Pluto invokes images of a shivering astronomer huddled over his instruments at the Lowell Observatory, moving the telescope to a new position listed in his detailed plan, focusing the telescope, putting a new plate in the camera, exposing the plate, using the blink comparator (the *mechanism*) to search for artifacts on the plate whose positions had moved relative to the previous plate, repeating this process night after night on the high windswept plateaus of Arizona. Such is science. Task methods are often numbingly repetitive and frustratingly slow.

Identify Collision Damage. The *procedure* for the insurance adjuster seems quite straightforward: inspect the car and fill out the standard damage-assessment form. Over the years, however, he has undoubtedly fleshed out this procedure, learning from experience, for example, how to differentiate between normal wear-and-tear and actual collision damage. The *mechanism,* which is the damage-assessment form, was probably devised long before the insurance adjuster was hired (perhaps born). It includes a detailed list of the types of damage and their associated costs, which is updated periodically.

Measure Fossil Age. The *mechanism* that the geologist uses for this task is presumably based on a well-accepted postulate that allows the age of a rock stratum to be accurately estimated by measuring the number and order of the rock strata that lie above it, as well as their thicknesses and mineral compositions. One *procedure* might be to measure the number, order, and thickness of the rock strata by photographing the exposed cliff face from a distance using optics calibrated with precise fiducials. Identifying the mineral composition of each stratum, however, might require that the geologist actually climb up (or down) the cliff face to obtain rock samples from each stratum, so that a detailed laboratory assay can be performed on each sample. Clearly, the procedure for this part of the solution entails some heavy labor and a real risk of physical injury.

Measure Drug Effectiveness. The *mechanism* of the solution for this task includes the dose ranges of the new drug, the physical parameters of the human subjects, the equipment required for administering and monitoring the progress of the subjects, *etc.* The *procedure* describes the step-by-step methods for selecting the subjects, for administering the test and control treatments, for tracking the cohorts after treatment, and so forth.

Compile Summaries. The *mechanism* for this task is the set of cognitive symbolic-language-processing tools that the doctoral student uses to summarize each source document and the format of the database that was designed to record these summaries. The *procedure* specifies how the student will access the documents, read and digest the material, enter the summary information in the form she has devised, and periodically review and revise the entries.

To summarize, the solution (mechanisms and procedures) provides the means for achieving the objective of the task, while the experiments (resources and protocol) measure the performance of the task unit when the solution is applied to it. These two components of the task method are discussed in more detail in Chapters 11 (Hypothesis) and 12 (Synthesis).

4.3 Task Range

> **Definition 4.12** The **task range** comprises all the products of the task, including knowledge, devices, and effects.

As was pointed out in Chapter 1, the purpose of a research task is to *acquire knowledge,* represented by its **results** and **conclusions**, while the purpose of a development task is to *apply knowledge* to create devices or effects. Although most of our running examples are research tasks whose products are knowledge, three are development tasks: *Remove Screws*, *Launch Spacecraft into Orbit*, and *Compile Summaries* (See Table 4.1). The objective of the development task *Remove Screws* is to generate an *effect,* namely the removal of the screws from the hinge plates on the door and doorframe. The objective of the development task *Launch Spacecraft into Orbit* is to generate an *effect,* namely the placement of the Mercury spacecraft in low Earth orbit. The objective of the development task *Compile Summaries* is to create a *device,* namely a database of summaries of published documents about Multiple Personality Disorder. This last example might appear to be a research task, because it deals with the compilation of intellectual knowledge in a clearly academic venue. However, no new knowledge is *acquired* by this task, just a distillation of existing knowledge. In fact, the process of summarization necessarily results in an inevitable *loss* of knowledge, which is intentional in this case.

No further explicit guidelines will be offered about the kinds of devices or effects that could be included in the range of a *development* task. In general, these items are highly specific to the task and are well defined and understood by those who mount the task, which is motivated by some clear and abiding requirement for the products.

On the other hand, **range knowledge**, which is generated by research tasks, deserves considerable development and discussion. For this reason, the explication of the fundamental concepts for knowledge representation and the specific roles of knowledge in R&D projects is the primary emphasis of Part III (Knowledge Representation).

Knowledge Representation

CHAPTER 5

An Epistemological Journey

Although the evidence is sparse, many physical anthropologists have tentatively concluded that the species homo sapiens has undergone no significant physical evolution over the past 40,000 years. Our bodies and genetic structure are no different today than those of our Cro-Magnon ancestors who lived in caves in the Paleolithic Era. This also implies that our brains have not changed for at least 40 millennia; we are no more or less intelligent now than we were then.

Why, then, have the vast majority of our technological advances apparently taken place in the last few centuries, less than one percent of the total period? Why hasn't the development of technology followed a more gradual and uniform growth pattern? Why didn't we have the steam engine in 5000 BC and land on the Moon a couple of thousand years later?

Certainly, part of the answer is that knowledge builds on itself exponentially, and exponential growth is not particularly dramatic until quite a few doubling periods have taken place. After all, if we are told that $1000 invested at 15% interest compounded annually will double in value to $2000 in 5 years, this does not seem very impressive. But if we extrapolate further, we discover that this $1000 investment will grow to a million dollars in 50 years, plenty for a comfortable retirement. And, of course, half of the final value of the investment was earned in the last 5 years. So, perhaps this helps to explain why the advances in technology seem rather paltry before the European Renaissance.

It is also likely that these advances in technology were very precarious in the beginning, and small hard-won gains were often obliterated by the calamities of human history, such as wars, disease, famine, natural disasters, and the unrelenting attrition of day-to-day survival. For most of human history knowledge was passed on from generation to generation by word of mouth, and whenever more pressing topics of conversation came up, which happened all the time, a new and better idea for building a house took a back seat. Often, in the light of immediate survival needs, a valuable bit of knowledge was lost entirely and had to be reinvented later on. This is like the investor who is too busy one day to check on the current value of his $1000 investment in a highly volatile stock, only to discover the next day that he has lost all his money. However, because the amount he lost wasn't very much anyway, he experiences no immediate hardship and simply starts over. Of course, had he been able to pay strict attention every day, he would have eventually accumulated a small fortune, instead of being periodically wiped out.

If this is the case, then what happened to trigger the enormous explosion of technology after the European Renaissance? In fact, this explosion was foreshadowed by an earlier rapid expansion of knowledge that began in ancient Egypt about 4000 BC and lasted until the beginning of the Christian era, with a particular flurry of activity at the height of the Greek civilization during the Age of Pericles. As always, most of the material innovations of this period were periodically wiped out, but the single critical invention that ushered in and fueled this rapid expansion of knowledge was preserved: the symbolic recording of knowledge, *i.e.,* written language.

Suddenly, with the advent of written records, it was possible for new ideas to survive their inventors without depending on word of mouth generation after generation, a process which is profoundly vulnerable to distortion, corruption, and selective memory. With the invention of written language, for the first time knowledge could be reliably accumulated and used as a springboard for the acquisition of new knowledge, which is, of course, **research**.

But the maintenance of written knowledge bases was still precarious. Books were extremely rare and not widely disseminated. Instead they were locked up in vaults, jealously guarded, and available only to their immediate custodians and the elite. Copying documents was laborious and expensive. Enough leaked out, however, to nourish a rapid expansion of knowledge over the next 4000 years. Mathematics was born. Civil and mechanical engineering flourished. Architecture thrived.

Then, in the middle of the first millennium of the Christian era, the Roman Empire collapsed under its own decadent weight, and the western world was plunged into darkness for a thousand years. Nevertheless, even though many manuscripts were lost or destroyed, enough survived to keep the spark of knowledge alive. Slowly the fight against ignorance regained ground. Land trade routes were established between Europe and Asia, and ship captains began to venture farther from home, although still forced to hug the coasts to navigate reliably. The world view slowly expanded. Manuscripts were written on many subjects, and monks spent their lives copying (and illuminating!) them, including the epistemological remnants of the classical era. But as before, few had access to these documents and their precious knowledge, and all but a few of the clerics and rulers were illiterate and ignorant.

And then, a single invention changed all that almost overnight. In Germany in 1454, a goldsmith named Johannes Gutenberg built the world's first printing press with movable type. It was slow and cumbersome, but the cat was out of the bag. Suddenly it was possible to make copies of the written word thousands of times faster than before. Now manuscripts could be printed and disseminated cheaply and quickly, and the written knowledge base of human civilization exploded. And as technology improved by leaps and bounds, the knowledge explosion, ignited by the written word and fueled by the printing press, was supercharged by the emergence of cheap technologies for communication, and sound and image recording. In just five centuries, the accumulated written knowledge base of human civilization mushroomed by a factor of about 100,000, an annual exponential growth rate of about 2.5%. The Library of Congress of the United States of America now stores about 10^{15} bits of knowledge. Ironically, this staggering number is just about equal to the intelligence capacity of a *single* human brain. Today, the Internet can instantly disseminate much of the accumulated knowledge of human civilization to billions of people around the world on demand. We are awash in knowledge.

In Chapter 1, it was stated that research seeks truth. But knowledge, *per se,* is not necessarily truth. Knowledge is simply *organized* information, not necessarily *correct* organized information. The search for truth can be very chaotic! For example, the Earth-centered universe proposed by the Greek astronomer Claudius Ptolemy in his book *Almagest* (*The Greatest*) in the second century AD held sway until this knowledge was challenged by Nicolaus Copernicus more than 1000 years later. Shortly thereafter, Galileo Galilei came out in support of the new Sun-centered model proposed by Copernicus, but was forced to recant by the Catholic Church, which found this notion heretical. Despite incessant intimidation by the Church, however, the scientific community quickly accepted the Sun-centered model, but it was not until the early twentieth century that the Church began to back down in the face of seemingly indisputable evidence. Then, just as religion and science were busy congratulating themselves on a possible meeting of their minds, Albert Einstein quietly interjected that it really doesn't matter which object you choose as the center of a gravitational system; it really just depends on your point of view, and the laws that govern the system's behavior are the same either way. How utterly confusing!

In fact, it is not possible to ascertain the *truth* about anything, because there is no independent agent who knows the truth first hand and may be routinely asked to corroborate our conclusions. All we can do is gather evidence, apply logic, and try to attach confidences to the conclusions we derive from this evidence. Theorems in mathematics, for example, are based on existing propositions, such as axioms, which are universally *presumed* to be true, but have never been (and often cannot be) proved. This is the heart of the deductive method of knowledge acquisition: a logical derivation based on a set of premises that are presumed to be true.

Scientists have long been aware of this vulnerability. Based on the inspired guidance of Galileo Galilei, Francis Bacon, and René Descartes in the early seventeenth century, scientists have evolved three strict criteria for *provisionally* accepting the knowledge gained from a research task as valid: reproducibility; completeness; and objectivity. It is important to remember that the acceptance of knowledge as valid is *always* provisional, because scientists must *always* be ready to change their minds, or at least reserve judgment, in the face of new contradictory evidence.

The criterion of **reproducibility** requires that all independent attempts to accomplish a task under the same conditions yield the same results. This helps to ensure that the results are not just flukes, but derive from a consistent and reliable method. The phrase "same results" is clearly subject to definition. For instance, if three people measure the length of a table, it is highly likely that they will come up with three slightly different answers. Moreover, because the world does not stand still, the conditions under which the length of the table is measured will not be *exactly* the same. It is this difficulty that motivated the development of the area of mathematics known as statistics, which is used to estimate the confidence that differences in measurements are attributable to chance, or alternatively, are real and significant.

The criterion of **completeness** requires that the propositions and methods used to accomplish a task be completely and clearly disclosed using standard terminology. This criterion is analogous to what lawyers call "due diligence." It also helps to satisfy the criterion of reproducibility. After all, if even seemingly minor details of the supporting knowledge and processes of a task are lacking, it may not be possible to reproduce the results faithfully. Insufficient documentation of a project is a very common and enormously frustrating violation of the completeness criterion in many research and development projects today. Project reports are woefully incomplete, confusing, and poorly organized, making reproduction of the results impossible. This is particularly apparent in software development. When programs written months or years ago must be resurrected, it is often discovered that insufficient documentation of the software and its operation was recorded, and the programs are largely incomprehensible — even when attempted by the original programmer (who has forgotten almost everything about his code!). Worse yet, investigators sometimes leave important details out of their research reports *intentionally* to "protect" their ideas. No doubt they are worried about others stealing their ideas, and perhaps justifiably, but combating such improprieties through purposeful obfuscation simply exacerbates the problem and stifles the overall advancement of knowledge, and in the long run, everyone loses. There are very effective ways to protect intellectual property without violating the criterion of completeness.

The criterion of **objectivity** requires that the governing propositions for a task be free of personal bias and universally acceptable as reasonable and valid. This requirement stresses the importance of basing the task on a carefully researched and well-documented framework of existing knowledge that is accepted by members of the scientific community in related fields. This does not mean that the task may not depart significantly from current paradigms. However, if it does, the logical progression of knowledge up to the departure point must be pedagogically correct, free of personal opinion and dogma, and clearly documented. There is no general rule that defines how far back toward fundamental axioms this train of logic must extend, but suffice it to say that any lack of rigor noted during peer review must be fully addressed and resolved.

The requirement for objectivity is often where science and religion part ways. Basing a *high-level* assumption on faith without building a pedagogically rigorous justification of its validity is not acceptable in science or engineering tasks. For example, an argument and its subsequent conclusions may not be based on an alleged revelation from a supreme being. Personal bias can too easily exert a strong influence on such knowledge, tainting its objectivity. On the other hand, in science and engineering some *low-level* propositions *must* be accepted on faith, such as the axiom "the shortest distance between two points is a straight line." Distinguishing between anecdotal speculation and universally accepted fundamental propositions is a challenge that has faced scientists for thousand of years.

One way to help ensure that these three criteria are satisfied is to establish a standard framework for the representation and roles of knowledge in research and development tasks for any field of science and engineering. The next chapter is devoted to the definition of this framework.

CHAPTER 6

Categories and Types of Knowledge

All knowledge belongs to one of four categories: **speculative**, **presumptive**, **stipulative**, or **conclusive** knowledge. Each category represents a different level of certainty with which the validity of the knowledge may be provisionally accepted — speculative knowledge is the least certain, and conclusive knowledge is the most certain.

Each item of knowledge is represented as a statement, called a **proposition**. A proposition is a single complete sentence, although it may be accompanied by a list, a diagram, a table, or a set of formulas that provide specific details.

Each category of knowledge comprises several different types of propositions, each of which has a different informational purpose, as summarized in Figure 6.1. The next four sections define and discuss each category and type of knowledge.

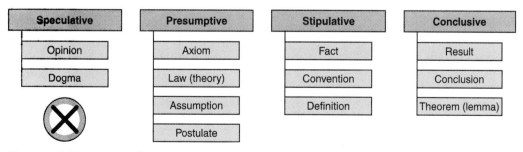

Figure 6.1 Taxonomy of categories and types of knowledge propositions

6.1 Speculative Knowledge

Speculative knowledge comprises propositions that rely solely on unsubstantiated belief or authoritarian ontological or moral statements. Such propositions have not and often cannot be validated by logical proof or statistical inference, and thus are not acceptable for scientific or engineering tasks. There are two types of speculative propositions, opinion and dogma.

Opinion is an unsubstantiated belief about the worth or value of something or someone. It generally expresses a personal feeling or a prevailing sentiment among a group of people:

> Slavery is immoral.
> Generosity is its own reward.
> Money is the root of all evil.
> The paintings of Auguste Renoir are magnificent.
> Cigar smoking is offensive.
> Violence is never justified.

Dogma is a rule or statement of causality or existence that is promulgated by a recognized authority, usually ecclesiastical, cultural, or political, and is intended to be accepted without question:

> All men are created equal . . .
> God created the heaven and the earth . . .
> Children should be seen and not heard.
> Thou shalt not covet thy neighbor's wife.
> The king rules by divine right.
> Only human beings are truly self-aware.

It is sometimes difficult to persuade those who are training to be scientists and engineers that neither opinion nor dogma may be expressed or applied in research tasks, except perhaps in the conclusions of a research report, and then only if these propositions are carefully acknowledged as opinions, as discussed in Chapter 13 (Validation). When such weak propositions are found in their work and they are asked to remove them, researchers may find it difficult to rebuild the necessary support for their methods and conclusions using more rigorous propositions. Such an eventuality should be regarded as a strong signal that the entire approach may be based on faulty premises; like a building whose foundation is weak and unstable, it may be necessary to tear it down and start over. This can be demoralizing for the researcher, but if speculative knowledge is assiduously avoided from the outset, the chance of this kind of mishap will be reduced.

6.2 Presumptive Knowledge

Presumptive knowledge comprises propositions that are assumed to be valid, but are always open to question and are explicitly qualified by an acknowledgment that the results and conclusions of the study depend on the validity of the propositions, which cannot be guaranteed. The investigator who uses presumptive knowledge does not bear the burden of proof. He is, however, expected either: 1) to provide convincing justification for the acceptance of a presumptive proposition; or 2) to demonstrate that a presumptive proposition is widely recognized by experts as reasonable.

There are four types of presumptive propositions: axioms, laws, assumptions, and postulates.

> **Definition 6.1** An **axiom** is a fundamental concept or property that is universally accepted as valid everywhere in the universe.

Axioms are presumptive propositions that have withstood the test of time and are universally accepted as valid. There are very few of them. Of all presumptive types of knowledge, axioms are the most rigorous. Most of the time they appear to be so obviously true and are so widely accepted, that few objections are raised when they are used, and they may be applied without any loss of rigor. Often they need not even be explicitly stated; their validity may be taken for granted. In fact, for tasks where such axioms would normally be expected to be applicable, but are *not* required to accept the results and conclusions, this is quite interesting and should be pointed out.

Most axioms are found in the field of mathematics, although there are also a few in the fields of physics and philosophy:

> The shortest distance between two points in Euclidean space is a straight line.
> The cardinality of the set of all prime numbers is infinite.
> If A implies B, and B implies C, then A implies C.
> No two objects can occupy the same space at the same time.
> Humans are conscious beings.
> Deduction and induction are equally valid methods of logical proof.

An investigator is, of course, free to state a new axiom. However, the strict requirements for justifying an axiom make this very difficult: the investigator must convince his audience that the new axiom is *universally* valid. And in mathematics and physics, at least, *universally* literally means everywhere in the universe. To satisfy that restriction will usually be quite difficult. Over the long history of mathematics and philosophy, axioms have emerged gradually from extended discourse and rigorous scrutiny over many years to confirm that the proposition appears to be consistently and universally true, regardless of the application domain.

> **Definition 6.2** A **law** (or **theory**) is a fundamental relationship or process that is universally accepted as valid everywhere in the universe.

A law is similar to an axiom, except that it specifies a *relationship or process,* rather than a *concept or property.* The same strict requirements that apply to axioms, also apply to laws. Like axioms, there are very few laws, and most are in the physical sciences and economics:

> Evolution is driven by the survival of the fittest. (Darwin's Law of
>   Evolution)
> Energy is conserved. (Law of Conservation of Energy)
> The product of gas pressure and volume is constant at constant
>   temperature. (Boyle's Law)
> Every action generates an equal and opposite reaction. (Newton's
>   Third Law of Motion)
> Bad money drives good money out of circulation. (Gresham's Law)
> The entropy of the universe is always increasing. (Second Law of
>   Thermodynamics)

Laws are usually not as obviously valid as axioms, because processes are almost always much more complicated to describe than properties, just as the hunting habits of a wolf are more complicated to describe than the color of its fur. The result is that laws typically have shorter lifetimes than axioms. Their inherent complexity makes them brittle in the face of new discoveries. When they cease to work, they must be repaired, or failing that, discarded. History is replete with accounts of laws being periodically modified to account for new observations.

The historical progression of the laws describing the planetary motion of the Solar System, from Ptolemy to Copernicus to Kepler to Galileo to Newton to Einstein, is a case in point. The famous Universal Law of Gravitation proposed by Isaac Newton in 1687 remained valid until Albert Einstein published his General Theory of Relativity in 1916, which not only yields Newton's law under appropriately simplified conditions, but also predicts how light bends as it skirts large masses in the universe, and how a mass is really nothing more than a local distortion in the space-time continuum. And this process is not finished (which Einstein knew full well). Physicists are still looking for the so-called Unified Field Theory or Theory of Everything.

Until the late nineteenth century, the laws of heredity presumed that an offspring inherited a weighted average of the traits of its progenitors. If the mother had blue eyes and the father had green eyes, it was possible that the child would have blue-green eyes. Then Gregor Mendel demonstrated clearly that heredity was not an averaging process, but a "voting" process in which one set of traits won out over another (although not necessarily those *expressed* by either of the parents). Then in the 1950s the laws of heredity were clarified even further when James Watson and Francis Crick discovered the helical structure of the DNA molecule, which encodes the genetic instructions for the individual. Today, as we attempt to clone living animals (and soon perhaps humans), new details of the laws that govern heredity are being revealed every day.

It is unlikely that the typical R&D project will come up with a new law or axiom (although it certainly is not impossible!). Nevertheless, it is important to appreciate the high level of rigor associated with existing axioms and laws, so that, when appropriate, they can provide highly rigorous support for the objectives and methods of an R&D task. That's why scientists and engineers study them in such detail in school.

On the other hand, every R&D project needs to build a complete foundation of knowledge to support the project, and in general it will not be possible for all (or even most) of the propositions to be axioms or laws. Less rigorous presumptive propositions will have to be accepted. These presumptive propositions need not, of course, be proved (because they are *presumptive*), but *they must be carefully stated, explained, and justified.* After all, why should those who review the project take the investigator's word that the presumptive propositions are valid? They will understand that such propositions cannot be proved, but they certainly have a right to demand that the investigator make a strong case for their validity. Such justification can be based on informal empirical evidence and/or the experience of qualified experts, usually via citations of their published research: "Because previous research has shown that process XYZ almost always yields normal distributions of the dependent variable [Smith 1996] [Jones 1994], it is assumed that the results of Test Q, which used this process, are normally distributed."

Supported by sufficient, clear, and strong justification, the two remaining types of presumptive propositions, the **assumption** and the **postulate**, are especially useful for building the knowledge framework necessary to support the objectives and methods of an R&D project. Unlike axioms and laws, assumptions and postulates need only be valid for the field of the investigation.

> **Definition 6.3** An **assumption** expresses an attribute or value of a property or a process that is required to be valid for the application domain of the task.

Assumptions are most frequently used to limit or restrict the task objective and, thereby, the domain and/or range of a proposed solution. For example, when an airframe development team is asked to design a new passenger airplane, they must make certain assumptions that will govern their design: the maximum wind shear forces that will occur; the maximum stress that may be safely imposed on the airframe; the maximum impact that runways will safely handle during landings; the fire resistance specifications of certain fabrics to be used for the airplane's interior; and so on. Some of these assumptions will be justified by the manufacturers' specifications, while others will be accepted largely on the basis of the extensive experience of the design engineers themselves. Initial tests of the prototype aircraft will often reveal whether or not these assumptions were valid. Sometimes, unfortunately, the lives of test pilots are lost before an assumption is discredited. Simulation can sometimes help limit such risks, but the complex behavior of airframes moving through the whimsical atmosphere is difficult to capture accurately in a computer model.

Like all presumptive propositions, assumptions need only be justified, not proved, although it is often the case that an assumption *could be proved*, if enough time and resources were available. Below is a list of examples to illustrate the form that assumptions may take; some of them are far less weighty than those given for the airplane development project just described:

The fabric Duratex IIa will not ignite during a cabin fire in the airplane.
The diesel-engine piston diameters are normally distributed.
Latex paint will cover the existing surface sufficiently well with one coat.
The prognosis for subject 27 is significantly worse than for subject 15.
$A > B > C > D.$
Depression is more difficult to deal with than anger in interpersonal relationships.
Car buyers regard fuel-consumption rate as the most important performance metric.
The difference between the different types of presumptive propositions is well understood.

To increase the likelihood that this last assumption is valid, it would probably be appropriate to reinvoke the running example tasks presented in Part II (See Table 4.1) and speculate about some of the assumptions that might well be required for each of them. But before we do so, let us introduce the last presumptive proposition, the **postulate**, which is a close companion to the assumption. Just as a law expresses a process and an axiom expresses a value, the postulate has the same relationship to an assumption, although they both demand less rigor than axioms or laws, because they need be valid only for the application domain of the investigation.

> **Definition 6.4** A **postulate** expresses a process, function, rule, or mechanism that is required to be valid for the application domain of the task.

A common use of the postulate in R&D projects is to formulate **performance metrics** by which the success or failure of tasks is evaluated. Clearly, the performance metrics must be acceptable to those who review and use the results of the task, and therefore these metrics must accurately reflect the task objective. For example, if the engineers who are designing a new passenger airplane decide to postulate a performance metric that only considers payload (number of passengers or amount of freight carried), they may well be criticized for having neglected many other factors that determine success, such as production cost, fuel consumption, comfort, safety, and so forth. Therefore, a task often postulates several performance metrics to measure how well the objective of each task has been met (**benefits**) and any associated disadvantages (**costs**), each from a different perspective. In addition, an overall **quality** metric may combine the benefit values, and an overall **payoff** metric may combine the benefit and cost values. These options are discussed further in Section 7.3 (Range Knowledge) and in Chapter 12 (Synthesis).

Postulates are also commonly used to transform a quantity from one representation into another by applying some kind of functions, rules, or tables. For example, it may be necessary for a task to employ a postulate to round a voltage to the nearest tenth of a volt, or to allocate a set of voltages to bins in a histogram whose class marks have been previously defined. Similarly, when a survey asks their age, people are generally intended to respond with an integer, like 35. When they do this, they have applied a postulate that (usually) truncates the true answer, which might be 35 years, 1 month, 25 days, 3 hours, and 6 minutes (expressed to the nearest minute). Such postulates often cause a loss of information and *simultaneously* carry an implicit assumption that the incurred error is not significant and may be neglected. If there is any reasonable doubt about this assumption, this qualification must be included *explicitly* in the postulate, so that the reviewers of the project are alerted to carefully consider the validity of this postulate.

The list of examples that follows illustrates the form that postulates may take. Just as laws are generally more complicated than axioms, postulates are generally more complicated than assumptions, because they must convey more information. Note that several of the listed postulates include the mechanisms that are used to generate the postulated knowledge:

Without significant loss of information, the average image-pixel intensity is linearly quantized into 10 equal-sized bins in the interval [0,1].

Engine performance is the fuel consumption rate, expressed in miles per gallon.

Insulation efficiency I at the end of each experiment trial is given by:

$$I = \log(T_i) - \log(T_e)$$

where T_i = absolute temperature inside the vessel (°K)

T_e = absolute temperature outside the vessel (°K)

The semester grade for an elementary school student is determined by the following:

SCHOLARSHIP \ EFFORT	*A*	*B*	*C*	*D*	*F*
A	A+	A	A	A	A
B	B+	B	B	B	B
C	B	C+	C	C	C
D	C+	C	C–	D	D
F	C	C–	D	D	F

The attraction of the beach for tourists is inversely proportional to the amount of cloud cover, and directly proportional to the air and water temperatures.

Well-formed dictionary definitions of English words obey the following grammar:

{A complete grammar for English dictionary definitions in Backus-Naur Form (BNF)}

Note that the postulate in the first example also states an assumption in the introductory phrase: "Without significant loss of information. . . ." The inclusion of an assumption in a postulate (or in other types of knowledge propositions) is a useful and perfectly acceptable way to combine two propositions that are closely related. Knowledge propositions may also state the **definition** of a special term, such as the name of a performance metric, a computational method, or a theory.

The complete grammar mentioned for the last example postulate, consisting of 425 BNF productions, has actually been developed. As partial justification of the validity of this postulate, the grammar has been tested on the definitions for more than 50,000 English words without failure. This, of course, is not conclusive proof of the validity of the grammar, nor is any proof required; it is, after all, a **postulate**. However, this record of consistent success lends credence to its reproducibility, a necessary (albeit not sufficient) argument for its acceptance.

Table 6.1 presents a few reasonable axioms, laws, assumptions, and postulates that might well be part of the knowledge requirements for the running example tasks introduced in Part II in Table 4.1. Certainly not all the necessary presumptive propositions have been included, and for purposes of illustration the statements of most of the propositions have been greatly abbreviated and often over-generalized.

To summarize, then, presumptive propositions (axioms, laws or theories, assumptions, and postulates) are useful for assembling the presumptive knowledge necessary to support a project. They do not require proof or conclusive evidence to be considered valid, although in some cases their validity could be proved or conclusively demonstrated if sufficient time and resources were available. Each presumptive proposition requires enough strong and clear justification to keep the reviewers from walking out of the room in utter disbelief. This constraint

Table 6.1 Abbreviated samples of presumptive propositions for the running example tasks

		PRESUMPTIVE PROPOSITION
Task Name	*Type*	*Statement*
Remove Screws	assumption	The screwdriver is large enough to loosen and remove all the screws.
	assumption	New hinges can be purchased at the only local hardware store.
Launch Spacecraft	law	Newton's Law of Gravitation (The force of gravity is proportional to . . .)
into Orbit	theory	Model of atmospheric interaction with the launch vehicle
	assumption	The astronaut will be in emotional control during inflight emergencies.
Identify Faster	assumption	The distributions of the speed samples are normal.
Traffic	assumption	A sample size of 100 is sufficient to provide accurate statistical conclusions.
Prove Fermat's	axioms	Many from mathematics
Last Conjecture	laws	Many from mathematics
Find New Planet	assumption	The missing mass is concentrated predominantly in one body orbiting the Sun.
	law	Newton's Law of Gravitation
Identify Collision	assumption	All damage caused by normal wear-and-tear can be visually recognized.
Damage	assumption	There is no hidden damage that could increase repair costs significantly.
Measure Fossil Age	postulate	The order, thickness, and mineral composition of strata identify their ages.
Measure Drug	assumption	A subject's unrestricted activities have no significant effect on drug benefit.
Effectiveness	assumption	Subject's awareness of his own health does not violate the experiment blinding.
	assumption	Reasonable drug doses are selected from the set. {5-, 10-, 15-, 20 mg}.
Evaluate	assumption	The room temperature has no significant effect on the performance of the user.
GUI	assumption	All evaluations by the user are honest and sincere.
Effectiveness	postulate	The complexity of the exercise is computed using the following formula: . . .
Compile	assumption	The internet will provide access to all of the important references.
Summaries	postulate	The following categories summarize the content of the articles adequately: . . .
Measure Liquid	assumption	Heat loss from the thermocouple connection cables may be neglected.
Temperature	assumption	The liquid temperature is the same throughout the volume of liquid.
Variation	law	The product of gas pressure and volume is constant at constant temperature.
Measure Age of	assumption	The effect of background radiation on the results may be neglected.
Organic Sample	postulate	The age of an organic sample is computed using the following formula: . . .

must be strictly enforced, because the confidence in the knowledge gained in a research or development task is only as strong as the confidence in its governing propositions.

6.3 Stipulative Knowledge

Stipulative knowledge comprises propositions that are considered valid for the task by reason of consensus among the investigators. This means that all those who participate in the task must simply agree or *stipulate* that the propositions are valid, acknowledging that their consensus is probably all that is necessary to be confident that others will accept the validity of the propositions. Lawyers at both tables in a court of law often *stipulate* something about an object or a piece of information so that it may be offered in evidence without further argument, *e.g.*, both sides stipulate that a particular handgun submitted as Exhibit A was, in fact, the murder weapon. Stipulative propositions, then, do not require justification or proof, but simply agreement among the investigators that the propositions are valid for the task.

There are three types of stipulative propositions: definitions, facts, and conventions. For a typical task there are usually more propositions in this category than in all other categories combined. For most tasks, a great deal of the routine stipulative knowledge needed to support a task is not at issue, will not be questioned, and requires no "suspension of disbelief" as presumptive knowledge often may. Stipulative knowledge is usually easy to accept at face value. Having said that, it is essential to remember that the criterion of completeness still applies: stipulative propositions must be clearly, unambiguously, and completely stated, so that peers of the investigators can easily and quickly understand them and, hopefully, accept their validity.

Definition 6.5 A **fact** is a statement of objective reality.

Facts are values that are readily accepted and can often be checked: "There are 3 apples on the table"; or "Each sample consisted of 100 measurements"; or "Washington, DC is the capital of the United States of America." Interestingly enough, the last example was originally a matter of *definition,* not of fact; in 1800 the federal government of the United States redefined the capital city and moved itself from New York City to Washington, DC. Thus, if the focus of a research project was a political analysis of the politics behind the transfer of the US capital in 1800, then the statement would be regarded as a definition. On the other hand, if the research involved an analysis of the geographical distances between the capital cities of the countries of the world, then the statement would simply be a fact.

Note that a knowledge proposition *derived from* an observed fact is not a *fact;* it is a **result**, which is a type of knowledge in the **conclusive** knowledge category (see Section 6.4). Facts are limited to raw observations that have not been processed in any significant way. For example, if three tables are observed to have 5, 6, and 7 apples on them, these observations are facts. However, the average number of apples on the tables (6) is *not* a fact, but a result, because it is conclusive knowledge that is the result of a (very short) descendant task called *Compute Average Number of Apples.*

Definition 6.6 A **convention** is a rule or statement governed by an accepted standard.

Conventions are stipulative propositions that acknowledge a widely accepted value or standard. The technical world is full of conventions, such as electronic communication standards, measurement standards, and environmental protection standards. For the most part, such conventions are helpful, but sometimes they create artificial and unnecessary barriers. For example, television transmissions in the United States conform to one standard (NTSC), in western Europe to another (PAL), and in Russia and the eastern European nations to yet another (SECAM). The result is that we cannot view each other's video tapes without a good deal of electronic contortionism, which is both inconvenient and expensive. Another well-known example is the confusion between the two major measurement-unit standards of the international community; MKS (meter-kilogram-second) units are used by the entire world *except* the United States, which uses FPS (foot-pound-second) units. This is a constant and irritating source of confusion for engineers and scientists. Unfortunately, converting the entire American society from FPS to MKS is a gargantuan task, which has been attempted on several occasions with little success. With some Americans it actually seems to be a matter of national pride, which is rather ironic, because the use of FPS units is the last vestige of the domination of America by England, who defined the FPS units centuries ago, but who have now themselves converted much of their culture and industry to MKS units.

Here are some examples of propositions that might be stipulated as conventions for a task:

All measurements are in MKS units.

All images will be made available in GIF format.

Angles are measured in the counterclockwise direction from the positive abscissa.

The symbol π represents the ratio of the circumference of a circle to its diameter.

All software for the project is to be written in C++.

The syntax of the dictionary definitions is defined in Backus-Naur Form.

Investigators do not generally invent conventions, although they certainly may *suggest* their introduction. Most official conventions have emerged from common usage over a long period of time, during which they became *de facto* standards. The process of establishing official *de jure* conventions is usually undertaken by government agencies such as the U.S. National Institute of Standards and Technology (NIST), international agencies such as the International Standards Organization (ISO), or major scientific and engineering societies such as IEEE and ASME.

Definition 6.7 A **definition** is a statement of meaning or membership.

Investigators have the freedom to define anything any way they want, *as long as the item to be defined is free of constraints imposed by any other widely accepted knowledge propositions.* They are free to define the value and role of a variable x in a mathematical equation, but *not* free to define the capitol of the United States as Boston (or anywhere else), because the definition of this term has already been stipulated as Washington, DC and is now a *fact*. Investigators are free to define the concept of intelligence, but are *not* free to define cigarette smoking as harmless, because there is conclusive evidence to the contrary. Sometimes these constraints on definitions are a bit fuzzy. For example, even though the *convention* of using the symbol π to represent the ratio of the circumference of a circle to its diameter was established long ago, mathematicians may choose to define the symbol π completely differently. Most mathematicians would ardently defend their right to use any symbol at any time for anything, almost as a matter of principle (a kind of Mathematicians' Bill of Rights). Nonetheless, conflicting definitions for π and other common symbols such as e and $+$ and Σ are confusing and should be avoided for practical reasons, unless there is an especially abiding reason for them.

Definitions are most commonly used to assign values to concepts or members to sets. For example, an investigator may wish to define "education level" as the number of years of education the subject has had, as opposed to, say, the highest degree the subject has earned. Or the investigator may find it convenient to define the term "highway vehicles" as cars, trucks, and buses, but not motorcycles and bicycles. Having once defined a term, then the investigator may use that term repeatedly throughout the investigation and expect the audience to remember and understand its meaning (although an occasional reminder never hurts). It becomes a kind of shorthand notation. Note that if the term "education level" were a performance metric instead of a demographic descriptor, then it would not be appropriate to *define* it; it must be *postulated*. A performance metric is a presumptive proposition and therefore requires more than merely the stipulation of its meaning; it requires a strong and clear justification that the metric is a useful and appropriate measure of performance, which is much more demanding.

6.4 Conclusive Knowledge

Conclusive propositions express knowledge obtained by statistical inference or formal proof. A conclusive proposition, subject to the accompanying fixed conditions and parameters (**governing propositions**) and the confidence with which it was validated, may be considered a *bona fide* addition to the accumulated knowledge of the human species. This is a profound statement, and the implied pedagogical responsibility should serve as a strong motivation for investigators to observe the rigorous methodology of the **Scientific Method**, all the way from a clear formulation of the problem to the final comprehensive peer review, as explained in Part IV.

There are two types of conclusive knowledge: results and conclusions.

Definition 6.8 A **result** is an intermediate measure of performance for a task.

Definition 6.9 A **conclusion** is the final result of a task that states the extent to which the task objective has been achieved.

Often the **conclusion** of a task is a decision that is required by the task objective. In our running list of example tasks, for example, the task *Identify*

Faster Traffic must ultimately decide which highway has the faster traffic traveling on it. The domain for this task includes the two lists of traffic speeds, which presumably were the conclusions of a previous task called something like *Obtain Samples of Traffic Speeds*. After receiving these two lists of speeds, the traffic engineer computes an average and a standard deviation for each list. From these four numbers, she applies the **performance metric** she has postulated for the task, which computes the standard difference between the averages of the two samples. Assuming the samples are normally distributed, she converts this standard difference into the statistical confidence that the two average speeds are different (*i.e.*, the confidence with which the null hypothesis of no difference may be rejected). This confidence is an *intermediate* outcome of the task, *i.e.*, a **result** of the task. It is not the final outcome, because no decision has yet been made about which highway has the faster traffic on it. When the engineer (or someone higher up the management ladder) uses this result (the statistical confidence) to *decide* which traffic is faster, this will be the *final* outcome or **conclusion** of the task. Furthermore, as you may remember, the introduction to this running example task in Chapter 4 suggested that its conclusion would become an input to another higher-level task that would decide which highway will be widened.

Even within a single task, it is often a long journey from the initial results (the raw sensor data) to the final conclusion of the task. Many sequential steps compute a chain of intermediate results that slowly reduces the raw data down to a form from which the required conclusions may be drawn. Each one of the steps can, of course, be considered a task unto itself, which is completely consistent with the recursive task hierarchy described in Chapter 3, and the iterative application of the Scientific Method which will be described in Part IV.

Not all conclusive knowledge is the result of statistical inference. Sometimes we can *prove* things. Not too often, but sometimes. When a task proves something, the conclusion has a special name: a **theorem** or **lemma**.

> **Definition 6.10** A **theorem** (or **lemma**[1]) is a conclusion that has been formally proved.

In contrast with empirical methods, theorems and lemmas are proved using the enormously reliable formal method of mathematical deduction, which was first postulated by Thales in 600 BC, and mathematical induction, which was bequeathed to us by Francis Bacon and Rene Descartes in the seventeenth century. Mathematics is the *language* of science, a linguistic framework that is welded together by a consistent and highly versatile syntax, as well as the semantic machinery necessary to manipulate instances of this language to discover new methods of expression. The mechanisms of deductive and inductive proof are the arbiters of this language, the gatekeepers for the evolving lexicon of mathematics.

---

[1]Lemmas are mini-theorems, also called "helping theorems," which are used as stepping stones along the journey undertaken to prove a theorem — critical oases in the desert of logic, as it were. While theorems are *conclusions* of wide interest, usually lemmas are important only as intermediate *results* in the context of the specific task.

Very few tasks, however, are fortunate enough to be able to achieve their task objectives by proving theorems, and when this does occur, usually the task domain is mathematics itself or a field intimately related to mathematics, such as physics, statistics, information theory, and so forth. Because they also have strong mathematical and logical underpinnings, even the fields of philosophy and linguistics occasionally offer opportunities for closed-form proofs of fundamental concepts. However, most of the social sciences still do not lend themselves well to the application of formal logic to prove theorems, simply because the fundamental processes that rule human behavior are enormously complex and just not well enough understood yet. They must still be approached empirically, hoping that a stochastic understanding of the underlying engines of personal and societal behavior will be sufficient to capture the general trends, to catch a statistical glimpse of macroscopic behavior. Someday, perhaps, our understanding of human reasoning and learning will be sufficient to construct highly accurate models, the strong tools of formal mathematics will be brought to bear, and the smallest details of human behavior will surrender to the microscope of logic. Who knows?

In the meantime, even though the preponderance of R&D tasks in science and engineering cannot be expected to introduce many *new* theorems, most of these tasks can make excellent use of *existing* theorems to construct their solutions, to build their supporting knowledge structures, and to reduce the results of their tasks. There has been a rich harvest of theorems and lemmas over the last three millennia, and they are ripe for the picking. In fact, every time you compute the area of a circle, the distance between two points in a Euclidean space, the strength of an energy field, or the variance of the probability of an independent random event, you are probably using theorems that were proved hundreds, if not thousands of years ago.

Even though most problems do not yield to closed-form solutions, it is important to note that empirical methods are not inherently less rigorous than mathematical proofs. Both can be weak or strong, depending on the weaknesses or strengths of the fundamental knowledge propositions on which they depend. Modern science recognizes the validity of both approaches, as long as they are rigorously conducted within the strict confines of the Scientific Method. Because of the enormous importance of this formal methodology for the conduct of R&D tasks, the Scientific Method is the singular focus of Part IV of this book.

Table 6.2 adds some stipulative and conclusive propositions to the sample presumptive propositions listed in Table 6.1 for our running example tasks. Once again, the lists are very incomplete, representing only abbreviated samples of propositions that might reasonably be necessary for these tasks.

Exercise 6.1
Select a task you have recently completed or are now working on. List the knowledge propositions that are required for this task. Label each proposition with its category and type.

Table 6.2 Abbreviated samples of knowledge propositions for the running example tasks

		KNOWLEDGE PROPOSITION	
Task Name	*Type and Role*		*Statement*
Remove Screws	assumption	gcs	The screwdriver is large enough to loosen and remove all the screws.
	assumption	gcr	New hinges can be purchased at the only local hardware store.
	fact	gcs	The only local hardware store is open on Sundays from noon until 400 pm.
Launch Spacecraft into Orbit	law	gcr	Newton's Law of Gravitation (The force of gravity is proportional to . . .)
	theory	gcr	Model of atmospheric interaction with the launch vehicle
	assumption	gpcE	The astronaut will be in emotional control during inflight emergencies.
	conclusion	gcr	The thrust of the Atlas rocket is computed by the following formula: . . .
	definition	gpcL	The following telemetry data are transmitted from the Mercury capsule: . . .
	definition	gpcL	The physical state of the astronaut includes the following parameters: . . .
Identify Faster Traffic	assumption	gpcE	The distributions of the speed samples are normal.
	assumption	gpcL	The sample size of 100 is sufficient to provide accurate statistical conclusions.
	result	gpcL	The samples of 100 speeds of vehicles from both highways
	definition	gpcR	Only cars, trucks, and buses were included in the vehicle speed samples.
	theorem	gps	The proportion variance for a binomial experiment is given by: . . .
Prove Fermat's Last Conjecture	axioms	gc??	Many from mathematics
	laws	gc??	Many from mathematics
	theorems	gc??	Many from mathematics
	definitions	gc??	Many mathematical symbols (Let $\mathbf{\Phi}$ be the matrix of probabilities . . .)
Find New Planet	assumption	gpcE	The missing mass is located predominantly in one body orbiting the Sun.
	law	gcr	Newton's Law of Gravitation (The force of gravity is proportional to . . .)
	conclusion	gcsL	At 10^{10} km the Lowell telescope can resolve objects with diameters > 800 km.
Identify Collision Damage	assumption	gpsE	All damage caused by normal wear-and-tear can be visually recognized.
	assumption	gpcE	There is no hidden damage that could increase repair costs significantly.
	convention	gcsL	All cost estimates are rounded off to the nearest 10 dollars.
Measure Fossil Age	postulate	gcr	The order, thickness, and mineral composition of strata identify their ages.
	fact	gpc	The fossil was found 29.4 meters below the edge of the cliff.
Measure Drug Effectiveness	assumption	gpsE	A subject's unrestricted activities have no significant effect on drug benefit.
	assumption	gpsE	Subject's awareness of his own health does not violate the experiment blinding.
	assumption	fpvL	Reasonable drug doses are selected from the set {5-, 10-, 15-, 20mg}
	convention	gpc	The placebo for the control cohort is a 1% saline solution.
	conclusion	rcb	There is a 99% confidence that the drug cured at least 25% of the subjects.
Evaluate GUI Effectiveness	assumption	gcsE	The room temperature has no significant effect on user performance.
	assumption	gcs	All evaluations by the user are honest and sincere.
	postulate	gcr	The complexity of the exercise is computed with the following formula: . . .
	definition	gcr	The goal of the exercise is to find the best price for an Acme 920 VCR.
	result	gcsL	The brightness of all display screens is maintained within 0.5%.
Compile Summaries	assumption	gcsL	The internet will provide access to all of the important references.
	postulate	gppL	The following categories summarize the content of the articles adequately: . . .
	result	gcs	The earliest reference to Multiple-Personality Disorder was published in 1850. . . .
Measure Liquid Temperature Variation	assumption	gcrE	Heat loss from the thermocouple connection cables may be neglected.
	assumption	gpsE	The liquid temperature is the same throughout the volume of liquid.
	law	gps	The product of gas pressure and volume is constant at constant temperature.
	definition	fcp	Three temperature sensors are used: thermocouple; IR sensor; thermometer.
	fact	gcsL	The accuracy of the thermocouple is rated at ± 0.1 Centigrade degrees.
Measure Age of Organic Sample	assumption	gcsE	The effect of background radiation on the results may be neglected.
	postulate	gcr	The age of an organic sample is computed using the following formula: . . .
	conclusion	gcs	The radioactive decay products of carbon-14 include the following: . . .
	result	rrc	The average background radiation level is 1.76 ± 0.23 microrads.

Role Codes   *(optional)*

g = governing proposition	+	{ p = parameter + s = scheme or c = constant }	{ L = limitation
		{ c = condition + r = regime or s = situation	{ R = restriction
f = factor	+	p = parameter or c = condition + p = policy or v = variable	{ E = exclusion
r = range proposition	+	r = result or c = conclusion + (b = benefit and/or c = cost) or p = payoff	

CHAPTER 7

Roles of Knowledge Propositions

Except for speculative knowledge (opinion and dogma), each type of knowledge proposition defined in the previous chapter may play a variety of roles in the development of the knowledge framework for a research or development task. These different **roles** of knowledge propositions are summarized in Figure 7.1 and explained in the remainder of this chapter.

One way to figure out the roles of knowledge propositions for a task is to pose exploratory questions. As we shall see in Chapter 10 (Analysis), posing a **research question** often helps to clarify the issues and is an effective way to kick-start the process of designing a task.

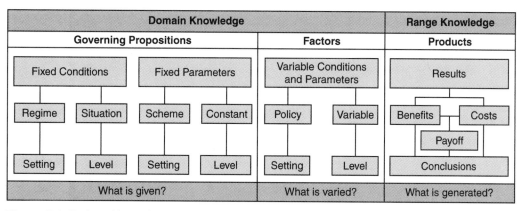

Figure 7.1 Roles of knowledge propositions for an R&D task

The first obvious question is: What knowledge is brought to the task to construct its fixed framework of information? What propositions are accepted as the foundation of knowledge for the task, even though many of them may lack conclusive evidence of their validity, such as assumptions and postulates? In short, what is given? Those knowledge propositions that remain constant during the execution of the task are referred to as **governing propositions**, because the task's reliance on the validity of this basic knowledge *governs* the entire task, in the same way that the basic premises of a deductive argument that are accepted without proof (such as axioms and postulates) govern the validity of the argument and its conclusions. The acceptance of a governing proposition, as has been pointed out repeatedly, carries an important admonition with it: a chain is only as strong as its weakest link.

This begs another question: What propositions are *not* fixed, but will, instead, be intentionally changed to explore their effects on the achievement of the task objective? In short, what is varied? Those propositions that the investigator intentionally selects to be varied to observe their effects on performance are called the **factors** of the task.

Which leaves the final question: What knowledge must the task generate in its range to achieve the task objective? How is the performance of the task to be measured? What are the results of the task? What kind of conclusions are to be drawn from the results? In short, what is generated? The results and conclusions generated by a task are the **range knowledge** of the task.

7.1 Governing Propositions

Governing propositions circumscribe the task by setting, limiting, restricting, or excluding certain **conditions** of the task environment and **parameters** of the task system. In so doing, they specify the bounds within which the results and conclusions of the task may be accepted as valid. Some of the governing propositions are imposed because of practical time and cost restrictions on the task, while others are necessary because of fundamental laws of the universe that cannot be abrogated or circumvented.

> **Definition 7.1** A **condition** is a specification of the task environment, whose value is either fixed or varied as a factor of the task.

> **Definition 7.2** A **parameter** is a specification of the task system, whose value is either fixed or varied as a factor of the task.

Once the environment and the system for a task domain have been identified, any knowledge proposition that specifies something about the environment is called a **condition** of the task, and any knowledge proposition that specifies something about the system is called a **parameter** of the task. Those conditions and parameters that are not intentionally varied during the execution of the task are called **fixed conditions** and **fixed parameters**. Even though most tasks have an enormous number of conditions and parameters that could be intentionally varied, typically most of them will not be varied, either because the effects on the objectives of the task are of no interest or irrelevant, or because the investigators simply do not have enough time or money to include so many conditions and parameters as factors. At some point during the design of the task, the investigators must decide which conditions and parameters will be varied and which will not, *i.e.,* which will be accepted as fixed governing propositions.

The number of governing propositions for a task is generally quite large, because they specify everything about the task that will be held at some fixed value during the task. Because there are typically so many governing propositions for a task, it is useful to define some names for the different kinds of governing propositions. This will establish a consistent and precise shorthand notation that can be used during the planning, execution, and documentation of the task (which, as was stated previously, is one of the major purposes of definitions).

> **Definition 7.3** A **regime** or a **scheme** is a process, function, rule, or mechanism that is a condition or a parameter, respectively; its value is called a setting.

> **Definition 7.4** A **situation** or a **constant** is a numerical or categorical quantity that is a condition or a parameter, respectively; its value is called a level.

We have defined a lot of new words here, but things are not really as complicated as they may seem to be. Examining the relative placement of these words in Figure 7.10 under the heading of governing propositions should help to clarify their roles and relationships in the task. Because regimes and schemes are processes, functions, rules, or mechanisms, one might well conclude that the types of proposition most commonly used to specify them are the postulate and the law, because the same words are used in their definitions (Definitions 6.2 and 6.4). This conclusion is correct. Similarly, the types of proposition most commonly used to specify situations and constants are the assumption and, although less often, the axiom; their definitions also use similar words (Definitions 6.3 and 6.1). The existence of the symmetries and commonalities between the types and roles of knowledge is not a coincidence; the definitions were intentionally set up with this parallelism in mind.

It is also appropriate to specify a governing proposition as a definition, fact, or convention if the knowledge can be stipulated, or as a conclusion (including the theorem) if the knowledge has been validated by a prior research task. Definitions are most appropriately used to assign arbitrary values to free variables or organize concepts into sets and thus establish concise references to these values and sets, just as we are now using them to organize the various categories, types, and roles of knowledge. In fact, the collection of knowledge propositions listed in this book are, of course, the governing propositions for the task undertaken by this book: *Establish a Formal Structure and Methodology for Conducting R&D Tasks*.

As shown in Figure 7.1, the names that refer to the assigned values of the governing propositions also exhibit a consistent symmetry: the values assigned to regimes and schemes are called **settings**, while the values assigned to situations and constant are called **levels**. A level is usually a simple assignment of a numerical or categorical quantity or set of quantities. A setting is generally more complicated, specifying a complete process or mechanism, such as formulas, blueprints, algorithms, instructions, flow charts, pseudo-code, and so forth.

Some governing propositions may be additionally characterized as **biased propositions**, reflecting some limitation, restriction, or exclusion that has been imposed on the task as a result of accepting the proposition. Many governing propositions, on the other hand, are not biased, because they do not impose a limitation, restriction, or exclusion. Called **neutral propositions**, they simply set an unbiased value of a condition or parameter.

> **Definition 7.5** A **limitation** is a fixed condition or fixed parameter that constrains the precision or amount of one of the task resources.

For example, perhaps an investigator must recognize the fact that the video camera used for an image-processing task provides only eight bits of grayscale intensity resolution, which will, in turn, limit the precision of the results and conclusions. Similarly, if a computer simulation of the motion of the planets of the Solar System must be limited by the simplifying assumption that the planets are represented as point masses, the predicted orbits may wander far from reality. If an experiment with a human cohort must be conducted with a very small number of subjects (which is very often the case), the lack of sufficient data can severely weaken the statistical confidence in the results. Each one of these example tasks accepts a biased proposition that is a **limitation** that could decrease the accuracy and increase the uncertainty of the results and conclusions.

> **Definition 7.6** A **restriction** is a fixed condition or fixed parameter that constrains the scope of the utility or validity of the products of a task.

It is often important for the investigator to identify how a governing proposition can restrict the utility of the results and conclusions of the task. For example, a lack of large-scale test equipment may restrict the conclusions of a vehicle-crash-test experiment to vehicles weighing less than 6000 pounds. Similarly, the lack of a sophisticated environmental test chamber may restrict the results of the quality-assurance tests of an electronic instrument to temperature ranges from −20 to +100 °F. If management is unwilling to spend the money necessary to develop a piece of software for a variety of operating systems, the package may be restricted to platforms running Unix. Very often a lack of resources or funds necessitates imposing biased propositions that place **restrictions** on the venues in which the products of the task are useful or valid.

> **Definition 7.7** An **exclusion** is a fixed condition or fixed parameter that avoids, minimizes, or ignores an unwanted influence on the task.

In the process of designing or modifying a device, engineers often take measures to forestall expected but unwanted effects that could disturb the operation of the device. For example, the casing of an electronic device might be constructed of stainless steel to minimize the influence of stray magnetic fields. Before delivering a tank whose engine must operate immediately and reliably in the field under any conditions, the production engineers decide to run the engine for several hours at the factory to avoid a low-speed break-in period in the field. Researchers also often apply exclusions to their tasks, such as intentionally hiding the type of experiment trial (control or test) from both the task team and the subjects before and during the experiment (double-blinding). **Exclusions** are biased propositions that recognize and avoid or minimize effects that could degrade the efficiency or reliability of the products of the task.

It may not be obvious at this point how the roles of governing propositions are assigned, but the process is fairly straightforward.

Step 1: Decide whether a given proposition is a governing proposition. This can be done by answering the question: Does this proposition describe knowledge that is used *during* the task, or is the knowledge generated *by* the task? If it is used during the task, then answer the question: Should the value of this proposition be intentionally varied during the task, or should it be kept fixed throughout the task? If the proposition is used *during* the task, but will retain a *fixed* value, then it is a *governing proposition*.

Step 2: Decide whether the governing proposition is a condition or parameter of the task. If it applies to the environment, then it is a *fixed condition*; if it applies to the system (which includes the task unit, of course), then it is a *fixed parameter*. Note that a governing proposition used to reduce the results of a task, *i.e.,* a performance metric, is a fixed condition, because the results derive from the output of the sensors, which are in the task environment. The performance metric was introduced previously in the discussion of the postulate.

Step 3: Narrow down the kind of governing proposition. If it specifies a process, function, rule, or mechanism, then the fixed condition is a *regime,* or the fixed parameter is a *scheme*. If the governing proposition specifies a numerical or categorical quantity, then the fixed condition is a *situation,* or the fixed parameter is a *constant*.

Step 4: Decide if the governing proposition is a biased proposition, *i.e.,* a *limitation,* a *restriction,* or an *exclusion*. To do this, ask the question: In the best of all possible worlds, would it be better if this governing proposition did not have to be imposed at all, *i.e.,* does it imply a possible weakness or uncertainty? If the answer is yes, then it is probably a biased proposition.

All the other outcomes of the four steps of this logical process designate the role of the proposition as either a *factor* or a **range proposition** of the task, which will be discussed shortly.

Let us examine a few of the governing propositions in Table 6.2 and use the logical process just outlined to determine their roles in the knowledge framework for the associated task. Note that the role of each proposition has been given a code in the third column of the table, which is explained in the role code footnotes at the bottom of the table. The role codes for factors (f) and range propositions (r) have not yet been explained, but will be shortly.

Launch Spacecraft into Orbit

assumption: The astronaut will be in emotional control during inflight emergencies.

role (gpcE):   g = governing proposition (fixed presumptive knowledge)
p = fixed parameter (in system: astronaut is part of task unit)
c = constant ("emotional control" is a categorical quantity)
E = exclusion (ignores an unwanted risk)

conclusion: The thrust of the Atlas rocket is computed by the following formula: . . .

role (gcr):   g = governing proposition (fixed presumptive knowledge)
c = fixed condition (in environment: rocket is an inducer)
r = regime (a function that models the rocket thrust)
(No fourth character in the code: not a biased proposition.)

Identify Faster Traffic

result: The samples of 100 speeds of vehicles from both highways.

role (gpcL):   g = governing proposition (fixed prior conclusive knowledge)
p = fixed parameter (in system: task unit is the speed samples)
c = constant (numerical results from a previous task)
L = limitation (ignores inaccuracy of samples with only 100 data points)

definition: Only cars, trucks, and buses were included in the vehicle speed samples.

role (gpcR):   g = governing proposition (fixed stipulative knowledge)
p = fixed parameter (in system: task unit is the speed samples)
c = constant (set of qualified vehicles is a categorical quantity)
R = restriction (limits the scope of the results and conclusions)

theorem: Proportion variance for a binomial distribution is given by: . . .

role (gps): g = governing proposition (fixed prior conclusive knowledge)

p = fixed parameter (in system: task unit is the speed samples)

s = scheme (theorem is expressed as a computational
function)

(No fourth character in the code: not a biased proposition.)

Find New Planet

law: Newton's Law of Gravitation

role (gcr): g = governing proposition (fixed presumptive knowledge)

c = fixed condition (in environment: gravity is the inducer)

r = regime (law is a function that models the force of gravity)

(No fourth character in the code: not a biased proposition.)

conclusion: At 10^{10} km the Lowell telescope can resolve objects with
diameters > 800 km.

role (gcsL): g = governing proposition (fixed stipulative knowledge)

c = fixed condition (in environment: telescope is the sensor)

s = situation (object diameter is a numerical quantity)

L = limitation (sensor precision could be too low to find
planet)

Evaluate GUI Effectiveness

assumption: The room temperature has no significant effect on user
performance.

role (gcsE): g = governing proposition (fixed presumptive knowledge)

c = fixed condition (in environment: laboratory)

s = situation (temperature of the room is a numerical quantity)

E = exclusion (neglects effects of changes in room temperature)

It may already have been noted that the naming conventions for governing
propositions is redundant to some extent. After all, if we know that a governing
proposition is a constant, then also stating that it is a parameter is unnecessary,
strictly speaking, because a constant is a parameter by definition. However, this
redundancy allows us to refer to a knowledge proposition in a number of ways at
various levels of abstraction. For example, in summary briefings for management
it may be enough to refer to a proposition simply as a *governing proposition*. If we
wish to be more specific without increasing the number of words significantly,
we may refer to the knowledge proposition as a *fixed parameter* or a *fixed condition*,
which automatically implies that it is a governing proposition. Finally, if we wish
to be very precise, then we may refer to the knowledge proposition as a *scheme*,
using a single word to provide all the important information about the type
and role of the knowledge proposition (in this case a governing proposition, a

parameter, and a scheme). The kind of bias, if any, can also be appended to any one of these references, if this level of detail is required.

7.2 Factors

Not all of the knowledge propositions for the task are fixed and unchanging. Almost always a few of them are intentionally varied to explore the impact of different settings or levels on the results of the task. Such a proposition is called a **factor**, which is analogous to an independent variable in the domain of a mathematical function.

> **Definition 7.8** A **factor** is a condition or parameter of a task whose value is intentionally varied to measure its impact on the results of the task.

The values that are applied to factors follow a structure parallel to that for regimes and situations, and for schemes and constants, an extension of the symmetry previously noted in Figure 7.1.

> **Definition 7.9** A **policy** is a process, function, rule, or mechanism that is a factor of a task; its value is called a **setting**.

> **Definition 7.10** A **variable** is a numerical or categorical quantity that is a factor of a task; its value is called a **level**.

In general, policies are different methods or techniques that an investigator wants to investigate as alternative approaches to achieving the task objective. For instance, engineers designing a new automobile fuel pump (the task unit) may want to consider both an electrical and a mechanical version. The version is one of the factors of the task, and each version is a different *policy,* because each represents a different mechanism. For each policy (version), the engineers may then want to vary several additional factors: the inlet aperture size (a parameter), the temperature of the working fluid (a condition), the viscosity of the working fluid (a condition), and for the electrical pump, the horsepower of the electric motor (parameter). Each of these factors is a *variable,* because it is expressed as a numerical quantity.

As another example, consider a research project that is investigating how the total amount of income tax paid varies with the demography of the taxpayer (task unit). Factors in this research might include the number of dependents, the income level, and the age of the taxpayer, all of which are parameters and variables. In addition the investigator may want to apply several different tax computation methods, factors that are conditions and policies.

Only two factors have been included in Table 6.2: the drug dose for the task *Measure Drug Effectiveness* (assumption → parameter, variable, limitation); and the type of temperature sensor for the task *Measure Liquid Temperature Varia-*

tion (definition → condition, policy). In addition, we may speculate briefly about some of the factors for several other running example tasks.

It will be recalled that the task **Remove Screws** is a descendant of the home-owner's primary task, **Fix Squeaky Hinges** (see Figure 3.2), in which the home-owner decided to try two different repair methods for the hinges: straightening them or replacing them. These were the two different settings for the policy (factor) which might have been called "repair method."

Andrew Wiles undoubtedly tried many different approaches to **Prove Fermat's Last Conjecture**, exploring a very long list of factors, including both *policies* and *variables,* drawn from the vast array of tools offered by classical and modern mathematics. Many of them were abandoned when they failed to yield useful results, but that does not refute the fact that they were formal factors at one time or another during the task. After all, that is often exactly what factors are for: to search through lots of possibilities to identify the optimum and discard the rest.

Probably the insurance adjuster who performs the task **Identify Collision Damage** will apply several different mechanisms to obtain an objective and complete appraisal of the damage to the car. First, he will inspect the car visually, taking photographs as needed. Second, if there is a chance that the engine, suspension, or brakes of the car were affected by the accident, he may drive the car himself to estimate the extent of the damage to these systems. Third, if the extent of the damage is unclear, he may ask a mechanic to help him diagnose the problems. All three *policies,* and probably several more, are possible factors for this task. The results of all of them will help him make his final damage estimate.

For the task **Evaluate GUI Effectiveness**, there are undoubtedly several factors the users will want to vary systematically to measure their impact on the effectiveness of the interface. For example, they may want to vary the background color of the graphics, the size and face of the text fonts, or the speed of sprite animation. These are all *variables,* whose levels are either categorical quantities selected from available sets of colors or font faces, or numerical quantities selected from available ranges of font sizes and animation speeds.

The distinction between policies and variables, schemes and constants, and regimes and situations is sometimes a bit fuzzy. It may not always be obvious whether the propositions are functions or quantities. This is not an issue of great concern, however, because deciding whether a factor is a policy or a variable is, in and of itself, not of great importance. Both kinds of factors are varied during the task, and as long as the investigator has clearly identified *what* must be varied, then what these factors are *called* is somewhat academic. Having said that, it is nonetheless important to sort the factors into the two piles of policies and variables as reasonably as possible, and to plan a precise sequence of the combinations of their values. Each combination of policy settings and variable levels is called a **treatment**. A set of treatments applied to a task unit during the course of an experiment is called a **trial**. These terms are formally defined and discussed in Section 12.2.2 of Chapter 12 (Synthesis).

The successful outcome of a task sometimes hinges on the sequence of treatments in a trial, *i.e.,* the order in which the different combinations of factor settings and levels are to be applied to the task unit. If the sequence of treatments in a trial is poorly planned, the application of one treatment might influence the results of a subsequent treatment. This makes the trial an order-dependent process, when in fact it is usually very important to be sure that results of individual treatments are as independent as possible from each other. As a simple example, consider how important it is to plan the order of a sequence of stress tests on a machine whose parts are expected to degrade during normal use, like a brake shoe.[2] Should the trial begin with the highest stress levels when the machine is in the best condition, or should it begin with the lowest stress levels so as to degrade the task unit as little as possible for as long as possible? There is no general right or wrong answer to this question, but a thoughtful decision must be made for each task. Of course, it might be best to apply each stress level to several machines, each at a different point in its life cycle, but such an ideal strategy is usually far too expensive.

The running examples in Table 6.2 include two examples of propositions that are specified as factors for their tasks. Their types and roles are summarized in what follows.

Measure Drug Effectiveness

assumption: Reasonable drug doses are selected from the set {5-, 10-, 15-, 20 mg}.

role (fpvL): f = factor (presumptive knowledge varied during experiments)

p = parameter (in system: cohort is the task unit)

v = variable (drug dose is a numerical quantity)

L = limitation (dose set may be too restricted)

Measure Liquid Temperature Variation

definition: Three temperature sensors are used: thermocouple, IR sensor, and thermometer.

role (fcp): f = factor (stipulative knowledge varied during experiments)

c = condition (in environment: temperature sensors)

p = policy (sensors are mechanisms)

(No fourth character in the code: not a biased proposition.)

Finally, although an investigator would prefer to have direct control over the conditions and parameters of the task, sometimes this is just not possible. For example, suppose an investigator wants to collect a set of questionnaire answers from people on the street, and then correlate these answers with their heights

---

[2]This is known as **partial-destructive testing**. **Full-destructive testing** causes total failure or destruction of the task unit, such as occurs when testing a flashbulb or a match. Clearly, most investigators try to avoid destructive testing of either kind, unless samples of the task unit are very cheap to acquire and test, or unless very accurate reliability estimates are essential, regardless of expense. If there is little or no cost involved, every task unit under test should eventually be driven into failure. See Chapter 12 (Synthesis).

(a variable factor). It would be impossible to interview the people in an orderly fashion by height. The responses (questionnaire answers and heights) must be collected on a first-come first-served basis as the people walk by. Then later the answers may be sorted by height, and a representative sample selected. Likewise, if an investigator wants to measure the behavior of a cohort when it is snowing (a fixed condition and situation), the supervisor must simply wait until Mother Nature decides to dish out a snow storm. Governing propositions that cannot be directly controlled by the supervisor are called **indirect conditions** or **indirect parameters**, and if the indirect parameter or condition is to be intentionally varied, then it is called an **indirect factor**. Thus, in the first example, the height of the respondent was an indirect factor, and in the second example the snowstorm requirement was an indirect condition.

   In summary, factors are knowledge propositions whose values are intentionally and systematically varied during the execution of the task, directly or indirectly. They are selected by the investigators from the entire set of knowledge propositions for the task and are assigned their values in some sequence during the execution of the task, while the values of the governing propositions remain fixed with their initially assigned values. Each combination of policy settings and variable levels is called a treatment. To minimize unwanted interactions among the treatments, their sequence in a trial must be very carefully designed. Guidelines for the design of **protocols** for experiment trials are given in Chapter 12 (Synthesis).

7.3 Range Knowledge

Range knowledge comprises the knowledge propositions that are the products of a task. These propositions may be the sole product of the task, or they may accompany other products of the task, such as devices and effects. Unlike governing propositions and factors, range propositions have their own special knowledge category, conclusive knowledge, and their own special types of knowledge propositions, results and conclusions. The theorem is a special kind of conclusion that refers to the final result of a mathematical proof. The types of conclusive knowledge have already been defined in Definitions 6.8, 6.9. and 6.10, and explained in Section 6.4.

   All range propositions are types of conclusive knowledge, but the role of conclusive knowledge propositions is not limited to range propositions. The results and conclusions of a previous task may be assigned the role of governing propositions or factors in the current task. This inheritance of information, passed forward from one investigation to the next, is exactly how human civilization has accumulated knowledge throughout the epistemological journey described in Chapter 5. Knowledge begets knowledge.

In Table 6.2 there are six examples of range propositions from a previous task used as governing propositions for the current task. Two of them are specified for the task *Identify Faster Traffic* and were explained in some detail at the end of Section 7.1.

It is often convenient to differentiate between two kinds of results of a task: benefits and costs, as well as two kinds of summary metrics, quality and payoff.

Definition 7.11 A **benefit** is a performance metric that specifies an advantageous result of applying the solution to the task unit.

Definition 7.12 A **cost** is a performance metric that specifies an immediate or long-term disadvantage of an associated benefit.

Definition 7.13 A **quality** metric is an arithmetic combination of several benefits that summarizes the extent to which the task objective has been achieved.

Definition 7.14 A **payoff** is an arithmetic combination of one or more benefits and costs that summarizes the impact of their interaction by increasing in value with increasing benefit and decreasing in value with increasing cost.

Ideally, every solution would completely repair or eradicate the associated problem. Reality, however, reminds us that solutions are rarely perfect and sometimes yield only marginal improvements. In fact, many tasks result in dismal failure. The benefit of a result or conclusion from a task can run the gamut from total failure to perfect success (although both extremes are highly unlikely). Thus, it makes sense to characterize the proportional **benefit** of a solution, rather than a binary conclusion that states either "it worked" or "it didn't work." Think how grateful we are for the many pharmaceutical products that have been developed to treat diseases and ease pain, even though very few actually *cure* the disease or kill the pain entirely. In truth, we marvel at a new drug that reduces the death rate from a serious disease by a mere 10%. We are not holding our collective breath for a breakthrough that increases efficiency of the family car by 1000%, even though that might be theoretically possible. Engineers are realists; even if the product of a task is far less than perfect, it may be good enough to acknowledge that the objective of the task was achieved to an acceptable extent.

By the same token, it is unreasonable to assume that a benefit can be realized without some associated costs, both direct and indirect. Every scientist and engineer learns that bringing order to one part of the universe must *inevitably* bring disorder to another part of the universe. While our internal combustion engines solve many of our transportation problems for us, they foul our atmosphere, and it is theoretically impossible to clean them up completely. The dream of unlimited, almost cost-free nuclear energy in the 1950s has turned into a nightmare of nuclear waste, whose disposal has presented us with a staggering, new problem. The prospect of genetic engineering techniques that can extend the human lifespan by a factor of three (or more) compels many to wonder how we will deal with the catastrophic population explosion that will inevitably result.

We try to train our engineers and scientists today to consider both sides of the coin: the **benefits** and the **costs**. And not just the immediate costs, but the long-term costs. And not just the energy or financial costs, but the environmental and societal costs as well. It is unfortunately an uphill battle, because most people do not want to waste the time or money educating our engineers and scientists in the subjects necessary to give them the sufficient vision to foresee nonobvious long-term costs: history; social science; and, yes, the humanities. This objection to giving engineering and science students a broader background is short-sighted. Feeding students a lean diet of technical information may make them smart enough, but it doesn't make them wise enough.

Although there is a strong moral imperative for understanding and measuring (or predicting) the long-term indirect costs of technological progress, a more prosaic example will effectively illustrate the purpose of postulating a comprehensive cost model and carefully segregating the results and conclusions of a task into benefits and costs. Suppose the government institutes a tax credit for people who use less home-heating fuel in winter by lowering their thermostats in the interest of conserving our finite reserves of fossil fuels. The policy designers, relying on narrowly educated engineers, assume that the benefit of this policy is the amount of fuel saved, and the cost is the revenues lost because of the tax credits. In our not-so-hypothetical example, however, another less obvious cost is unexpectedly incurred, which turns out to be even larger than the loss of tax revenues: an enormous increase in medical expenses for the treatment of pneumonia and respiratory infections, particularly among elderly and poor people, who found they could earn large and badly needed tax credits by turning their thermostats down to dangerously low values. This potential cost was totally overlooked when the new policy was designed; it never occurred to the policy makers to try to predict the social consequences of the new policy by gathering demographic information about those who would use the tax credit. Only later, when the increase in sickness and death became evident as a result of longitudinal studies, was the correlation made, and by then it was too late. A pilot study designed by more broadly educated engineers *in cooperation with health-care professionals and social scientists* might well have revealed the hidden cost before the policy was implemented on a large scale.

There are two important lessons here. First, although the potential benefits of a task are usually easy to identify from the outset of a task, the potential costs are much more difficult to predict. Therefore, it behooves us to pay more attention to the identification of the immediate and long-term, direct and indirect costs, and unintended side effects of a technological solution, which are often very nonobvious and elusive. Second, the results and conclusions of a task should always be carefully segregated into benefits and costs and analyzed separately. Merging them into a single performance metric may hide the real impact of the costs and obscure meaningful information as a result of arithmetic operations (*e.g.,* statistical moments) that ignore the details of the underlying distributions. It may also make the inclusion of new costs much more difficult, necessitating a change in a postulated performance metric, rather than simply in the list of results.

> **TIP 7**
> Combining the individual benefits into a summary metric (quality) or combining the benefits and cost into a single summary metric (payoff) is always possible, but has the potential of hiding meaningful information.

Finally, it is entirely possible that a result of a task is both a benefit *and* a cost. While one aphorism suggests: "Every cloud has a silver lining," another warns: "Beware of Greeks bearing gifts." Consider, for example, an amazing result of research in genetic engineering that will probably be achieved in the not-too-distant future: a drug that confers immortality, or at least that lengthens the human life expectancy to many hundreds of years. How would you characterize such a breakthrough: a benefit or a cost? For industrial R&D tasks, anything that keeps the stockholders happy may be considered a benefit, but in academic research, scientists have the freedom to carefully consider the impact of their discoveries and inventions. If not they, who?

The two example tasks in Table 6.2 for which a result or conclusion has been listed will serve to illustrate how results and conclusions are stated as (new) knowledge propositions, which can also be characterized as benefits or costs.

Measure Drug Effectiveness

conclusion: There is a 99% confidence that the drug cured at least 25% of the subjects.

role (rcb):   r = range proposition (knowledge product of the task)
c = conclusion (final outcome of the task)
b = benefit (indicates the extent to which the task objective has been achieved)

Measure Age of Organic Sample

result: The average background radiation level is 1.76 ± 0.23 micro-rads (± 1 standard deviation).

role: r = range proposition (knowledge product of the task)

(rrc) r = result (intermediate knowledge by the task)

c = cost (noise that degrades the significant sample radiation)

This concludes our discussion of the roles of knowledge propositions. It is important for project team members to maintain a comprehensive list of all knowledge propositions and their roles for the task. The construction of this list is always a work in progress; it will continue to evolve as the task is planned and executed.

It is not difficult to formulate a knowledge proposition for a task once the need for one has been recognized. The trick is recognizing the need for it in the first place. This recognition usually comes during the planning process, like a beacon from a distant shore. However, as you navigate the labyrinth of creativity, focused squarely on design tasks, the last thing you need is to be sidetracked by the recognition of some academic knowledge proposition. It is much easier to make a quick mental note to yourself and push on to "more important" things in the rush of inspiration. While that maintains the *creative* course, when you eventually get around to the task of codifying the knowledge propositions, you will have to wrack your brain trying to recollect those mental notes, probably with little success. So a compromise is recommended. When a knowledge proposition pops into your head, to avoid breaking your train of thought completely, just stop for just a moment, jot down a brief note in your research notebook (which is right beside you!), and then return to your creative task. Don't worry about using the right words or spelling or phraseology for these notes. Just get the basic information down in shorthand form. Later, when you find a stopping place, you can go back to these notes, clean them up, organize them, and steadily build a solid and complete set of knowledge propositions.

> **TIP 8**
> As each task is planned, make a brief note of each knowledge proposition in your research book as it occurs to you. Periodically collect the notes into a master list, wordsmith them into *bona fide* knowledge propositions, organize them by category, type, and role, and make sure there are no conflicts.

> **Exercise 7.1**
> For each knowledge proposition you listed in Exercise 6.1, give a complete codified description of its role in the task.

CHAPTER 8

Limits of Knowledge

A major focus of physics research in the nineteenth century was the origin and behavior of electricity and magnetism, which eventually led to the electromagnetic field theory published by James Clerk Maxwell in 1873. In 1905 Einstein unified the behavior of mechanical systems and Maxwell's theory of electromagnetism, publishing his work in a series of papers in the prestigious journal *Annalen der Physik,* including his famous Special Theory of Relativity. Although the brilliance of Einstein and this insight cannot be overstated,[3] it must be remembered that other very bright minds in the physics community at that time were also on the brink of this breakthrough. If Einstein had somehow been deprived of the opportunity of publishing his work (which he almost was, being a mere patent examiner), someone else, albeit not as brilliant, would have probably figured it out a few years later. To physicists of the late nineteenth century, a unified theory of mechanics and electromagnetism was their Holy Grail.

Research on electromagnetic radiation was begun in earnest in the mid-1660s by Isaac Newton, who spent his undergraduate years at Cambridge University using telescopes and prisms to investigate the nature of light, the most obvious form of electromagnetic radiation. His book *Optiks,* which was published in English in 1704, launched the passionate crusade to unlock the secrets of light, which culminated in the theories of Maxwell and Einstein 200 years later.

---

[3]What is perhaps an equally astonishing testament to Einstein's genius is that his famous 1905 paper was based on an essay called *On the Electrodynamics of Moving Bodies,* which he had written 10 years earlier when he was 16 years old! What were you doing when you were 16 years old?

Light was a tricky medium to investigate. Everyone could see it, but no one could feel it. Anyone could stop it, but no one could trap it. It was a mystery worthy of the best minds of the Scientific Revolution. Slowly but surely, scientists added to Newton's primer. One important research question was raised very early in this crusade: how fast is light propagated through space? It was evident that it moved very quickly, because no one had yet been able to observe any delay. Thus, one of the early challenges that scientists took up was to measure the speed of light.

There is a poorly documented (and perhaps apocryphal) account of two French scientists in the late eighteenth century who designed and conducted a simple experiment to measure the speed of light. The protocol of the experiment was as follows. Scientist A stood on the top of hill A with a lantern and the eighteenth-century equivalent of a stopwatch, while Scientist B stood on top of hill B with another lantern. The distance between the two lanterns on hill A and hill B, which the scientists had carefully measured, was 2640 meters (expressed in modern units[4]). Both lanterns had shutters, which were closed at the outset of the experiment. To start a measurement, Scientist A opened the shutter on his lantern and started his stopwatch simultaneously. When Scientist B saw the light from hill A, he opened the shutter on his lantern. When Scientist A on Hill A saw the light coming from the lantern on Hill B, he stopped the stopwatch and recorded the elapsed time. The two scientists repeated this experiment 100 times. They then went back to the local bistro and reduced their results, which they announced to the world (but never formally published) in a series of lectures during the following year. Based on a large sample of the time delay and the distance between the lanterns, the scientists concluded that the speed of light was 3040 meters per second with a standard deviation of 32.3 meters per second.

---

[4]To unify the many inconsistent measurement systems in Europe at the time, the meter-kilogram-second (MKS) system of units was introduced in 1795 by Lagrange, Laplace, and Lavoisier at the request of the new government of the French Revolution. It is now the standard everywhere in the world except the United States.

Which, of course, is wrong. We now know the speed of light is about 300 million meters per second, so their result was off by a factor of 100,000, an accuracy of about 0.001%. However, although the result was highly inaccurate, their measurements were quite precise; they were able to measure the distance to the nearest 10 meters and the time to the nearest hundredth of a second. Moreover, the ratio of the standard deviation to the average, a measure of fractional uncertainty called the coefficient of variation, was about 1%, indicating very little uncertainty. Although it was not the speed of light, clearly they were measuring *something* with high precision and accuracy and little uncertainty.

As you have probably deduced, what these two scientists had actually measured was their combined physiological reaction times, *i.e.,* the time delay between seeing a stimulus (light from the lantern) and acting in response (stopping the stopwatch). Even though they were saddled with the primitive technology of the late eighteenth century, they really should have known better. Their experiment protocol had a serious fault, one that Galileo, Descartes, and Bacon had warned about almost two centuries earlier.

The task method used by the two scientists was based on two assumptions: 1) Their timing device was sufficiently accurate and precise (condition, regime, limitation); and 2) their physiological reaction times were significantly shorter than the light delay and could therefore be neglected (condition, situation, exclusion). The first assumption was necessary; presumably they were using the best timing mechanisms available in the late eighteenth century. The second assumption, however, was false and unnecessary. Moreover, it was probably completely overlooked by the two scientists, because to check the reasonableness of this assumption, all the scientists needed to do was to conduct a **control task**. By repeating the experiment while standing just one meter apart, they would have obtained almost exactly the same time delay, a dead giveaway that their results were being swamped by a large and constant **source of error**. Had they discussed their experiment protocol and their results with their peers before embarking on their lecture tour, almost certainly someone would have warned them of the critical deficiency in their protocol. That's what Galileo, Descartes, and Bacon had insisted must be a required and routine phase in every research task: **validation** through **peer review**. This topic is discussed in detail in Chapter 13 (Validation).

Could the scientists have increased the distance between the lanterns and improved the accuracy of their results? Despite the reasonably high precision of their measurements for this period in history, they would have had to use a distance of at least *4000 kilometers* to discriminate between the light delay and their combined physiological reaction times with a statistical confidence greater than 90%. Clearly that was impossible. No, in addition to a rigorous control task, they simply needed a much more precise timing device. It was more than a century later in 1923 before advances in the technology of timing devices allowed Albert Michelson at the University of Chicago to obtain the first accurate and highly precise measurement of the speed of light with very little uncertainty: 299,798,000 meters per second.

> **Exercise 8.1**
> Confirm that the two scientists would have needed a distance between the lanterns of at least 4000 kilometers to discriminate between the light delay and their combined physiological reaction times (say 1.0 second) with at least a 90% confidence. Assume the coefficient of variation of the delay times is 0.01. Be sure to define all the task components and state all your governing propositions explicitly.

Three words appeared several times in the previous story: **precision, accuracy**, and **uncertainty**. Although the words *precision* and *accuracy* are often thought to be interchangeable, they have very different meanings. Ironically, such misuse happens more frequently in technical contexts than in literary contexts or everyday conversation. When you tell someone something during a normal conversation, and then they ask you to be more *precise*, they are probably not doubting the *accuracy* of your statement, but simply want additional descriptive information to help clarify their understanding. For example, if you tell someone you saw a very tall building, their request to be more precise probably implies that they want more details, such as the number of stories. Similarly, when you admire someone for speaking precisely, you are probably not referring to the *accuracy* of what they are saying, but rather their careful enunciation. On the other hand, when you characterize the statement "*Wuthering Heights* was written by Henry James in 1847" as inaccurate, you are disputing its accuracy, not the level of detail. All these conversational examples use the words *precision* and *accuracy* correctly (*i.e.,* accurately).

Unfortunately, in technical contexts the terms *precision* and *accuracy* are used incorrectly much more frequently. Often we describe a measurement as very precise, when we really mean that it is very accurate, *i.e.,* very close to the correct answer (assuming we know the correct answer).

The word *uncertainty* is usually applied accurately in both conversational and technical contexts. It indicates the level of confidence in some knowledge, from a conversational statement such as "I think her name is Kathy, but I'm not sure," to a technical statement such as "There is at least a 95% confidence that a measurement of the length of the room will be between 9.9 and 10.1 meters," or "The standard deviation of the sample of measured vehicle speeds is 7.4 miles per hour." Uncertainty can arise from the variance of a source of information, like a sensor, or from the propagation and accumulation of uncertainty along a chain of arithmetic operations.

It is important to emphasize the differences among the three concepts of precision, accuracy, and uncertainty. Consider the following thought experiment. Suppose we ask four different groups of people to answer the question: "What is the sidereal period of the orbit of the planet Mars?" The members of each group are allowed to discuss and refine their answers among themselves before the group submits its final result, which is the average and standard deviation of their individual answers.

The average answer for the group of *astronomers* is 686.98 days with a standard deviation of 0.05 days, which is very precise, highly accurate (correct to two decimal places), and extremely certain (very low coefficient of variation). We would expect no less of astronomers.

The average answer for the group of *politicians* is 584.1235 days with a standard deviation of 276.4531 days, which is highly precise, fairly accurate, and rather uncertain. (The coefficient of variation is 0.47, which is very large.) Obviously, the answers of the individual politicians ranged all over the place. Perhaps this is a key attribute of a representative democracy: although legislators seldom agree with each other, they often manage collectively to come up with a reasonable answer, albeit typically articulated with excessive precision.

The average answer for a group of *first-grade children* is 8 days with a standard deviation of 91 days, which is very imprecise, very inaccurate, and extremely uncertain. (The coefficient of variation is 11.3, which is enormous.) What else would we expect? Just imagine what the discussion among the children must have been like.

The average answer for a group of *teenagers* is 122.45 days with a standard deviation of 9.03 days, which is very precise, very inaccurate, and very certain. (The very low coefficient of variation of 0.073 indicates close agreement among the discussants.) Peer pressure can be a formidable force for uniformity. As George Bernard Shaw once said: "The trouble with youth is that it is wasted on the young."

From these brief examples, it may already be clear that *precision* expresses the level of detail provided by a value, *accuracy* expresses the relative correctness of a value, and *uncertainty* expresses the consistency of the result. To avoid ambiguity, however, we need to establish precise, accurate, and very certain definitions of these three important terms.

8.1 Accuracy and Error

> **Definition 8.1 Accuracy** is a measure of how close a result is to some baseline.

Tweedledum and Tweedledee stood facing each other on a hot and sultry afternoon. Tweedledee said, "Dear brother, how far apart are we standing?" Tweedledum, who'd had his eyes squeezed tightly shut, opened them wide, scanned the sky, squinted one eye, licked his thumb, and raised it into the air. "About five feet, I'd say, dear brother." Tweedledee rolled his eyes in scorn and said pontifically, "No, dear brother, that's not how *far apart* we are, that's how *close* we are! Please answer the question posed." Tweedledum stared unblinkingly at his brother for several hours while the sweat streamed down his face. Finally he pronounced, "I am no closer to being far apart from you than you are far apart from being close to me." And then they had a cup of tea.

Lewis Carroll did not write this. But, had he undertaken to philosophize about the concept of accuracy, he might well have. Furthermore, he might have suggested a tape measure with two sides to it: one for measuring how far apart things are, and the other for measuring how close things are. If he had labeled each side of the tape, the "how close" side might have been titled "Accuracy," and the "how far apart" side might have been titled "Error."

To measure accuracy and error the same way seems silly — at the very least a waste of two perfectly good words. Moreover, to measure accuracy, which is an optimistic metric, as the distance from a baseline seems inappropriate; as the distance gets smaller and smaller, the accuracy, which is improving, goes toward zero. This seems backwards. Golf scores notwithstanding, improving performance is usually reflected by *increasing* values of the performance metric. It makes good sense to represent high accuracy by high values, and poor accuracy by low values.

How can we postulate an accuracy metric whose value increases as the result gets closer to the baseline? The answer is to start with the other side of the tape measure: error. As the distance from the baseline gets larger and larger, the error, which is a pessimistic metric, also gets larger and larger, and *vice versa*. This seems appropriate. Once we have defined the term "error" and postulated some metrics for it, then perhaps we can postulate some metrics for accuracy that behave in the opposite way.

> **Definition 8.2 Error** is a measure of the distance of a result from some baseline.

Although unthinkable from a literary point of view, the question posed by Tweedledee might now be: "Dear brother, given that my position is correct, what is the error in your position?" Of course, Tweedledum might well take umbrage at the assumption made by Tweedledee that *his* position is "correct." Nonetheless, even with the question reversed, Tweedledum's answer would probably be the same as before. No wonder so few issues were ever put to rest in Wonderland.

Before we can postulate any metrics for error or accuracy, we must first define a concept that so far in this discussion has been taken for granted: **baseline**. A baseline may be defined as any one of several different reference values: the correct value, the desired value, the expected value, *etc.* In fact, several different baselines may be defined for a task, so that the error and accuracy of the results and conclusions can be appreciated from several different perspectives. There really are no constraints on the definition of a baseline except, of course, for the caveat that accompanies the validity of *all* stipulative knowledge: it must reflect the consensus of all those participating in the task, arrived at with the expectation that the process of peer review will confirm its validity, or at least not reject it outright.

Perhaps the purest definition for a baseline is the *correct value,* which has been established as conclusive knowledge or stipulated as fact. Unfortunately, the correct value for the result of a task is very often unknown. It would be safe to say that neither Tweedledum nor Tweedledee knew the correct distance between them. And certainly the two French scientists who tried to measure the speed of light were completely in the dark about its correct value; if they had had any notion at all, they would have quickly realized the immensity of their error (if not the source of it) and not set off on a self-congratulatory lecture tour.

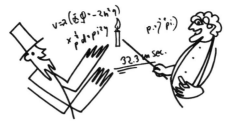

If we examine the example running tasks in Table 6.2, only the first two have conclusive knowledge of the correct values for the results of their tasks. By having counted them in a previous task, the homeowner performing the task *Remove Screws* knows that there are six screws to remove. Therefore, as he removes the screws one at a time, he can compute the error at the end of each step. At the end of the first step the error is five screws, and it decreases by one at the end of each step until the error is zero at the end of the task when all six screws have been removed. The task *Launch Spacecraft into Orbit* had access to considerable conclusive knowledge that permitted a highly accurate prediction of the final orbit into which the rocket would insert the spacecraft. Any deviation from this prediction could have been rigorously characterized as an error. None of the other example tasks in Table 4.1 has the luxury of knowing what the correct answer is.

Sometimes the correct value for a result of a task can be stipulated as *fact*. For example, if the task is to count the number of cups sitting on a table using an image-processing algorithm applied to a video image of the table, then certainly the correct number of cups on the table can be determined by simply having a human look at the table and count the cups. It is not unusual to test a task method (solution *and* experiments) by generating test data for which the correct solution is known. We can shake-test an aircraft navigation instrument and measure its performance under carefully controlled laboratory conditions before we expose the instrument to the violent conditions inside the avionics bay on a jet fighter aircraft in combat. Similarly, to test a clustering algorithm, we can generate probability density functions of known mixtures of parametric components, and measure the error of the clustering algorithm under these test conditions before applying it to unknown distributions acquired from real-life situations. Neither of these examples will give us measures of performance under actual operating conditions, but if our test conditions are sufficiently realistic, the assumptions about the estimates of the probable error will be accepted as valid. Considerable care must be taken to justify the credibility of the test conditions in a set of appropriate governing propositions.

When the correct answer is not available, the baseline may be defined by a *desired* value. Two of the example tasks in Table 4.1 have desired results that were probably defined by management. For the task *Measure Drug Effectiveness*, the marketing department may have determined that the drug will not be profitable unless it can cure at least 10% of the cohort. To justify further investment in the development and production of the software product, the management of the company conducting the task *Evaluate GUI Effectiveness* may have defined a minimum acceptable improvement in the performance of the cohort using the new GUI (test trials) over the performance of the cohort using the competitor's GUI (baseline trials). Comparing the benefits of two or more competing products is a common example of using a desired value as a baseline for error measurement and estimation. The process of defining desired values for the results of a task, also known as **performance criteria**, is discussed in Chapter 10 (Analysis).

Regardless of the basis for the baseline value, a word of caution must be repeated here. Great care must be taken to make sure that the people who are actually involved in designing and executing the task *do not know* what this baseline value is. They may quite naturally guess that one exists, but keeping its exact value secret from them until all the test and control data have been acquired is the only way to ensure that results have not been tainted by unconscious manipulations of the experiment or the data. Much more will be said about avoiding such "self-fulfilling prophecies" in Chapter 12 (Synthesis).

The **raw error** is a dimensional quantity that expresses the difference between the result and the baseline in the same units as the result itself. If the measurement of a voltage is 4.9 volts and the baseline (desired) value is 5.0 volts, then the absolute error is 0.1 volts. Comparing the raw errors from several different dynamic ranges of the measurements, however, may be misleading. Suppose, for example, that the measurement of the voltage just mentioned is taken from a power supply designed to provide three different supply voltages, and the objective of the task is to measure the error in these three supply voltages. Samples of voltages at the three settings yield the following average raw errors: 0.12 volts for the 1.0-volt supply, 0.96 volts for the 8.0-volt supply, and 1.8 volts for the 15-volt supply. Although it is quite obvious that the raw error increases as the voltage increases, the *ratio* of each raw error to its baseline is constant at 0.12, or 12%. This dimensionless ratio, which is called the **fractional error**, facilitates the comparison of errors across different dynamic ranges. For example, although Australia suffered far fewer battle deaths in World War II (27,000) than the United States (292,000), the percentage of Australian soldiers who were killed (2.7%) was much greater than the percentage of US soldiers (1.8%).

Table 8.1 postulates several of the most common type of metrics used to represent raw and fractional errors. Note that some metrics use the signed value of the error, while others use the absolute value; the best choice is simply whichever is more appropriate to the task.

The *fractional* metrics postulated in Table 8.1 have a potential problem. If the denominator of a mathematical expression is zero, the operation of division is not defined, and neither is the quotient.[5] Thus, if the value of the baseline β is zero, the values of the fractional metrics cannot be computed. Unfortunately, the value of the baseline may well be zero. For example, if a weight scale is being calibrated with no weight on the pan, then the desired value of the result (the

---

[5]Unfortunately, many scientists and engineers have been taught that division by zero is infinity. This is not true. Division by zero is mathematically undefined. It may be possible to apply limit theory to show that the value of a certain function *approaches* infinity as the denominator approaches zero in the limit. However, to assume this is true in general without doing the math is inviting catastrophe — the airplane may well fall out of the sky.

Table 8.1 Common error metrics

Formula	Type	Formula	Type
$\varepsilon = x - \beta$	signed raw	$\varepsilon = \dfrac{\Sigma(x - \beta)}{n|\beta|}$	average signed fractional
$\varepsilon = \dfrac{x - \beta}{|\beta|}$	signed fractional	$\varepsilon = \dfrac{\Sigma|x - \beta|}{n}$	average absolute raw
$\varepsilon = |x - \beta|$	absolute raw	$\varepsilon = \dfrac{\Sigma|x - \beta|}{n|\beta|}$	average absolute fractional L_1 norm
$\varepsilon = \dfrac{x - \beta}{|\beta|}$	absolute fractional	$\varepsilon = \sqrt{\dfrac{\Sigma(x - \beta)^2}{n}}$	root–mean–square (rms) L_2 norm
$\varepsilon = \dfrac{\Sigma(x - \beta)}{n}$	average signed raw	$\varepsilon = \dfrac{1}{|\beta|}\sqrt{\dfrac{\Sigma(x - \beta)^2}{n}}$	fractional root–mean–square (rms) L_∞ norm

x = measured value β = baseline value n = number of measurements ε = error value

baseline) is zero. Under such conditions, the fractional error metrics cannot be used.

It is important to note that none of the fractional errors necessarily normalizes the raw error into the bounded range $[-1, +1]$ for signed errors or $[0, +1]$ for absolute errors. The dynamic range can be unbounded. For example, if the baseline is 4.0 volts and the measured voltage is 6.0 volts, then the raw error is 2.0 volts and the fractional error is 0.50 or 5$\underline{0}$%. On the other hand, if the measured voltage is −6.0 volts, then the raw error is −1$\underline{0}$ volts and the fractional error is −2.5 or −250%. (The underlined zeros will be explained in Section 8.3 on Precision.)

If a **normalized error** in the range $[-1, +1]$ is needed, then the minimum possible result x_{min} and the maximum possible result x_{max} must be known. Given these extrema, the normalized error can be computed from the result x and the baseline β:

$$\widetilde{\varepsilon} = \frac{x - \beta}{x_{max} - \beta} \text{ if } x > \beta \quad \text{or} \quad \widetilde{\varepsilon} = \frac{x - \beta}{\beta - x_{min}} \text{ if } x < \beta$$

Suppose, for example, that the maximum range of voltages that can be supplied by the power supply in the previous examples is from −2$\underline{0}$ to +12 volts. If the baseline (desired voltage) for one of the settings of the power supply is 4.0 volts and the measured voltage is 6.0 volts, then the raw error is 2.0 volts, the fractional error is 0.50 or 5$\underline{0}$% as before. The normalized error, on the other hand, is the ratio of the raw error of 2.0 volts to the difference between the maximum supply voltage of 12 volts (x_{max}) and the baseline of 4.00 volts, or 0.25 or 25%. In a similar fashion, if the measured voltage is −6.0 volts, then the raw error is −1$\underline{0}$ volts, and the fractional error is −2.5 or −250% as before, but the normalized error is the ratio of the raw error −1$\underline{0}$ volts to the difference between the baseline of 4.0 volts and the minimum supply voltage of −2$\underline{0}$ volts (x_{min}), or −0.42 or −42%. The normalized error can be no less than −100%, corresponding to the largest possible error in the negative direction, and no greater than +100%, corresponding to the largest possible error in the positive direction. Finally, the absolute normalized error is defined simply as the absolute value of the normalized error. Figure 8.1 illustrates these relationships.

Note that the behavior of the normalized error is only piecewise linear, which, of course, is a consequence of splitting the normalization process into two regions, one for results greater than the baseline, and the other for results less that the baseline.

Now that we have postulated a variety of error metrics, can we postulate some *accuracy* metrics, *i.e.,* metrics that represent how *close* a result is to the baseline, rather than how *far away* the result is from the baseline? The only reasonable possibility is to postulate the accuracy λ simply as the unit-complement of the absolute value of the normalized error, applying the same sign as the error:

$$\lambda = \left[1 - \left|\tilde{\varepsilon}\right|\right] \ \text{sign}(\tilde{\varepsilon})$$

Now we have several ways to measure the error of a result, and even a way to convert this error into a measure of the accuracy of the result, if the error can be normalized. However, the accuracy or error of a single result can be dangerously misrepresentative. Nature is noisy, and most human endeavors are deeply entangled in the thermodynamic machinery of the universe. A single highly

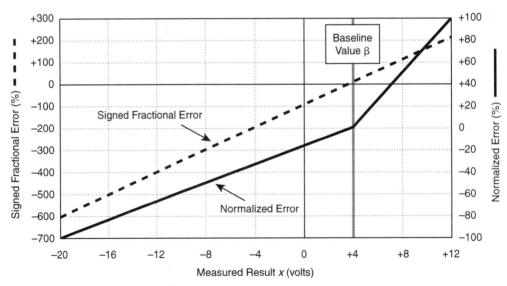

Figure 8.1 Behavior of two different error metrics

accurate result can be deceptively encouraging, and a single erroneous result can be unfairly discouraging. All measurement is fraught with uncertainty. Although a careful design of the solution and the experiments can help to reduce the uncertainty, it cannot be avoided entirely. That is a fundamental law of the universe — one of the few we think we know for certain.

8.2 Uncertainty

> **Definition 8.3 Uncertainty** is a measure of the inconsistency in a sample.

Tweedledee was very worried. The Queen of Hearts had just commanded him to count the number of rabbits in her North Pasture, but he was having great difficulty. For one thing, he couldn't quite remember whether 7 came after 8, or the other way around. Finally he gave up, and turned to Tweedledum, who was idly watching the rabbits cavort on the pasture. "Dear brother, don't just stand there!" he whined. "We desperately need a count of the number of rabbits. Can you please help?" Tweedledum sighed, and began to move his finger slowly from one rabbit to the next, mumbling a sequence of numbers. Tweedledee could not be sure it was correct, but was at least somewhat encouraged when Tweedledum didn't hesitate when he put 8 after 7. Tweedledum, he had to admit, had a much higher IQ than he. After an hour or so, Tweedledum turned to his brother and announced, "27." Tweedledee looked at the pasture. It seemed to him that there were many more rabbits than that. "Are you certain, dear brother? Just 27?" Tweedledum said, "No, not particularly. The rabbits keep moving around," and lay down to take a snooze. Tweedledee screamed in exasperation, jumping up and down, "Well, then count them again! You must count them again to be certain!" Tweedledum wearily hoisted himself back to his feet, and began moving his finger across the pasture again, while Tweedledee circled him shouting words of encouragement like "Atta boy!" and "Don't forget the little gray one!" Finally Tweedledum stopped counting and turned to his brother, who was standing on tip-toes in frenzied anticipation. "There are 62," he announced. Tweedledee's jaw dropped open. He might not have been able to count very well, but he certainly knew trouble when he heard it. "Before you said 27. Now it's 62?!" He fell to the ground in despair. "She'll have our heads, for sure. Oh my, oh my! What will we do?" Tweedledum bent over his weeping brother and said, "Well, maybe you could keep the rabbits from hopping around so much while I count them." "How, dear brother," cried Tweedledee. "How does one keep rabbits from hopping around. It's what they do!" Tweedledum thought for a moment and then said, "Why don't you make a very, very loud sound, and then the rabbits will all freeze with fright, and I can count them." Tweedledee jumped to his feet. "Brilliant suggestion! Brilliant, dear brother." With which he turned toward the pasture and let out such a loud scream that the trees bent over and the sun went down. The rabbits, although quite properly frightened, did not freeze, but vanished in a blur of hopping feet into the woods. The pasture was empty. "See, now it's quite certain," said Tweedledum. "No rabbits. Zero." And because it had gotten dark, they both went to sleep.

8.2.1 PARAMETRIC ESTIMATES OF UNCERTAINTY

Although one may be able to estimate the error and accuracy of a single data point, uncertainty can only be estimated from a **sample** of measurements or quantities. It makes no sense to estimate the consistency of a result consisting of a single data point.

Let us suppose that three groups of 6 students are each given the task of measuring the length of one of two different rooms: Room A or Room B. In each group each student makes one measurement, using a tape measure calibrated in meters and centimeters. The results of their tasks are in Table 8.2.

The average length of Room A measured by both Group 1 and Group 2 is 10.00 meters. But when asked to characterize the relative uncertainties of the measurements of Group 1 and Group 2, you probably feel much more uncertain about the measurements of Group 2, which appear much more inconsistent than the measurements of Group 1. (In fact, it is a bit difficult to imagine how the students in Group 2 could have obtained six such wildly different measurements, unless perhaps they were getting some help from Tweedledum and Tweedledee.) One way to quantify the uncertainties of these sets of measurements is the standard deviation of the sample. The sample for Group 1 has a standard deviation of 0.040 meter, while the sample for Group 2 has a standard deviation of 0.40 meter, ten times larger. These values seem to reflect the relative uncertainties of the corresponding samples.

Table 8.2 Small samples of room-length measurements

MEASURED LENGTHS OF ROOMS A AND B (IN METERS)		
ROOM A		ROOM B
Group 1	Group 2	Group 3
9.97	9.65	1.97
10.03	10.38	2.03
9.97	9.59	1.97
10.01	10.42	2.01
9.96	9.67	1.96
10.06	10.28	2.06

Postulate 8.1 The **uncertainty** u of a sample $\underline{x}$ is the standard deviation of the sample σ_x.

Note that the standard deviation is not *always* a measure of uncertainty; it is often used simply to represent the dispersion in a sample of results that are not *expected* to be consistent, such as student scores on an aptitude test, the age of people living in a retirement community, the distribution of per capita income in a city, or the speeds of vehicles on a highway. Each of these distributions is expected to have a fairly wide dispersion. The standard deviation may be used as a measure of uncertainty when the sample consists of repeated measurements of the *same* quantity, which are all expected to cluster around a single stationary value, such as measurements of the length of a room, the speed of light, or the height of a cliff. In such cases, the cause of any dispersion in the values is an unwanted disturbance, either in the object being measured, in the sensor making the measurements, or in the measurement procedure itself. The source of this unwanted disturbance can be random noise, an extraneous signal, or a systematic bias. This subject has already been briefly discussed in Section 4.1 (Task Domain) and is covered in greater detail in Chapter 12 (Synthesis).

Although the standard deviation can be used to indicate uncertainty, it has a major deficiency: it does not reflect the central tendency of the measurements themselves. This makes it impossible to know if a standard deviation is especially large or small with respect to the range of values of the measurements. For example, if all we knew was that the standard deviation of the sample for Group 1 was 0.040 meter, how could we judge whether that represented a lot of uncertainty or very little uncertainty without knowing something about the range of the measured values of the length of the room? Although an unexpected expense of several thousand dollars is a catastrophe to a family with an annual income of $30,000, it means little to the CEO of Microsoft.

To shed some light on this question, consider the sample of measurements of Room B made by Group 3, shown in the last column of the preceding table. The average of the measurements of the length of this tiny room is only 2.00 meters, but the standard deviation is exactly the same as that obtained by Group 1 for Room A: 0.040 meter. Does it makes sense to say that the uncertainty of the measurements of the shorter room is the same as the uncertainty of the measurements for the much longer room, just because they have the same standard deviation? It would probably be better if the uncertainty metric incorporated information about both the standard deviation *and* the range of values of the measurements.

A popular metric of uncertainty that meets this criterion is based on the coefficient of variation, which is the ratio of the standard deviation to the average of a sample:

Postulate 8.2 The **fractional uncertainty** δu of a sample of measurements is the absolute value of the coefficient of variation of the sample:

$$\delta u = \left| \frac{\sigma_x}{\bar{x}} \right| \text{ where } \bar{x} = \text{sample average}$$
$$\sigma_x = \text{sample standard deviation}$$

If the standard deviation is constant, the fractional uncertainty decreases as the average increases, because the standard deviation becomes a smaller and smaller fraction of the average. Thus, the fractional uncertainty of the Group 1 sample for Room A is 0.040/10.00, or 0.0040; and the fractional uncertainty of the Group 3 sample for Room B is 0.040/2.00, or 0.020, five times larger. The fractional uncertainty of the Group 2 sample for Room A is 0.40/10, or 0.04, which is the largest value among the three samples, as we would intuitively expect.

Although the average of a sample may well be negative, the *absolute value* of the coefficient of variation is specified for the metric, because a negative value for uncertainty has no reasonable interpretation.

Note that if the sample average is zero, the value of the coefficient of variation is not defined. Fortunately, this does not happen nearly as often as it does with the fractional error metrics postulated in the previous section. For the coefficient of variation to be undefined, the *average* of a sample (rather than a single value) must be exactly zero, which in turn means that sum of the values in the sample must be *exactly* zero, which is highly improbable. In fact, if you ever come across a sample of measurements whose average is *exactly* zero, a loud alarm should go off in your brain: check your data and your arithmetic!

Although the fractional uncertainty metric is well behaved — the higher the coefficient of variation, the higher the uncertainty about the consistency of the measurements — it might be useful to postulate a metric that puts a more positive spin on things, *i.e.,* the higher the value of the metric, the greater our **certainty** about the consistency of the measurements.

If we can normalize the fractional uncertainty into the interval [0, +1], then the unit-complement of this normalized fractional uncertainty will serve nicely as a certainty metric. To normalize the fractional uncertainty into the interval [0, +1] requires knowing the maximum possible fractional uncertainty that could occur. (The minimum possible fractional uncertainty is, of course, always zero.) As it turns out, if every nonzero value in the sample has the same sign, the maximum possible fractional uncertainty is simply the square root of the sample size. The reason for the restriction that the dynamic range of the sample may not span across the negative and positive domains, but must lie entirely in one or the other, is simple. In the worst case of a sample that *does* span across both domains, the average (which is the denominator of the fractional uncertainty) could theoretically be exactly zero, in which case the fractional uncertainty becomes undefined, obviating any possibility of normalization. And although this possibility is remote, the average can *approach* zero arbitrarily closely, allowing the fractional uncertainty to become arbitrarily large, once again obviating any possibility of normalization. But even if *all but one* of the values is zero, then the maximum possible fractional uncertainty (the absolute value of the coefficient of variation) is the square root of the sample size. (The proof of this lemma is fairly straightforward and is left to the reader. See Exercise 8.2.)

Fortunately, the requirement that the dynamic range of the sample lie entirely in either the positive or negative domain is not terribly restrictive. Almost always the dynamic ranges of measurements lie exclusively in one or the other of these domains; they do not include both negative and positive values. However, there is no law against it, and if the sample of measurements does span across both domains, then the fractional uncertainty simply cannot be normalized.

Given that this condition on the domain of the dynamic range of the sample is satisfied, we may immediately postulate a normalized fractional uncertainty:

> **Postulate 8.3** The **normalized fractional uncertainty** $\delta\hat{u}$ of a sample of n measurements that all lie in either the positive or negative domain is given by:
>
> $$\delta\hat{u} = \frac{\delta u}{\delta u_{max}} = \frac{\delta u}{\sqrt{n}}$$

Let's apply this metric to the small samples of the measured lengths of the two rooms obtained by all three groups of students. Table 8.3 presents the three measurement samples and the statistics necessary to compute the necessary intermediate metrics. The last two lines of the table are used to compute the normalized certainty, our current goal, which will be discussed shortly. But first, take a look at the actual values of the normalized fractional uncertainty values in the third to last row of the table. The largest value (0.0163) belongs to Group 2 and is 10 times the value for Group 1 (0.00163), as we would expect from the chaotic measurements made by Group 2. The value for Group 3 (0.00816) lies between the other two, which is also quite reasonable given the much shorter length of Room B. These values are more representative than the simple uncertainties (*i.e.,* the standard deviations), which report Groups 1 and 3 as having the *same* uncertainty, a direct consequence of the fact that the raw uncertainty ignores the ranges of the measurements

Table 8.3 Illustration of uncertainty metrics for student room-length measurements

Measurements and Statistics	Room A	Room A	Room B	Parametric Uncertainty Metrics
	Group 1	*Group 2*	*Group 3*	
Sample of measurements (meters)	9.97	9.65	1.97	
	10.03	10.38	2.03	
	9.97	9.59	1.97	
	10.01	10.42	2.01	
	9.96	9.67	1.96	
	10.06	10.28	2.06	
Minimum measurement	9.96	9.59	1.96	
Maximum measurement	10.06	10.42	2.06	
Dynamic range	0.10	0.83	0.10	
Average	**10.00**	**10.00**	**2.00**	
Standard deviation	**0.040**	**0.40**	**0.040**	uncertainty u
Absolute value of the coefficient of variation	0.00400	0.0400	0.0200	fractional uncertainty δu
Maximum possible absolute coefficient of variation	2.45	2.45	2.45	square root of sample size
Normalized absolute coefficient of variation	0.00163	0.0163	0.00816	normalized fractional uncertainty $\delta \tilde{u}$
Scaled normalized absolute coefficient of variation	0.0768	0.193	0.146	scaled normalized fractional uncertainty $\delta \tilde{u}$ (scaling factor = 0.4)
Unit complement of scaled normalized absolute coefficient of variation (%)	**92%**	**81%**	**85%**	certainty c

(the averages). Thus, as hoped, normalizing the fractional uncertainty seems to improve the relative values of the fractional uncertainties of the samples.

However, although the relative values of the normalized fractional uncertainties seem appropriate compared with each other, their *actual* values are problematic. The largest value, which belongs to Group 2, is 0.0163 or less than 2%. This value seems far too small to express the uncertainty of a sample that was previously characterized as consisting of "wildly different measurements." (The dynamic range of this sample is 83 centimeters!). The reason is because the coefficient of variation is a *very* small fraction of the maximum possible coefficient of variation (2.45). So, while the *relative* values of the three normalized fractional uncertainties are reasonable, the *actual* values themselves are much too small.

We can rectify this problem by applying a **scaling function** that amplifies the normalized fractional uncertainties by replacing the linear normalization function with a nonlinear transformation. The goal of this scaling function is to compensate for the sample size, which almost always imposes an extraordinarily large maximum possible coefficient of variation. We are free to do this as long as the scaling function is continuous, well-behaved, and does not violate the required bounds on the normalized value: [0, +1]. One way to do this is simply to raise the normalized fractional uncertainty to a fractional power. The choice of this exponent, called the **scaling factor**, is completely arbitrary, and its choice really just depends on how much you want to stretch the certainty scale. (It is necessary, of course, to use the same scaling factor with all the samples you intend to compare.)

> **Postulate 8.4** The **scaled normalized fractional uncertainty**
> $\delta\tilde{u}$ of a sample of measurements is given by:
> $\delta\tilde{u} = \delta\tilde{u}^s$ = the scaled normalized fractional
> uncertainty where s = scaling factor and $0 \le s \le 1$.

For the example of the student measurements of the lengths of the two rooms, a value of 0.4 has been selected (quite arbitrarily) for the scaling factor. The results of applying this scaling function are in the second to the last row in Table 8.2. As a result of the exponential scaling, the ratios among the three scaled normalized fractional uncertainties are not as large as those in the preceding row, but their actual values are much more realistic.

Finally, the last row in Table 8.2 presents the unit-complements of the scaled normalized fractional uncertainties in the preceding row, expressed as percentages. This final step achieves the goal of this short mathematical journey, *i.e.*, to convert the unnormalized fractional uncertainty into a **certainty** that lies in the interval [0, 100%]. The certainties for the three groups of student measurements are now quite reasonable. The Group 1 sample has a certainty of 92%, considerably higher than the certainty of the Group 3 sample (85%), which has the same standard deviation, but a much lower average. The sample for Group 2, which seems to have been very carelessly collected, has the worst certainty (81%).

> **Exercise 8.2**
> Prove the following lemma (hint: you'll need a bit of differential calculus):
> *The absolute value of the maximum possible coefficient of variation of any sample that lies entirely in the interval [0, +∞] or [−∞, 0] is the square root of the size of the sample.*

The field of statistics provides a completely different parametric approach to this challenge by providing a theoretically rigorous method to compute the **confidence interval** in a sample of measurements. (Confidence, after all, is just another word for certainty.) The metric will not yield a single number for the confidence, but instead will provide either 1) an estimate of the confidence that the value will lie between two specified bounds (the interval); or 2) the bounds of the interval that corresponds to a specified confidence. These two cases are the inverses of each other. Thus, one may pose a question about the confidence interval in one of two ways:

- What is the confidence that a measurement will lie in a specified interval (x_{low}, x_{high})?

or

- What are the bounds of the interval (x_{low}, x_{high}) that will yield a specified confidence?

The most obvious approach to answering these questions is to begin by computing the standard distance z for the interval, which is based on the average and standard deviation of the sample:

$$z_{high} = \frac{x_{high} - \bar{x}}{\sigma_x} \quad \text{and} \quad z_{low} = \frac{x_{low} - \bar{x}}{\sigma_x} \tag{8.1}$$

It should be clear from these two equations that a value of z represents the distance of either x_{low} or x_{high} from the sample average $\bar{x}$ in units of the standard deviation, and because the units of the standard deviation and the average are always the same (*e.g.,* meters), values of z are always dimensionless. That is why the z-distance is called the *standard* distance.

The values of z themselves do not solve our problem, because they are unbounded variables that do not directly reflect the confidence of the interval between them. To complete the process of computing the confidence, we must find the areas under the sampling distribution of the measurements between the two bounds z_{low} and z_{high}.

How easily that last phrase falls from the lips! In fact, this is the tricky part, because unless we know the shape of the associated sampling distribution, we cannot compute this area. After all, if you were asked to compute the area of plywood necessary to cover the side of a roller coaster, your first question would be: What is the shape of the roller-coaster track? Given either a sideview of the roller coaster or a mathematical function that described the trajectory of the track, you could compute this area using either geometric methods or the integral calculus, respectively. But without this information, estimating the area would be sheer guesswork.

From just six measurements of the length of each room in the previous example, there is no way in which we can compute the shape (parametric function) of the sampling distribution. There are not even enough data points to construct a histogram from which we might intelligently estimate the function. And unfortunately experiments that generate very little data are very common. If we *must* use a parametric method regardless of the dearth of data, there is only one possible approach (and it is not a very good one): *assume* a functional description for the distribution based on previous experience with similar tasks, and proceed with our fingers tightly crossed. Needless to say, because this assumption is based on knowledge whose relevance is highly questionable, it is necessary to be very clear about this weakness by stating it explicitly as part of the corresponding governing proposition:

> **Assumption:** Based on experience with measurements made under similar conditions, the distribution of the measured room lengths is normal (or some other parametric distribution function, such as Poisson, Weibull, binomial, *etc.*).

Having bitten the presumptive bullet, we can now estimate the confidence corresponding to a z-distance either by looking it up in a table of normal areas, by using the Excel function NORMSDIST,[6] or by applying a numerical approximation of the integral of the Gaussian probability density function. If you don't have a table of normal areas handy, or if you want to embed the computation of the single-tail normal area in a computer program, a convenient and fairly accurate approximation of the single-tail area of a standard normal distribution between the average and either plus or minus z standard deviations is given by:

$$\text{Area} = 0.5\sqrt{1 - e^{-\left(\frac{2z^2}{\pi}\right)}}$$

(8.2)

Let us consider a couple of examples to illustrate this process. Suppose we want to find out the confidence that a measured length of room A in the previous example will lie within ±0.5% of the average length obtained by the Group 1 students, *i.e.*, between x_{low} = 9.95 meters and x_{high} = 10.05 meters (±5.00 centimeters). First we calculate the two corresponding z-distances:

$$z_{high} = \frac{10.05 - 10.00}{0.040} = 1.25 \quad \text{and} \quad z_{low} = \frac{9.95 - 10.00}{0.040} = -1.25$$

Because the specified interval is split symmetrically about the average, the two z-distances have the same absolute value with opposite signs, and thus they represent two equal areas above and below the average. From Eq. (8.2), the area for an absolute z-distance of 1.25 is 0.397 (the corresponding value in a table of normal areas or from the Excel NORMSDIST function is 0.394). The total area between the bounds, then, is the sum of these two partial areas, 0.397 + 0.397 = 0.794. (Note that if the z-values have the *same* sign, then the two areas lie on the same side of the normal distribution, and the area for z_{low} must be *subtracted* from the area for z_{high} to find the net area between the bounds.)

Thus, if we assume that the measurements gathered by student Group 1 are normally distributed, we may conclude that there is a 79% confidence that measurements of the length of Room A made under the same conditions will lie within ±0.5% of the average length. (Note that statistical confidences are typically expressed as a percentage with no more than two significant digits.) This conclusion is, of course, mitigated by all of the governing propositions, including the shaky assumption of the normality of the underlying distribution of the measurements.

Nonetheless, pushing on, we can now compute the confidence for the same interval in the Group 3 sample for Room B: ±0.5% of the average length. Because the average length of Room B is much smaller (2.00 meters), the

---

[6]NORMSDIST takes a *signed* value of z as its input parameter and returns the single-tail area between −∞ and this input z value. Thus, to get the area between the average and z, you must subtract 0.5 from the area returned by NORMSDIST and take the absolute value of the result, because the area between −∞ and the average in a symmetrical density distribution is always 0.5.

bounds are much closer together: $x_{low} = 1.99$ meters and $x_{high} = 2.01$ meters (± 1.00 centimeters). As before, we calculate the two corresponding z-distances:

$$z_{high} = \frac{2.01 - 2.00}{0.040} = 0.25 \quad \text{and} \quad z_{low} = \frac{2.00 - 1.99}{0.040} = 0.25$$

Using Equation (8.2) as before to compute the single-tail area (0.098), under the assumption that the measurements gathered by student Group 3 are normally distributed, we may conclude that there is a 20% confidence that measurements of the length of Room B made under the same conditions will lie within $\pm 0.5\%$ of the average length. Quite appropriately, the confidence for Group 3 is much lower than the confidence for Group 1, despite the fact that both samples have the same standard deviation.

Now let's turn things around and consider the case when a bounded interval is to be found for a specified confidence. First we must solve Equation (8.2) for A_s to find the z-distance that corresponds to the given confidence (area):

$$z = \pm\sqrt{-\frac{\pi}{2}\ln\left(1 - 4A_s^2\right)} \quad \text{where } A_s = \text{the single-tail area } (<0.5) \qquad (8.3)$$

The positive and negative values of z are z_{high} and z_{low}, respectively, from which we can immediately compute the corresponding values of x_{high} and x_{low} by solving Eq. (8.1) for them:

$$x_{high} = \bar{x} + z_{high}\sigma_x \quad \text{and} \quad x_{low} = \bar{x} + z_{low}\sigma_x \qquad (8.4)$$

To illustrate this approach, let's compute the 92% confidence interval for the Group 1 sample of room-length measurements in Table 8.3. The single-tail area A_s corresponding to a confidence of 92% is $0.92 - 0.50 = 0.42$. From Eq. 8.3:

$$z = \pm\sqrt{-\frac{\pi}{2}\ln\left[1 - 4(0.42)^2\right]} = \pm 1.39 \quad \text{(Normal table area is 1.41)}$$

From these values for z_{high} and z_{low}, the bounds of the corresponding interval are computed using Equation 8.4:

$$x_{high} = 10.00 + (+1.39)(0.04) = 10.06 \text{ meters}$$

$$x_{low} = 10.00 + (-1.39)(0.04) = -9.94 \text{ meters}$$

Thus we may conclude that a confidence (or certainty) of 92% will include measurements within $\pm 0.6\%$ of the average for the Group 1 sample *under the assumption that the distribution of the measurements is normal.*

The reader may have noted that the confidence for this sample was set to the same value as the certainty as shown in Table 8.3. Although this selection was intentional, no comparison is really meaningful. *Confidence* is a metric that is computed or specified for a *bounded interval* within the sample, while the *certainty* metric attempts to characterize the *entire sample,* based on its coefficient of variation. They present different perspectives of the consistency of the sample.

Exercise 8.3

For each of the measurement samples in Table 8.2, calculate the intervals for each of the following certainties, or *vice versa:*

90%	[9.8, 10.0]	±0.75%
95%	[10.0, 10.5]	±3%
99%	> 10.0	[−0.5%, +1%]

If you are left feeling a bit uncertain about the usefulness of all these metrics for uncertainty, certainty, and confidence, given all the caveats and restrictions imposed on them, you are most certainly in good company. Fortunately, there is a completely different approach available to us, that makes no parametric assumptions whatsoever about the underlying distributions and is based on solid theoretical grounds.

8.2.2 GEOMETRIC ESTIMATES OF THE UNCERTAINTY OF MEASUREMENTS

More often than not, the distribution of the sample of measurements cannot be assumed to be normal. Equally often, the sample is just too small to construct a meaningful histogram that can be used to check whether the distribution of the measurements is normal. And even when there is sufficient data, the histogram often reveals that the distribution is significantly nonnormal, exhibiting heavy skew, kurtosis, or even multiple modes.

Fortunately, there is a very straightforward geometric method to estimate confidence intervals that requires no parametric assumptions about the underlying distribution. The general method is called **geometric confidence estimation**. A brief introduction of the method applied *specifically* to characterizing the consistency of a sample of measurements is presented here.

Before we introduce this method, it is important to understand the fundamental difference between parametric and nonparametric methods for estimating confidence intervals. Every sample of data has an underlying distribution. These data may be measurements, computer-generated values, counts, or values of a random variable in a theoretical function. The number of data points in the sample is called the **sample size**. Table 8.2 presented three examples of measurement samples, each of size six.

The distribution of the sample is a representation of the sample that describes how the individual data points are spread out (distributed) across the dynamic range of the sample. If we want to represent the distribution of a sample of measurements, we generally construct a **histogram**. The horizontal axis of a histogram is a set of **bins** spanning the entire dynamic range of measurement values, from the lowest possible value to the highest possible value. Each bin has a **class mark** that defines the central value of the bin, and the **width** of each bin is an interval that is centered on the class mark. Although the widths of all the bins in the histogram need not be the same, in general they are, *i.e.,* the class marks are uniformly distributed across the dynamic range and the bin width is constant. The most common exception to this convention of constant bin widths is with distributions that are (at least theoretically) unbounded, in which case the lowest bin includes the entire range below its upper bound to $-\infty$, and the highest bin includes the entire range above its lower bound to $+\infty$.

To illustrate the construction of a histogram, suppose we ask the Group 1 students who measured the length of Room A for the task described in the previous subsection to repeat their measurements 100 times to provide us with a larger and therefore more reliable sample. From the analysis presented in the previous section, this group of students had the least uncertainty in their results, so we expect fairly consistent results. Table 8.4 presents the set of measurements in their sample.

The order of the measurements in the table is simply the order in which they were acquired, top to bottom, left to right. Simply inspecting the sample gives us very little insight into its distribution. To help us visualize the distribution, let us construct a histogram using a bin width of one centimeter (0.01 meter), covering the entire dynamic range of the measurements from 9.90 meters to 10.10 meters. This histogram is shown in Figure 8.2.

Once again, the consistency of the measurements of Group 1 is evident. The distribution is very symmetrical, closely approximating a normal distribution, which is shown as the light dashed line. From this information, we are encouraged to apply the parametric metrics for uncertainty, certainty, and confidence that were postulated in the previous section (using the correct standard normal areas, not the approximate areas computed by Eq. 8.3):

Table 8.4 Large sample of room-length measurements (in meters) by Group 1 for Room A

9.94	10.02	9.98	9.95	9.99	10.04	10.03	10.03	10.01	9.94
9.94	9.98	9.97	9.96	9.99	10.05	10.01	10.06	9.96	9.95
10.02	9.97	9.99	9.98	10.00	10.01	9.99	10.04	10.03	9.96
9.92	9.97	9.97	9.99	10.04	9.99	9.95	10.05	10.03	10.06
9.93	9.97	10.02	10.00	10.02	10.00	9.94	10.05	10.06	10.09
10.03	9.97	9.98	10.01	9.95	10.00	10.03	10.02	10.04	9.96
10.00	9.96	9.98	10.01	10.00	10.07	10.04	9.95	9.98	10.07
10.02	10.02	9.99	10.00	10.02	10.05	10.05	10.01	10.04	9.96
9.97	10.00	9.97	9.99	9.99	10.04	10.00	10.03	9.93	10.07
10.01	9.98	10.00	9.98	9.91	10.08	10.01	10.01	10.08	9.92

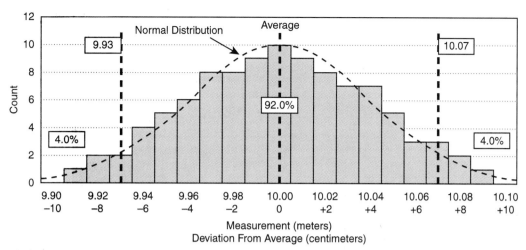

Figure 8.2 Histogram of a large sample of room-length measurements by Group 1 for Room A

$$\text{average} = 10.00 \text{ meters}$$
$$\text{uncertainty} = \text{standard deviation} = 0.040 \text{ meters}$$
$$\text{fractional uncertainty} = 0.0040 \text{ (coefficient of variation)}$$
$$\text{maximum possible fractional uncertainty} = 10 \text{ (square root of the sample size)}$$
$$\text{normalized fractional uncertainty} = 0.00040$$
$$\text{scaled normalized fractional uncertainty} = 0.00040^{0.4} = 0.044 \text{ (scaling factor} = 0.4)$$
$$\text{normalized fractional certainty} = 96\%$$

$$\text{interval for 50\% confidence} = \quad 0 \ \sigma = \pm 3 \text{ cm} = 9.97 \text{ to } 10.03 \text{ meter}$$
$$\text{interval for 92\% confidence} = \pm 1.28 \ \sigma = \pm 7 \text{ cm} = 9.93 \text{ to } 10.07 \text{ meter}$$
$$\text{interval for 99\% confidence} = \pm 2.33 \ \sigma = \pm 9 \text{ cm} = 9.90 \text{ to } 10.10 \text{ meter}$$

Take a moment to visualize (or even draw) the placement of the boundaries of these confidence intervals on the histogram in Figure 8.2, so that they define symmetrical deviations on either side of the average. The boundaries of the 92% confidence interval are shown in the figure.

Note that the current sample of 100 measurements for Group 1 yields a somewhat *higher* normalized fractional uncertainty (96%) than the initial small sample of 6 measurements (92%). Given the very small size of the sample in the initial task (6 data points), there really was no reason to expect that the value for the uncertainty would be very dependable. We are much better off relying on the values of the uncertainty, certainty, and confidence metrics calculated with the much larger sample of 100 data points.

The previous analysis depended completely on the validity of the assumption that the sample of measurements was normally distributed. For the large sample of 100 measurements, we may feel comfortable with that assumption, and it would not be difficult to justify to our peers. But now, let us give our strangely chaotic group of students (Group 2) the same task and see what they come up with. Their results are shown in Table 8.5.

Table 8.5 Large sample of room-length measurements (in meters) by Group 2 for Room A

9.89	10.10	10.14	9.93	9.88	10.08	10.09	9.87	10.11	9.90
10.09	10.15	9.89	9.91	10.14	10.12	10.13	10.15	10.11	10.07
10.11	10.06	9.87	10.09	9.89	9.93	9.83	9.86	9.94	10.07
9.88	10.09	9.90	9.90	9.92	9.90	10.14	9.95	10.15	10.10
9.87	10.06	10.08	10.09	9.89	10.06	10.19	10.11	9.93	9.95
9.89	10.09	9.91	10.11	9.80	9.91	9.94	10.10	10.18	10.04
10.11	10.06	9.84	10.12	9.89	9.89	9.88	10.16	9.93	9.85
9.91	9.91	10.02	9.91	10.11	9.90	10.13	10.07	10.15	9.90
9.81	9.88	10.08	9.93	10.08	10.11	10.08	10.13	10.12	9.90
9.88	10.06	9.89	9.90	10.11	10.10	9.97	9.90	9.92	9.89

As before, simply inspecting the data points in the table gives us very little information about the distribution of the sample, so we construct a histogram, which is shown in Figure 8.3.

My, my!! What have we here?! The distribution is strongly bimodal, and although the average of the sample is the same as the average of the Group 1 sample (10.00 meters), there are *no* data points at the average. In fact, there are no measurements within 1 centimeter of the average, and very few are even close to the average.

We can only speculate why the sample has two modes. Perhaps the sub-group of the students responsible for the higher mode mistakenly picked a target point on the far wall that was not on a line normal to the point on the near wall where the tape measure was anchored. This would cause the slightly diagonal distance to be larger than the actual distance. This error, however, cannot explain

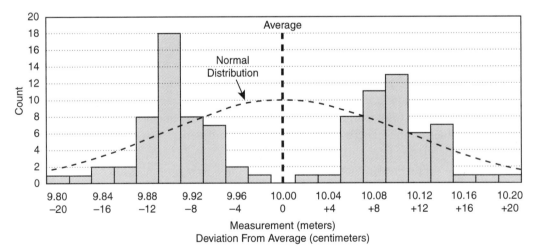

Figure 8.3 Histogram of a large sample of room-length measurements by Group 2 for Room A.

the *lower* mode. Perhaps the two subgroups measured the distance at widely separated points along the two walls, assuming mistakenly that the walls were parallel. However, a range error of 24 centimeters (the distance between the modes) along a wall of any reasonable length is very large and highly unlikely. Then again, perhaps Tweedledum and Tweedledee had somehow imposed their scatterbrained methodologies on the two subgroups, a chaotic battle of nitwits, so to speak. Whatever happened, the two subgroups clearly were not measuring the same thing and/or were using vastly different experiment protocols. Clearly, they were not playing by the same rules, *i.e.,* not abiding by the same governing propositions, thus creating an enormous discrepancy in the results.

Nonetheless, let us apply the methods described in the previous section to compute the uncertainty, certainty, and confidence of the sample:

$$average = 10.00 \text{ cm}$$
$$uncertainty = standard\ deviation = 0.11 \text{ cm}$$
$$fractional\ uncertainty = 0.011 \text{ (coefficient of variation)}$$
$$maximum\ possible\ fraction\ uncertainty = 1\underline{0} \text{ (square root of the sample size)}$$
$$normalized\ fractional\ uncertainty = 0.0011$$
$$scaled\ normalized\ fractional\ uncertainty = 0.0040^{0.4} = 0.065 \text{ (scaling factor} = 0.4)$$
$$normalized\ fractional\ certainty = 93\%$$

This value for the certainty, which is only 3% lower than the value for the well-behaved Group 1 sample, probably strikes you as way too high, and rightly so. After all, the distribution in Figure 8.3 is a hallmark of confusion and inconsistency. The cause of this unreasonably high value is the inappropriate use of methods that derive the normalized fractional certainty from the coefficient of variation. While not dependent on knowledge of the distribution, these methods *are* dependent on the *unimodality* of the distribution, requiring that the bin counts decrease in a well-behaved and reasonably monotonic manner as the absolute deviation from the average increases. To top it off, it must be obvious from the dashed line in Figure 8.3 that there is no way we can accept the assumption that the sample is normally distributed, so calculating the confidence of an interval parametrically is out of the question. We desperately need another method to estimate the consistency of a sample that casts off these parametric chains.

Let us go back to basics. Whenever we compute a confidence interval, what we are really attempting to do is to measure the *area* under a section of the histogram, which is an estimate of the probability density function that underlies the sample, and presumably of the *population* from which the sample was taken. Therefore, because we know the total count in the histogram, all we really need to do is to sum up the bin counts between two interval boundaries (the area between the boundaries) and divide this sum by the total count in the histogram (the total area). The quotient is then an estimate of the probability that a random measurement of the same process made under the same conditions will fall somewhere in this interval. It really is as simple as that.

Of course, if there are very few total counts (data points) in the histogram, the histogram will only represent a very rough approximation. But then again, even if one knew that the small sample was normally distributed, a parametric estimate of the confidence would be just as rough, based on correspondingly unreliable estimates of the average and standard deviation. The reliability and accuracy of the two methods are comparable, but the geometric approach does not rely at all on *a priori* knowledge of the distribution. It will work just as well for the crazy sample acquired by the Group 2 students as it does for the well-behaved sample acquired by the Group 1 students.

Let us try it out. As before, we must first specify either: 1) a confidence for which we would like to know the corresponding bounded interval; or conversely 2) a symmetrical interval about the average for which we would like to know the corresponding confidence. In the first case the confidence is expressed as a percent; in the second case the symmetrical interval is expressed as a deviation from the average, expressed as a fraction of the average (dimensionless or percent).

To illustrate the first case, let us find the 50% confidence interval for the nearly normal sample acquired by the Group 1 students; the histogram for this sample is shown in Figure 8.4.

Previously, when we accepted the parametric assumption that the sample was normally distributed, we computed the high and low bounds (x_{high} and x_{low}) by solving Equation (8.1) using the values of z_{high} and z_{low} that corresponded to a

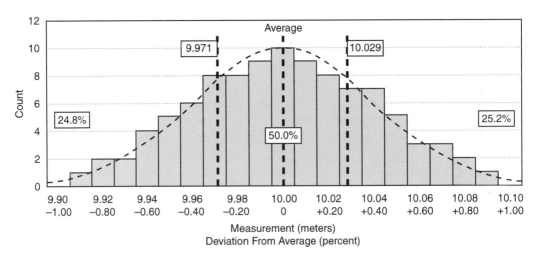

Figure 8.4 Histogram of the Group 1 sample showing the 50% confidence intervals

specified confidence interval. With this new geometric estimation method, however, we start at the average measurement and increase the deviation symmetrically in both directions, adding up the area of the included columns in the histogram until the sum is 0.50 or 50%, interpolating as necessary. As shown in Figure 8.4, this sum is reached at a deviation of ±0.029 meters. We often express this deviation as the **fractional deviation**, which is the ratio of the deviation (0.029 meters) to the average measurement (10.00 meters) or 0.0029 or 0.29%. Note that this fractional deviation is a bit smaller than the **fractional uncertainty** reported in Table 8.3 (0.0040 or 0.40%).

Because the distribution is not perfectly symmetrical (although very nearly), the areas of the two tails below and above the two symmetrically displaced bounds are not quite the same. The low tail has an area of 24.8%, while the high tail has an area of 25.2%. Likewise, the low and high sections of the central area do not have the same area; the low central section has an area of 25.2%, while the high central section has an area of 24.8%. This asymmetry in the central area is a direct result of our decision to sum the areas by keeping the *deviation* symmetrical, not the areas. We could have just as easily found the low and high *asymmetrical* deviations by keeping the low and high areas equal. (Note that statistical confidences are rarely expressed to the nearest 0.1%. We have done so here only to emphasize the slight variation from the symmetry of a normal distribution.)

Based on this geometric analysis of the histogram, the students in Group 1 may come to several conclusions concerning the certainty of their sample of measurements:

1) The average of the sample of measurements is 10.00 meters with a standard deviation of 0.04 meters.
2) For a measurement made under the same conditions, there is a 50% confidence it will lie in the interval ±2.9 centimeters, yielding a fractional deviation of 0.0029 or 0.29%.
3) For a measurement made under the same conditions, there is a 24.8% confidence that it will be less than 9.971 meters and a 25.2% confidence that it will be greater than 10.029 meters. (Once again, the excessive precision is only for purposes of illustration.)

Because the histogram of the Group 1 measurements closely matches a normal distribution, the parametric method previously presented in Section 8.2.1 should yield very similar results. From a table of normal areas (or the Excel NORMSINV function) we find that the 50% confidence interval about the average corresponds to symmetrical z-distances ($z_{low} = -0.674$ and $z_{high} = +0.674$), from which we can immediately compute the corresponding values of x_{low} and x_{high} using Equation 8.1:

$$x_{low} = 10.00 + (-0.674)(0.04) = 9.973 \text{ meters}$$
$$x_{high} = 10.00 + (+0.674)(0.04) = 10.027 \text{ meters}$$

As expected, these results are in very close agreement with the deviation bounds found by the geometric method (9.97 and 10.03).

Now let us apply this same geometric method to the strange sample of measurements made by Group 2. The histogram of this sample of measurements is shown in Figure 8.5. As witnessed before, the superimposed normal distribution (dashed line) illustrates the absurdity of making such an assumption about this distribution.

Once again, to find the 50% confidence interval we start at the average and increase the deviation symmetrically in both directions, adding up the area of the columns in the histogram until the sum is 0.50 or 50%, interpolating as necessary. This sum is reached at a deviation of ±0.097 meters, yielding a fractional deviation of 0.0097. This deviation is more than *three times* that of the Group 1 sample, a clear indication of the greater uncertainty of the Group 2 sample. The low tail has an area of 26%, while the high tail has an area of 24%; likewise, the low central section has an area of 24%, while the high central section has an area of 26%.

Based on this geometric analysis of the histogram, the students in Group 2 may come to several conclusions concerning the certainty of their sample of measurements:

1) The average of the sample of measurements is 10.00 meters with a standard deviation of 0.11 meters (almost useless statistics with the bimodal distribution).

2) For a measurement made under the same conditions, there is a 50% confidence that it will lie in the interval ±9.7 centimeters, yielding a fractional deviation of 0.0097 or 0.97%.

3) For a measurement made under the same conditions, there is a 25.6% confidence it will be less than 9.903 meters and a 24.4% confidence that it will be greater than 10.097 meters.

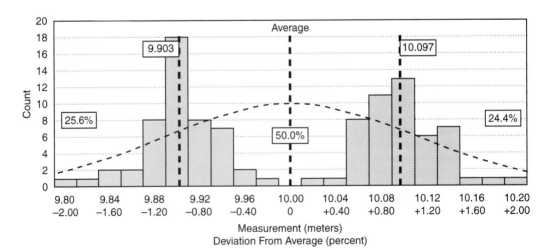

Figure 8.5 Histogram of the Group 2 sample showing the 50% confidence interval

Based on the fractional deviation, note that the uncertainty of the Group 2 measurements is 3.3 times the uncertainty of the Group 1 measurements for the 50% confidence interval, *i.e.,* the ratio of 0.0097 to 0.0029. This ratio is called the ***K% uncertainty ratio*** and is useful for comparing the uncertainties of different samples. Unfortunately, the value of the uncertainty ratio changes with the value of K, the specified confidence. For example, the 90% uncertainty ratio for the two samples is 2.1, which is the ratio of 0.0143 to 0.0067. As the specified confidence is increased, more and more of the structure of the distribution is included in the confidence interval. For the 100% confidence interval, which by definition includes the entire distribution, the 100% uncertainty ratio for the two samples is 2.2. How, then, should the confidence for an uncertainty ratio be selected? The smaller the confidence interval, the less likely that critical distribution structures will participate in the computation, reducing the reliability of the result. The larger the confidence interval, the more likely that critical structures (such as multimodalities) will be merged into the result and not isolated. The 50% uncertainty ratio is often a good compromise. Another approach is to compute the uncertainty ratios for all integer confidences, reporting the largest (worst-case), the average, and the smallest (best-case) values.

Exercise 8.4

Using the data samples in Tables 8.4 and 8.5, construct a spreadsheet to compute all the integer uncertainty ratios for the confidence intervals from 0% to 100%. Construct a histogram of the results.

The geometric method just described presents clear advantages over parametric methods. No information about the shape of the distribution is needed, and the search for deviation bounds that correspond to confidence intervals and *vice versa* can be easily automated using spreadsheets or computer programs. Constructing the histogram necessary for this method, however, is often problematic; the data must be quantized into an arbitrary number of bins whose class marks are usually (but not necessarily) linearly distributed. The number of bins and the distribution of the class marks can have a profound effect on the shape of the histogram, regardless of the fact that the underlying data are the same. A poor choice of the number and position of the bins can introduce significant **quantization errors**, especially if the sample is small.

Fortunately, there is no need to construct a histogram at all. Instead, we can use a geometric method for computing confidence intervals that is completely free of governing propositions and whose accuracy is degraded only by an insufficiency of data (something which no statistical method can overcome). Clearly, any sample of data can be sorted in ascending order based on the value of the individual data points. Such a sorted list is called a Pareto plot, named for Vilfredo Pareto, a European economist and sociologist in the early twentieth century.[7]

---

[7] I reference Pareto's work with some antipathy, because he used his mathematical ideas to develop a theory of social elitism that anticipated some of the principles of fascism.

This Pareto plot can be used very easily to compute confidence intervals anywhere throughout the sample. Although this method will work with any sample size, for purposes of illustration, let us once again apply it to the room-length measurement samples acquired by student Groups 1 and 2 as given in Tables 8.4 and 8.5. Figure 8.6 shows the Pareto plots of both groups.

The first thing to notice about the Pareto plot in Figure 8.6 is that the measurements are now plotted along the *vertical* axis, while in the histograms they were plotted along the horizontal axis. The horizontal axis is simply an index into the sorted list of data points. The line representing the average of the measurements is now a horizontal line, while before in the histograms it was a vertical line. (Of course, you are free to interchange the axes of a Pareto plot, but traditionally the horizontal axis is the data point index.)

Because the measurement values have been sorted in ascending order, the lines for the Group 1 and Group 2 measurements are necessarily monotonically increasing, although not necessarily *strictly* monotonically because there may well be repeated values in the samples. Looking at Tables 8.4 and 8.5, you will find many such repeated values, which account for the many short horizontal line segments in the Pareto plots in Figure 8.6.

The line for the Group 1 measurements is a smooth and symmetrical S-curve that is characteristic of a normal or nearly normal distribution. The line for the Group 2 measurements has a steep vertical section around the average, indicating a bimodal distribution that has very few data points in the middle of the dynamic range. A uniform (flat) distribution (not shown) would appear as a straight line of constant slope (although perhaps staircased with many short horizontal line segments representing repeated values) connecting the minimum

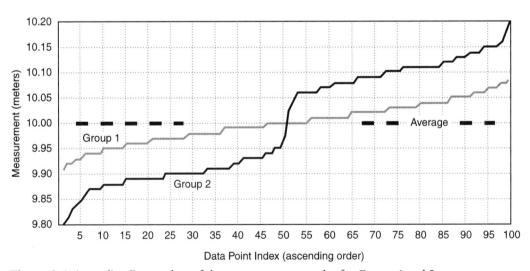

Figure 8.6 Ascending Pareto plots of the measurement samples for Groups 1 and 2

measurement at the left end of the plot with the maximum measurement at the right end of the plot.

Exercise 8.5

Construct a spreadsheet to simultaneously display several Pareto plots for normal distributions of samples with the same average, but different standard deviations. Observe how the shape of the Pareto plot changes as the standard deviation of the sample is varied. By applying a rounding function, observe the effect of staircasing as the precision of the data points is varied.

Construct a spreadsheet to display a Pareto plot of a distribution consisting of a mixture of two normal components with different averages and standard deviations. Observe how the shape of the Pareto plot changes as the averages and standard deviations of the two normal components are varied.

The procedure for finding the deviation bounds that include a specified confidence interval, or *vice versa*, is similar to the procedure just described for histograms. Figure 8.7 shows the Pareto plot for the Group 1 room-length measurements. To find the confidence interval for a specified deviation, simply project the two specified deviation bounds horizontally over to the Pareto curve and then down to the data-point-index axis, and then count the number of included data points, linearly interpolating across flat plateaus of repeated values as needed. This count divided by the total count (expressed as a percent) is the confidence that a measurement will lie between these bounds. Alternatively, to find the bounds for a

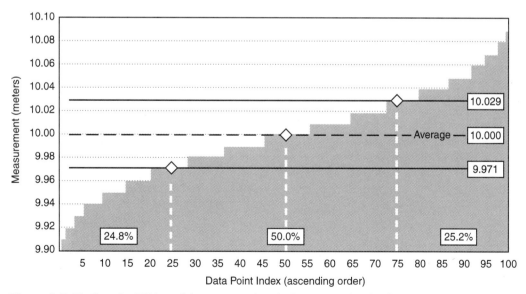

Figure 8.7 Finding the 50% confidence interval with the Pareto plot for the Group 1 sample

specified confidence interval, find the data point index on the horizontal axis that projects to the intersection of the average measurement on the Pareto curve, and then move away from this central index in both directions until you have included the required number of data points, that is, the specified (fractional) confidence times the total number of data points. The example in Figure 8.7 shows the symmetrical deviation bounds around the average for the 50% confidence interval. This approach yields exactly the same results as the histogram method (±0.029 meters; see Figure 8.4).

Figure 8.8 shows a similar analysis of the bimodal Group 2 sample of room-length measurements. Once again the results are very close to, *but more accurate* than those obtained with the histogram method (±0.103 meters *vs.* ±0.097 meters; see Figure 8.5).

The major advantage of the Pareto method over the histogram method is the avoidance of the need to distribute the measurements into an arbitrary number of bins with arbitrary class marks, which can lead to significant quantization error due to aliasing. Histogram binning is an inherently chaotic and brittle process. In contrast, the Pareto method uses all of the data points individually to find the area (confidence) between two specified measurement bounds, or alternatively to find the measurement bounds that include a specified area (confidence). Moreover, when the sample is quite small, it is often impossible to construct a meaningful histogram, because either the number of bins is too small (underfitting the distribution) or the number of counts in the bins is too small (degrading the statistical reliability of each bin count). The Pareto method, on the other hand, minimizes these problems by using all of the unquantized information in a data sample, and thus yields more accurate confidence estimates.

The real power of **geometric confidence estimation** lies in its usefulness for testing hypotheses about data samples *without introducing quantization error*

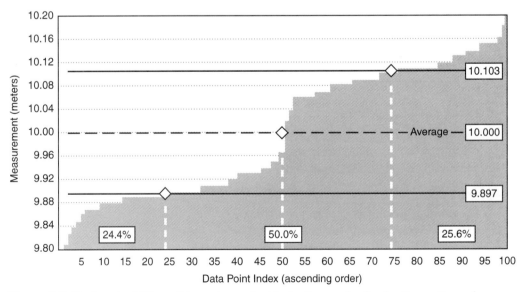

Figure 8.8 Finding the 50% confidence interval with the Pareto plot for the Group 2 sample

and without making risky parametric distribution assumptions, e.g., to decide whether two or more samples are drawn from the same population or whether two or more samples have the same average and/or standard deviation. The reader is encouraged to explore these flexible applications of the Pareto plot.

Before we conclude our discussion of the various methods for measuring uncertainty, certainty, and confidence, Table 8.6 summarizes the uncertainty metrics postulated in this section.

Table 8.6 Summary of uncertainty metrics

---

1) If there is no need to normalize the measure of uncertainty of a sample to allow comparison with other samples with different averages, then the **uncertainty** and **fractional uncertainty** are useful metrics.

uncertainty = u = standard deviation   (Postulate 8.1)

fractional uncertainty = δu = absolute value of coefficient of variation   (Postulate 8.2)

Note: If the average is exactly zero, the coefficient of variation is undefined.

2) If a normalized measure of uncertainty is required to allow comparison of two or more samples with different averages, then the **normalized fractional uncertainty**, **scaled normalized fractional uncertainty**, and **certainty** are useful metrics.

normalized fractional uncertainty   = $\delta\hat{u}$ (Postulate 8.3)

                                         = ratio of the fractional uncertainty to the square root of the sample size

scaled normalized fractional uncertainty   = $\delta\tilde{u}$ (Postulate 8.4)

                                         = normalized fractional uncertainty raised to the s power where s = scaling factor $(0,1)$

certainty = c = unit complement of the scaled normalized fractional uncertainty

Note: The value of the scaling factor k is arbitrary, but constant across all comparisons.

3) The **confidence** that a measurement will lie in a specified confidence interval and the **confidence interval** that will include measurements with a specified confidence are useful uncertainty metrics.

Parametric methods (*e.g.,* the standard distance) may be used to compute the values of these metrics *if (and only if) the distribution is known, e.g.,* Gaussian (normal), Poisson, Binomial, *etc.*

Nonparametric geometric estimation methods can be used to compute the values of these metrics by constructing and analyzing histograms and/or Pareto plots of the samples. Histogram binning will introduce quantization error. For Pareto plots, accuracy of the estimates is only a function of sample size (data sufficiency).

4) To compare the uncertainties of two samples, the $K\%$ uncertainty ratio is a useful metric.

$K\%$ uncertainty ratio = r_K = ratio of the fractional deviations for two samples

fractional deviation = ratio of width of the $K\%$ confidence interval to the sample average

Note: 50% is a common value for K.

    The minimum, average, or maximum r_K over all $K[0, 100\%]$ may also be useful.

8.2.3 PROPAGATION OF UNCERTAINTY

You may recall that the students measured the length of the room with a tape measure. Although it was not explicitly stated, you probably assumed that they were using a tape measure with a length of at least 10 meters, so that they could obtain the length of the room with a single measurement. But suppose that were not the case. Suppose instead that the students were using a tape measure with a maximum length of only 3 meters, and that they actually had to make four individual measurements each time the room length was measured. First they held one end of the tape against the near wall and marked a spot on the floor 3 meters away toward the far wall. Then they moved the end of the tape to this mark, held it firmly in place, and once again made a mark 3 meters farther on. They repeated this process one more time, and then measured the length of the final distance (about 1 meter) from the third and last mark to the far wall. The four measurements were then added up, and the sum was recorded in the result table. This protocol was repeated 100 times to gather the complete sample.

It seems intuitively obvious that the method requiring four measurements would be fraught with much more uncertainty than the method requiring only one measurement. It might be difficult for one student to hold the near end of the tape accurately on each mark, while another student stretched the tape and made a new mark farther along. The marks themselves would have a physical dimension, making accurate placement of the tape difficult. Keeping the measurements running along a straight line normal to the walls could easily degenerate into a zigzag path. Thus, the uncertainty of the sum of the four measurements really must be some combination of the uncertainties of the four individual measurements.

This raises an important research question: How is uncertainty propagated in arithmetic combinations of measurements? Do they add? Do they multiply? Fortunately, this question has been answered very comprehensively in an elegant short book by John R. Taylor, who is a professor of physics at the University of Colorado: *An Introduction to Error Analysis: The Study of Uncertainties in Physical Measurements*. While most of the information in his book is beyond the scope of this book (albeit extremely valuable), Table 8.7 presents a summary of the formulas for computing the uncertainty propagated through a sequence of measurements.

Using the formulas in Table 8.7 requires that you know whether your measurements are random and independent, or dependent. How do you decide this? Measurements are members of an independent and random sample if: 1) the *order* of the values is uncorrelated with the *values* themselves; *and* 2) the measurements (or sensors) are statistically (or functionally) unrelated to each other. Failure of either of these criteria designates the measurements as dependent to some degree. These criteria can be checked by several statistical tests, including the correlation coefficient, but a complete discussion of such tests is beyond the scope of this book. Any comprehensive book on applied statistics will offer several very practical statistical approaches. However, in addition to the quantitative tests for independence, it is important for you to ask yourself some important *qualitative* questions about the way in which the measurements were acquired, *i.e.,* the experiment **protocol**. Very often the task procedure is tainted by some subtle influence that results in a sample of measurements that is neither random nor independent. This subject is treated more thoroughly in Chapter 12 (Synthesis).

If you are able to assert that your measurements are random and independent, the formulas on the left side of Table 8.7 can be used to compute the total uncertainty that accumulates as a result of some arithmetic combination of intermediate measurements. If you are not sure if your samples are random and/or independent, or if you know they are not, then the formulas on the right side of Table 8.7 can at least provide an upper-bound (worst-case) estimate of the total uncertainty that will be propagated through the process of arithmetic combination.

Table 8.7 Propagation of uncertainty in arithmetic combinations of sample results

Random and Independent Samples	Dependent Samples
Given a set of measurement sample results $u, v, \ldots, z$ with uncertainties $\delta u, \delta v, \ldots, \delta z$:	
IF the results are combined as: $q = u \pm v \pm \ldots \pm z$	**IF** the results are combined as: $q = u \pm v \pm \ldots \pm z$
THEN the uncertainties add in quadrature: $\delta q = \sqrt{(\delta u)^2 + (\delta v)^2 + \ldots + (\delta z)^2}$	**THEN** the uncertainties add: $\delta q \le \delta u + \delta v + \ldots + \delta z$
IF the results are combined as: $q = \dfrac{x\,y\,\ldots\,z}{u\,v\,\ldots\,w}$	**IF** the results are combined as: $q = \dfrac{x\,y\,\ldots\,z}{u\,v\,\ldots\,w}$
THEN the fractional uncertainties add in quadrature: $\dfrac{\delta q}{q} = \sqrt{\left(\dfrac{\delta u}{u}\right)^2 + \left(\dfrac{\delta v}{v}\right)^2 + \ldots + \left(\dfrac{\delta z}{z}\right)^2}$	**THEN** the fractional uncertainties add: $\dfrac{\delta q}{q} \le \dfrac{\delta u}{u} + \dfrac{\delta v}{v} + \ldots + \dfrac{\delta z}{z}$
In general, given the function $q(u, v, \ldots, z)$ for combining the measurement sample results:	
$\delta q = \sqrt{\left(\dfrac{\partial q}{\partial u}\delta u\right)^2 + \ldots + \left(\dfrac{\partial q}{\partial z}\delta z\right)^2}$	$\delta q \le \left|\dfrac{\partial q}{\partial u}\right|\delta u + \ldots + \left|\dfrac{\partial q}{\partial z}\right|\delta z$
	EXAMPLE: for $q = x^n$, $\dfrac{\delta q}{|q|} \le |n|\dfrac{\delta x}{|x|}$

> **Exercise 8.6**
> Prove the formula given for the uncertainty in the example in Table 8.7:
> $$q = x^n$$
> Derive the formulas for the uncertainty of some other common expressions of arithmetic combination for both dependent and independent samples.
> (You will need some knowledge of differential calculus to do this exercise.)

Computing the average of the 100 measurements of the room length made by the Group 1 students using a long tape measure propagates no uncertainty, because the uncertainty of each individual measurement is zero. It is a single number, duly measured and recorded, without any inherent uncertainty. However, as was discussed previously, the uncertainty of the entire sample may be adequately represented by the standard deviation of the sample, as well as the values of the other metrics postulated in the previous sections.

On the other hand, as was just suggested, if the Group 1 students had been forced to use a shorter 3-meter tape measure, they would have been well advised to compute the averages and standard deviations of the four subsamples of 100 measurements individually and then apply the formulas in Table 8.7 to estimate the *propagated* uncertainty of their measurements. Using this approach they would obtain a larger, but *more accurate* estimate of the total uncertainty incurred in this piecemeal process than they would obtain by simply adding up the four measurements of each subsample, repeating this 100 times, and then computing the uncertainty of the grand totals.

Table 8.8 illustrates the hypothetical results of this procedure using both measurement methods. It assumes that the Group 1 students would incur about the same standard deviation for each of the four subsamples using the short tape measure, as they did for the single sample using the long tape measure; we made the same assumption about the Group 3 measurements of the short room (See Table 8.3).

Table 8.8 Uncertainties propagated when combining four room-length measurements

		Average Length (meters)	Uncertainty (standard deviation) (meters)	
Short tape measure	subsample 1	3.01	0.041	
	subsample 2	3.04	0.042	
	subsample 3	2.96	0.039	
	subsample 4	0.99	0.038	
			0.080	sum in quadrature (independent)
			≤0.160	sum (dependent)
Long tape measure	single sample	10.00	0.040	

If the samples are independent and random, Table 8.8 reveals that the propagated uncertainty of 100 samples acquired with four measurements using the short tape measure (0.080 meters) is about twice the uncertainty of 100 samples acquired with a single measurement using the long tape measure (0.040 meters). That's the good news. The bad news is that if the samples are dependent to some degree, this ratio could be as large as four (0.160 *vs.* 0.040 meters).

Consider another example. Suppose our task objective is to measure the average volume of a sample of (supposedly) identical aluminum rods. It is assumed that each rod is a right cylinder. We design two solutions. The first solution measures the diameter and length of each rod using a vernier caliper. The second solution fully immerses each rod in water in a volumetric flask and measures the height of the water level before and after immersion.

The procedure for the first solution divides the average measured diameter by two to obtain the average radius R. (This step propagates no uncertainty, because the denominator of the operation is the exact number 2, which has no uncertainty.) The average volume of a rod is then the product of the average measured length of the rods L and their average cross-sectional area πR^2. We assume that the fractional uncertainties of the radius and length measurement samples are the same. Table 8.9 is a summary of the hypothetical (and admittedly a bit contrived) measured values and computed results with their corresponding uncertainties. It assumes the measurement samples are independent and random, which means the fractional uncertainties add in quadrature.

The second solution uses a glass volumetric flask. The marks on the calibrated section of the flask have a granularity of 0.5 cubic centimeters. The flask is filled with water to the low end of the calibrated section of the flask. When a rod is lowered into the flask and fully immersed, a volume of water equal to the volume of the rod is displaced. Therefore, the volume of the rod in cubic centimeters is simply $0.5n$, where n is the number of marks that the water level rises. Because the meniscus on the surface of the water makes it difficult to count the marks accurately, we have assumed that the fractional uncertainty of the sample of mark counts is *twice* that of the length measurements of the first protocol. Table 8.10 is a summary of the hypothetical measured values and computed results with their corresponding uncertainties, assuming once again that the measurement samples are independent and random.

Table 8.9 Propagation of uncertainty of the volume measurement: first protocol

Quantity	Average	Standard Deviation	Fractional Uncertainty
Radius R	1.02 cm	0.0551 cm	0.0540 = 0.0551/1.02
Radius-squared R^2	1.04 cm^2		0.108 = (2) (0.0540)
Length L	3.21 cm	0.173 cm	0.0540 = 0.173/3.21
Volume = $\pi R^2 L$	10.5 cm^3	1.27 cm^3 $\leftarrow$	$0.121 = \sqrt{0.108^2 + 0.0540^2}$

Table 8.10 Propagation of uncertainty of the volume measurement: second protocol

Quantity	Average	Standard Deviation		Fractional Uncertainty
Displacement *n*	21.0 marks	2.27 marks		0.108 = 2.27/21.0
Volume = 0.5n	10.5 cm^3	1.13 cm^3	←	0.108

Despite the fact that the fractional uncertainty of the sample of mark counts for the second protocol (0.108) is *twice* that of the length measurements of the first protocol (0.0540), the fractional uncertainty of the volume measured using the second protocol (0.108) is actually *less than* that of the first protocol (0.121). Clearly a more direct method of measurement yields lower uncertainty.

> **TIP 9**
> To minimize the net uncertainty of the results of a measurement task: 1) Measure things as directly as possible, minimizing the number of arithmetic combinations of intermediate results; and 2) ensure that samples of measurements are as independent and random as possible.

8.3 Precision

Definition 8.4 Precision is the number of states that spans the dynamic range of a quantity, which is a measure of the maximum amount of information that can be expressed by the quantity.

The morning had begun very badly. Punctuating each word with a sharp rap of her scepter on Tweedledee's head, the Queen of Hearts had declared in no uncertain terms that she would have a *precise* estimate of the number of rabbits on the North Pasture by noon, or the heads of the two brothers would come off forthwith. Rubbing his head forlornly, Tweedledee collapsed next to Tweedledum and pleaded, "We desperately need a count of the number of rabbits out there, dear Brother. Please help me!" Tweedledum, who was inspecting his thumbnail with ferocious intensity, just kept mumbling to himself, "Where does it all end?" "Please, dear brother," moaned Tweedledee, "We need a precise count of the rabbits, and we need it right now!" Tweedledum harrumphed and glanced up at the pasture for a moment. "Many," he said, and went back to his thumb. "No, no, no!" cried Tweedledee, the tears of frustration streaming from his eyes. "We need a much more precise answer than that!" With a heavy sigh, Tweedledum glanced up again at the pasture. "There are 1,372," he said, and went back to his thumb. "Really?! 1,372?! Is that the right number?" shouted

Tweedledee gleefully. "How should I know?" muttered Tweedledum, touching his thumbnail briefly with the tip of his tongue. "You didn't ask me for the *right* number, just a *precise* number. 1,372 is a very precise number." With which Tweedledee began kicking his brother's shins as hard as he could, which didn't seem to disturb Tweedledum at all.

As defined, precision is a dimensionless number: the total number of different states that a quantity can assume over its entire dynamic range. For that reason, it is sometimes referred to as the **absolute precision**. If a voltmeter can display voltages from 0 to 10 volts in increments of 0.01 volts, the absolute precision of any measurement made with this instrument is 1000, which is the dynamic range of 10 volts divided by the increment 0.01 volts. Regardless of whether a measurement is very small, such as 0.17 volts, or very large, such as 9.56 volts, the absolute precision is constant at 1000. As another example, a computer monitor usually has a different horizontal and vertical absolute spatial precision, say 1024 by 768 pixels, respectively.

The base-2 logarithm of the number of states may also be used to express the absolute precision in bits, which is the fundamental unit of information. This is especially appropriate when the device is a digital instrument whose components have dynamic ranges that are whole numbers when expressed as a power of two. For example, a typical precision for a word of computer memory is 32 bits, which is considerably easier to express than the equivalent decimal precision of 4,294,967,296, which is the total number of unique states that this word can assume. Similarly, the pixel intensity in an image format is usually limited to a maximum number of gray levels or colors, an absolute precision that is often expressed as both decimal integers and bits, say 256 gray levels (8 bits) or 16777216 colors (24 bits).

As an alternative to absolute precision, the *relative* precision of a sensor or a quantity may be specified. A relative precision is expressed in the units of the associated number. There are two kinds of relative precision: **granularity** and **resolution**.

Granularity is the smallest measurement increment specified for a device or quantity. For example, the calibration marks on the side of a measuring cup indicate the volume of liquid in increments of one fluid ounce, this being the smallest distance between physical marks that can be reliably distinguished by the human eye, given the uncertainty caused by the meniscus on the surface of the liquid and the instability of the liquid itself. Regardless of how full the flask is, the relative precision of the measured volume is one fluid ounce. If you need to know a volume more precisely, you must find a measuring instrument with a smaller granularity. Similarly, a vernier caliper with a granularity of 0.1 millimeter means that all measurements made with this caliper, regardless of the size of the object to be measured (within the limits of the dynamic range of the caliper, *i.e.,* the maximum distance between its jaws), will have a relative precision of 0.1 millimeter. The granularity of a common hypodermic syringe is 0.01 cc; if a more precise dose must be injected, a more sophisticated infusion device with a smaller granularity must be used.

Resolution is the reciprocal of granularity. If the granularity of a pixel or dot on a computer display screen is 0.25 millimeter (sometimes called the *dot pitch*), then the resolution is 4<u>0</u> dots per centimeter or 102 dots per inch. The resolution of a magnetic disk might be 100<u>0</u> tracks per radial centimeter, which is a granularity of 0.01 millimeter per track. Ordinary photographic film has a resolution of about 50 lines per millimeter, which is a granularity of 0.02 millimeter per line. A parking garage might have a resolution of 363 parking places per acre, which is a granularity of 12<u>0</u> square feet for each parking place.

The preceding definitions for relative precision assume that the granularity or resolution is constant over the entire dynamic range of the quantity. In some cases, however, the relative precision of a quantity may vary with its value, *e.g.,* low values might have a higher relative precision than high values, or *vice versa*. If its distribution is not uniform, then the relative precision cannot be expressed by a single number, but instead must be expressed by a function or perhaps a table. An example of this is the manner in which grayscale brightness is displayed on computer monitors. The distribution of brightness is not uniform over the entire dynamic range, but follows a nonlinear function controlled by the *gamma* parameter to compensate for the nonlinear perception of the human vision system.

Sometimes the relative precision appears to vary as a function of some parameter, but can be transformed into a constant. Consider the knowledge proposition for the task *Find New Planet* in Table 6.2: "At 10^{11} km the Lowell telescope can resolve objects with diameters > 80<u>0</u> km." This statement seems to imply that the granularity of the telescope varies with distance, but this is only because the granularity is expressed in units of distance. If, instead, the granularity is expressed as the angle subtended by the object, then it is constant at about 10^{-6} degrees over all reasonable distances for objects of Pluto's brightness. For example, this telescope could resolve similarly illuminated objects with diameters as small as 5 meters on the surface of the Moon.

Just as a chain is only as strong as its weakest link, the results and conclusions of a task may not be expressed with more precision that the *least* precise item of knowledge that participates in the generation of the results and conclusions. If the results are expressed with any additional precision, the investigator has *invented* some information. Clearly, then, the precision of all the devices and quantities established by the governing propositions must be ascertained, because the minimum of these values determines the maximum allowable precision for the results and conclusions of the task, called the **precision limit**. The governing proposition that states the value of the precision limit is called the **limiting proposition**. Once the precision limit of the numeric value in the limiting proposition has been determined at the beginning of the task, the precision of the results and conclusions generated at the end of the task can be trimmed to the precision limit.

In some tasks, the chain of arithmetic operations that transforms the numeric inputs (domain knowledge) into the numeric results (range knowledge) does not cause a change of units. For example, if the task is to compute the average of a set of vehicle speeds, then the result has the same units as the inputs, *e.g.,* kilometers per hour. The precision limit for such cases is expressed as a *relative* precision, *i.e.,* the minimum resolution or granularity of the inputs. For the case of the average vehicle speed, the appropriate relative precision might be the granularity of the speed sensor, *e.g.,* 0.1 kilometers per hour. The average speed is expressed with the same granularity. Likewise, the average of a list of expenses expressed to the nearest dollar may also be expressed to the nearest dollar, but not to the nearest penny.

If the units of the results of a chain of arithmetic operations are *different* from the units of the inputs (which is very often the case), then the precision limit must be expressed as a dimensionless *absolute* precision, not a relative precision. Suppose, for example, your task is to compute the area of a dining table. One governing proposition states that the radius R of the table is 0.75 meters (a fact), a second characterizes the table as circular (an assumption), and a third states that the area of the table is πR^2 (a theorem). The raw area is computed to be 1.767145868 . . . square meters. However, before you can state your conclusion, you must trim the precision of the raw result to the precision limit. But you cannot set the precision of the result expressed in *square meters* based on a granularity expressed in *meters*. Thus, somehow the *absolute* precision of the measured radius must be ascertained. This dimensionless number will be defined as the precision limit, which is then used to set the precision of the final result.

8.3.1 ASCERTAINING THE PRECISION LIMIT

With the exception of quantities associated with digital devices, the absolute precision of a quantity in a governing proposition is seldom explicitly stated. When it is not, it must be calculated or estimated. If the dynamic range and either the granularity or the resolution of the quantity is provided by the governing proposition, the absolute precision may be immediately calculated by dividing the dynamic range by the granularity, or by multiplying the dynamic range by the resolution. For instance, if the granularity of a magnetic disk is 0.010 millimeter per track and the overall radius of the recording surface is 2.0 centimeters (2<u>0</u> millimeters), then the absolute precision is 2<u>0</u> divided by 0.010 or 20<u>0</u>0 (tracks). If the resolution of a monitor screen is 40.0 dots per centimeter and the width of the active portion of the screen is 32.0 centimeters, then the horizontal absolute precision is 32.0 times 40.0 or 1280 (dots).

If it is impossible to compute the absolute precision of a quantity in a governing proposition because the granularity or resolution and the dynamic range are not provided, which is very often the case, the last resort is to estimate it. Consider, for example, the proposition stated for the task *Measure Fossil Age* in Table 6.2: "The fossil was found 29.4 meters below the edge of the cliff." The fact that the measurement is expressed to the nearest 0.1 meter implies that the granularity of the measuring instrument was 0.1 meter. Now all we have to do is estimate the dynamic range of the measuring instrument, and then the corresponding absolute precision can be computed by dividing the estimated dynamic range by the granularity.

The most conservative estimate for the upper limit of the dynamic range of the measuring instrument is 29.4 meters, because we know it must be at least that large to have produced a measurement of the same value. This conservative upper limit would yield an absolute precision of 294 (29.4 meters divided by the granularity of 0.1 meter). But it is unlikely that a measuring instrument would have such a strange upper limit to its dynamic range. It might be more reasonable to assume that the measuring instrument could measure distances up to 30 meters or perhaps 100 meters without loss of precision. The corresponding absolute precisions would be 300 (30 meters divided by the granularity of 0.1 meter) or 1000 (100 meters divided by the granularity of 0.1 meter). Experience with measuring instruments suggests that even the higher of these possible precisions is not unreasonable, and because there is little else to go on, the investigator is probably on safe ground in assuming that the absolute precision of this quantity is 1000 (which, of course, becomes one of the governing propositions for this task). If this governing proposition turns out to be the *limiting* proposition for the task, then 1000 is the precision limit, and we are ready to set the precision of the results of the task *Measure Fossil Age*.

In general, then, whenever the dynamic range, granularity, and resolution are not given, the absolute precision of the number given in a governing proposition must be estimated from the number itself. To do this, we must first determine which digits in the number are significant, discounting those zeros that are simply placeholders. This was trivial in the previous example for the distance of 29.4 meters, because there are no zeros to complicate matters, implying that all the digits in the number 29.4 are significant. This, however, will not always be true.

In secondary school we learned how to isolate the significant digits in a number:

1) Sequences of zeros embedded between two nonzero digits are significant.
2) Sequences of zeros at the end of the fractional part of a number (if any) are significant.
3) Underlined zeros are significant. (This is what all those underlined zeroes mean.)

All other zeros are nonsignificant. Underlining a zero forces its significance when the first two rules would otherwise automatically designate the zero as nonsignificant. For example, suppose an investigator measures a distance and reports it as 350 meters. Without any further information about granularity or resolution, the rules just postulated would require us to assume that the zero is nonsignificant, implying that the measurement had a granularity of 10 meters. If, however, the distance is reported as 35<u>0</u> meters, the underlined zero forces its significance and indicates that the measurement had a granularity of 1 meter.

If an *entire* number is underlined, it is considered exact, *i.e.,* has unlimited precision. For example, the number <u>2700</u> is not limited to four significant digits, because the entire number is underlined, indicating that this number is exact. This is necessary when a number is a count or is derived from a count, rather than a measurement, such as a sample size, the number of categorical answers to a survey question, a deterministic probability, the width of an image in pixels, *etc.* For example, if we interview every fourth voter emerging from a polling place without fail, the probability that a voter was interviewed is exactly <u>0.25</u> (<u>1</u> divided by <u>4</u>). If an exact number has a repeating decimal, then the number is underlined and followed by an ellipsis (. . .). For example, if we develop every third picture on a roll of film, the probability that a picture was developed is <u>0.33</u> . . . (<u>1</u> divided by <u>3</u>). (If the number is *not* underlined, a trailing ellipsis simply means that the decimal fraction is longer, but has been truncated.)

Once all zeros have been designated as significant or nonsignificant, the significant digits can be isolated by deleting the non-significant zeroes, the decimal point, and the sign, if any. The length of the resulting string is the number of significant digits in the value, which is the basis for estimating its absolute precision. This process is illustrated in Table 8.11, applying the rules just given.

Deciding whether to use the minimum and most conservative estimate of the absolute precision, or one of the higher and perhaps more reasonable estimates of the absolute precisions, is difficult. Additional information about the quantity, such as how it was measured or generated, may help to decide this issue. Finally, when all the precisions for all the numbers in the governing propositions that participate in the computation of the results and conclusions have been estimated and listed, only one of them is chosen to specify the precision limit for the task. Although this is normally the minimum of the absolute precisions, the actual choice may well be influenced by the relative unambiguities and reliabilities

Table 8.11 Examples of the estimation absolute precision

Original Number	Rule 1	Rule 2	Rule 3	Significant String	Significant Digits	Most Conservative Absolute Precision	Reasonable Absolute Precisions
	(significant zeroes are **bold**)						
10.3002	10.3002	10.3002	10.3002	103002	6	103002	200000 or 1000000
−0.00980	−0.00980	−0.00980	−0.00980	980	3	980	1000 or 10000
67<u>000</u>	67<u>000</u>	67<u>000</u>	67**000**	6700	4	6700	7000 or 10000
67000	67000	67000	67000	67	2	67	70 or 100
+0.13010	+0.13010	+0.13010	+0.13010	13010	5	13010	20000 or 100000

of the items on the list. A near-minimum that is reliable and clear may be preferable to the true minimum.

8.3.2 SETTING THE PRECISION OF RESULTS AND CONCLUSIONS

The methods just described for ascertaining the precision limit of the quantities in governing propositions assume that the precision of the numbers has been accurately represented. This is not always the case. Often the numbers in governing propositions obtained from other tasks in the same project or even from completely different projects are poorly notated, casting doubt on the computation or estimation of their absolute precisions. One common error occurs when a measurement just happens to yield a whole number and is reported without any decimal fraction, despite the fact that far more precision might actually be justified. For example, a result that is reported as 25 meters may actually deserve to be reported as 25.00 meters. And the opposite is also often true. An inexperienced engineer may compute a result using a calculator or a spreadsheet, which routinely performs its calculations with 14 or more digits, and then reports this number including all these digits, far exceeding the precision than can actually be justified. An engineer who reports a distance of 25.739587665321 meters, lifted right off a spreadsheet, has clearly exaggerated the justifiable precision by many orders of magnitude. There is no instrument that can measure a distance of several meters to the nearest 100 millionth of a micron, which is smaller than the estimated diameter of the hydrogen atom.

All of this should serve to motivate the task personnel to express their results and conclusions with no more precision than is justified, and to carefully encode the number reflecting this precision according to the rules previously given. In addition, governing propositions must be included that state the absolute precision (or the dynamic range and the granularity or resolution) of each sensor and inducer used for the task and of each number that is brought to the current task from a previous task, if known. If the absolute precision of a number is not known, it must be estimated as was just described.

> **TIP 10**
> For each knowledge proposition that includes a number, state or estimate the precision of the number whenever possible, either directly in the proposition itself or in an accompanying proposition. For example:
> **Fact:** The video camera supports 256 gray levels.
> **Fact:** Speeds in the range 0 to 120 mph were measured to the nearest 0.5 mph.

Once you are satisfied that you have ascertained the precision limit from reliable estimates of the absolute precisions of the numbers in the governing propositions that participate directly in the generation of the results and conclusions of your task, this precision limit may be applied to each raw numeric result to set its precision appropriately.

The first step in this process is to multiply the absolute value of each raw numeric result $|r|$ by 10^n, so that the product is equal to the precision limit K.

$$|r|10^n = K$$

To determine the position of the least significant (rightmost) digit of the result, this equation may be solved for n:

$$n = \log_{10}\left(\frac{K}{|r|}\right)$$

(8.5)

The value of n computed by this equation is a real number in the range $(-\infty, +\infty)$. However, because the position of the least significant digit in the result must be expressed as an integer, the real value of n must be converted to an integer N, which is accomplished by setting N to the nearest integer less than n: (**IF** $n = 0$, **THEN** $N = \text{truncate}(n)$, **ELSE** $N = \text{truncate}(n) - 1$). The value of N is the position of the least (rightmost) significant digit in the result, measured on the scale shown in Table 8.12, in which d stands for any decimal digit.

For example, if a raw result $r = 0.00123581$ and the precision limit K is ascertained to be 1000, then from Eq. 8.5 $n = 5.908 \ldots$ and $N = +5$. Thus, from Table 8.12 the least significant digit in r is the fifth digit to the right of the decimal point. After trimming off the excessive precision, the value for r is appropriately reported as 0.00124 (the 3 is rounded to 4, because the digit to the right of 3 in r is 5). Similarly, using the same value of K, if the raw result $r = 123500$, then $n = -2.091 \ldots$, and $N = -3$, which means that the least significant digit in r is the third digit to the left of the decimal point. Thus the value of the result expressed with the correct amount of precision is 124000. (Notice that the zeroes are *not* underlined, because they are not significant.)

Now let us go back to the example of calculating the area of the circular table from the previous section. Based on the radius of 0.75 meters, we computed the raw area to be $1.767145868 \ldots$ square meters (πR^2). Lacking knowledge of the granularity of the measuring instrument, we must estimate the absolute precision of the measured radius from its expressed value. Based on the postulated rules, the number of significant digits in the measured value 0.75 is 2. The most conservative absolute precision is 75, but a more reasonable value is probably 100, so we set the precision limit K to 100. From Eq. 8.5 we compute the position of the least significant digit N to be $+2$, for which Table 8.12 specifies

Table 8.12 Determining the position of the least significant digit

	Value of N											
...	−5	−4	−3	−2	−1	0	+1	+2	+3	+4	+5	...
...	d	d	d	d	d	.	d	d	d	d	d	...

two decimal digits. Thus, after trimming off the excessive precision, the area of the circular table is appropriately reported as 1.77 square meters.

TIP 11

Below is an Excel formula that can be used to express a result with the correct precision, given the precision limit and the raw result:

=FIXED(rawResult, IF (LOG10(precisionLimit/ABS(rawResult)) < 0,
    TRUNC(LOG10(precisionLimit/ABS(rawResult))) – 1,
    TRUNC(LOG10(precisionLimit/ABS(rawResult)))), TRUE)

Note that the result of the FIXED Excel function is text, not a numeric value.

Granularity and resolution are also useful for characterizing the precision of the results and conclusions of a task. A histogram of the results of a task might have a resolution of 16 bins per pound, which is a granularity of 1.0 ounce per bin. The granularity of a table of the results of an economic study may limit the expression of amounts to the nearest million dollars, either because the data in the governing propositions for the study were similarly limited in precision, or because the investigator simply wants to hide the details and emphasize global behavior to enhance data visualization. In a similar fashion, to allow graphic results to be easily visualized, the precisions of the axes of a graph are often greatly limited to allow the principal relationships between a factor (independent variable) and a performance metric (dependent variable) to be grasped at a glance.

To make such histograms, tables, and graphs, high precision values must be mapped into a smaller number of low-precision categories, called **bins**, that are distributed over the dynamic range of the result, a process called **quantization**. Clearly, the many-to-few mapping of the quantization process destroys some information, and a task that employs quantization must assume either that the reduction of information has no significant effect on the achievement of the task objective (a limitation), and/or that the reduction in information is necessary to avoid the need for more resources than are available for the task (exclusion). For example, reducing a sample containing 10,000,000 high-precision data points to a histogram with only 100 bins may save scarce computational resources, while still maintaining an accurate representation of the distribution of the original sample. This histogram may help the investigator visualize the behavior of the task unit while different factors are varied. It may also allow the application of some geometric methods for testing hypotheses and estimating the statistical confidences of high-level conclusions. The geometric methods for doing this have been presented briefly in the preceding Section 8.2.2 (Uncertainty). Regardless of the purpose, assumptions that are limitations and/or exclusions imposed by quantization processes must be stated explicitly in governing propositions and accompanied by clear and credible justifications.

The running example task *Identify Faster Traffic* will serve nicely to illustrate the application of the principles just described to the analysis and presentation of the results and conclusions of a task. Assuming the traffic engineer has finished her investigation, she now must create the presentation of her results and conclusions. She decides, quite appropriately, on a top-down presentation, starting with the lowest-precision conclusion that may be drawn from the results:

Vehicles on Highway A travel significantly faster than vehicles on Highway B.

If she is lucky, her boss will thank her for this succinct answer to the research question for the task and send her back to work on a new project. But more likely he will want a few more details. So she prepares additional results of her study in order of increasing precision:

- There is a 9<u>0</u>% confidence that vehicles on Highway A travel faster than vehicles on highway B.
- The average speed of vehicles on Highway A is 66 mph; the average speed of vehicles on Highway B is 54 mph.
- The average speed of vehicles on Highway A is 66 mph with a standard deviation of 5.7 mph; the average speed of vehicles on Highway B is 54 mph with a standard deviation of 7.4 mph.
- The histogram of the vehicle speeds for both highways is shown in Figure 8.9.

Each succeeding example of the results of the task given above presents a greater amount of information, *i.e.,* a higher-precision result. The highest-precision

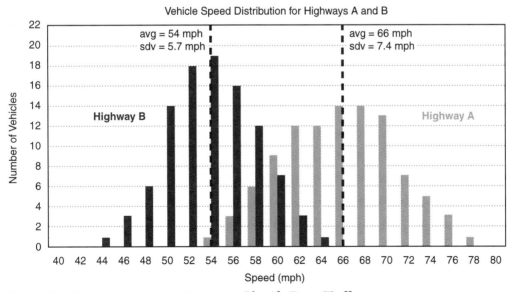

Figure 8.9 High-precision result for the task *Identify Faster Traffic*

result that the traffic engineer could present is a table of the original samples of 100 vehicle speeds acquired from each highway. Although such a lengthy table would probably be of little interest to most people, a Pareto plot of each sample might provide some visual insight into the distribution of the data.

It is important to remember that none of the results listed here says anything about accuracy. In fact, in the absence of some independent estimate of the true distribution of vehicle speeds on these two highways, there is no way to compute the accuracy of the results of this task. Only the traffic engineer's strict adherence to the Scientific Method and her carefully documented compliance with the three criteria of reproducibility, completeness, and objectivity can engender sufficient confidence in the validity of the results and conclusions of the task. The remaining chapters in Part IV of this book focus squarely on rigorous methods to justify such confidence.

Exercise 8.7
Some of the governing propositions you listed for Exercises 6.1 and 7.1 probably include numbers that will participate directly in the computation of numerical results for the task. For each of these numbers, notate its precision using the rules stated in this chapter. If known, include the actual values for the relative and absolute precision of each number in the same governing proposition or an accompanying governing proposition, as appropriate. If the absolute precision is not known, estimate it.

8.4 Knowledge, Truth, and Humility

All research and development tasks in science and engineering begin and end with knowledge. Even the devices and effects created by development tasks are often accompanied by intellectual insights that add to the archive of human knowledge, from glimpses into the fundamental processes of the universe (*e.g.,* atomic structure and genetic codes), to knowledge that liberates time and energy for pursuits less menial than raw survival (*e.g.,* digital logic and weather forecasting), to political and economic systems that recognize and celebrate the dignity of the human spirit (*e.g.,* democracy and socialism). It was John, the disciple of Jesus of Nazareth, who said: "The truth shall make you free"; and 1500 years later it was Francis Bacon who said: "Knowledge is power."

But truth in advertising requires us to fully accept and admit that knowledge and truth can be very different things, often at perilous odds with each other. While we grasp knowledge securely in our hands, truth may lie somewhere far over the horizon. Hoping against hope, we all-too-often cling fervently to what we know, confusing it with truth, and trusting time-honored tradition to carry the day and maintain our warmth and security. Too often truth has imprisoned the soothsayer, and knowledge has crippled the powerful, as Galileo Galilei and King Oedipus came to know full well.

Scientists and engineers enjoy no immunity to this. Some hedge their bets by demanding that new hypotheses be met with skepticism: better safe than sorry. Others protect their interests by claiming that pragmatism must accelerate the acceptance of new ideas: better first than second. Both are convenient biases, one motivated by fear of error, the other by fear of failure. Both think that truth is on their side. Both discourage thinking out of the box. But, of course, there is no real danger in *thinking,* only in *thinking you are in possession of the truth.*

The proper balance, it would seem, lies in intellectual neutrality, understanding that expectations, skeptical or pragmatic, cautious or hopeful, simply threaten objectivity and promote self-deception. Only formal evidence can tip the scales of neutrality and fill the vacuum of ignorance. And in the absence of such evidence for or against, the carefully protected *a priori* assumption of neutrality must be honored, allowing no other conclusion than to *reserve judgment.* Humility in science and engineering means learning to say: "I don't know." Or more completely: "Here is what others have learned, here is what I have learned, but about everything else I must reserve judgment."

This mantra, like all mantras, must be repeated again and again and again to infuse itself thoroughly into your being: wax on, wax off. The discipline is challenging, but attainable. Exercise your mind to its limits. Maintain eternal vigilance. Press for evidence, leach out every drop of information, digest every morsel of knowledge, separate each seed from the overwhelming chaff, knead the product to keep it soft and pliable, *but reserve judgment until the last possible moment.* As the popular jargon suggests: "Wait for it." Or, as we refer to it in the field of statistical learning: "Hard decisions late."

But, of course, not *too* late. That's the trick. Part IV of this book sheds some light on this trick.

The Scientific Method

CHAPTER 9

Overview

9.1 History of the Scientific Method

Imagine, if you will, the great philosopher Buddha (Prince Gautama Siddhartha) seated under a banyan tree in Nepal about 500 BC, surrounded by several of his devoted students. One of the more spunky students pipes up: "Master, how do we know if something is true or false? How do we know what to believe?" By way of an answer, Buddha quotes from the *Kalama Sutra,* which he has recently written:

> Do not believe in something simply because
> you have heard it.
> Do not believe in traditions simply because
> they have been handed down for many generations.
> Do not believe in something simply because
> it is spoken and rumored by many.
> Do not believe in something simply because
> it is found in your religious books.
> Do not believe in something simply on the
> authority of your teachers and elders.
> But after careful observation and rigorous analysis,
> when you find that something agrees with reason,
> and is conducive to the good and benefit of one and all,
> then accept it and live up to it.

It is rumored that the same student, on hearing this, responded: "So then, Master, does this mean that we should not necessarily believe this statement of yours either?" It is not known how Buddha reacted to this. Perhaps he smiled and enjoyed the irony with his student. Perhaps he banished the student to a corner facing the wall. It was a long time ago.

This statement by Buddha is the earliest extant example of a scholar's attempt to codify the process of scientific investigation. In addition to admonishing against accepting knowledge as a matter of faith or loyalty or popularity, he put his finger on a couple of important guiding principles for a reliable process of knowledge acquisition.

First, he suggests the task begin with a clear understanding of the problem that motivates the task. While this may seem an obvious place to start, his emphasis on "careful" observation and "rigorous" analysis implies that this should be a very thorough process, resulting in a precise statement of the objectives of the task and governed by a set of propositions that tightly constrain the form and method of any proposed solution. If the CEO of Boeing asks his engineers to design a new high-payload aircraft, a careful investigation of practical constraints, as well as a rigorous review of the state-of-the-art, will probably rule out a powerful catapult that flings a large load of anesthetized passengers from Paris to London.

Second, Buddha suggests that this painstaking process of analysis be followed by a search for a solution to the problem that fulfills the need. The only method proposed for this search is "reason," implying that process of intellectual deduction is sufficient unto the task; think about a problem long enough and deeply enough, and a reasonable answer, if one exists, will come to you. This is where the modern Scientific Method departs radically from the seminal model offered by Buddha. In classical times, the deductive constraint was quite natural; the *metaphysical* questions that the philosophers were pondering did not lend themselves well to an empirical approach, which requires that experiments be run and results be gathered and reduced to a conclusion. It was (and still is) extremely difficult to run experiments on the transcendental nature of existence. However, as the centuries progressed and the mysteries of the behavior of the *physical* universe began to capture the attention of philosophers and their intellectual progeny (scientists and engineers), experiments became possible, and much could be learned from trying solutions in the laboratory before developing them for application in the real world.

Finally, Buddha suggests that all proposed answers be validated based on the common advantage they can provide: "conducive to the good and benefit of one and all." Again, such egalitarian metrics do not survive well outside the metaphysical domain where all humans presumably share the same spiritual fate. Today, the "bottom line" is usually much more secular: "conducive to the good and benefit of the customers," or perhaps even more cynically: "conducive to the good and benefit of the shareholders." In spiritual matters, Buddha believed we are all equal shareholders. In today's crass and commercial world, the watchword is *caveat emptor.*

In the short quotation from Buddha, he does not suggest that validation include any kind of intellectual peer review of the task results and conclusions.

However, from his complete works we know that he encouraged frequent and open discussion of new ideas among his students and disciples, subjecting the ideas to healthy criticism and cross fertilization. This part of the validation process is a mainstay of the modern Scientific Method. Perhaps peer review seemed so obvious to Buddha, being so much a part of his everyday activities, that it did not warrant any special mention in his quotation. Ironically, it is this essential process of peer review that was totally missing from the evolution of the Scientific Method in Europe until the end of the Renaissance. Even to this day, many researchers stop their tasks short of this critical final phase. The reasons for this omission, now and in the past, are manifold and often rather self-serving.

Even the great classical Greek and Roman philosophers who followed closely on the intellectual heels of Buddha, albeit on the other side of the world and without any knowledge of his work, had little to say about the *process* of knowledge acquisition. They, too, were more concerned with the *nature and definition* of truth, rather than how to acquire it. They, too, were preoccupied with deductive rationalism, building an archive of philosophical knowledge on a foundation of axioms and definitions, for which, once again, the only reasonable platform for validation was intellectual contemplation and argumentation among a small group of disciples in close contact with each other. In all fairness, beyond their immediate working groups, little communication was possible, but such intellectual inbreeding stifled innovation.

Only a few sensors for experimentation were available to the classical scholars, mainly precision instruments for measuring length. Greek mathematicians proved many fundamental theorems in geometry and mathematics that could be tested and validated empirically with these instruments. Some even attempted to use these instruments to measure the positions and deduce the motions of the heavenly bodies, but only with marginal success. The Romans were less interested in such far-flung ideas, but achieved significant Earth-bound innovations in the fields of architecture, agriculture, and civil engineering, building massive and efficient transportation systems, including highways and aqueducts. However, despite these achievements, the Greeks and Romans gave little thought to understanding and codifying the *process* of scientific investigation.

With the decline of the Greek civilization, the advent of Christianity, and the fall of Rome, almost all intellectual thought in Europe ceased. The inquisitive nature of science was firmly shackled to the chains of religious dogma. Any intellectual thought that was not literally consistent with the text of the New and

Old Testaments, the axiomatic source of all fundamental truth, was proscribed and severely punished. Literacy in Europe dropped to almost zero while the feudal barons divided and redivided the continent and hacked each other to death in the name of religion under the righteous eye of the Church. No new intellectual ideas emerged from the Dark Ages. Almost a thousand years was lost to the epistemological maturation of the human species.

And then, as described in Chapter 5 (An Epistemological Journey), the Western world began to pull out of its slump. Trade routes were established to the East, and ships began to venture farther from the European continent. With the invention of the printing press, the curse of intellectual sterility was broken, and the exchange of information exploded. Shortly after this world-shattering invention, Phillippus Aureolus Theophrastus Bombastus von Hohenheim (1493–1541) was born in Switzerland. Wisely adopting the nickname Paracelsus, this highly intelligent and wild character divided his life between medicine and debauchery. Traveling all over Europe and the Middle East, he learned everything there was to know about classical medicine and alchemy, and ended up throwing it all out of the window. He was contemptuous of the reliance of physicians on the precepts of classical alchemy that had survived the Dark Ages in musty books, which were proclaimed to contain the inspired and unimpeachable wisdom of the Golden Age. Instead, Paracelsus insisted that science must begin with a deep and clear understanding of the problem domain, followed by an hypothesis about a possible solution that must be tested impartially under carefully controlled conditions. Only then could knowledge be accepted as valid. His perpetual challenge to the classical academicians and their irrefutable classical rules, coupled with his wild life style, could have easily marked him as a heretic in those dangerous days of the Holy Inquisition, and he and his ideas were regarded with much suspicion in Catholic Europe. However, in Northern Europe under the fortunate protection of the leaders of the Protestant Reformation, his profound genius, both in knowledge and method, prevailed, and he was eventually acknowledged as a brilliant diagnostician, a brand new specialty in the field of medicine. As an example of his rigorous application of empirical research methods, Paracelsus discovered a cure for syphilis in 1527.

The inductive genie was out of the bottle. Scholars were abandoning the classical traditions that could not be validated, and seeking new ways to understand the behavior of the universe based on the examination and support of empirical evidence. And their findings were being printed, published, and disseminated throughout Europe. Three scientists in particular were about to merge these lessons into a single guiding principle for the conduct of research and development in science and engineering.

In England, Francis Bacon (1561–1626) not only encouraged observation and experimentation with many cases before arriving at any conclusion, but also recognized the critical importance of noting the *failures* of a theory as well as the successes. This set the ground rules for explicitly formulating and stating all governing propositions, as well as testing the proposed solution under conditions that intentionally drive it to failure.

In France, profoundly impatient with the arcane presumptive propositions that were the ironclad foundation of medieval scholasticism, René Descartes (1596–1650) argued vehemently that the process of intellectual reasoning must be founded on one, and only one, fundamental axiom, namely the acknowledgement of one's own consciousness (*cogito, ergo sum; je pense, donc je suis;* I think, therefore I am.) Everything else must derive from that, either deductively or inductively. His principles were published in 1637 in his controversial book *Discourse on the Method for Reasoning Correctly and Searching for Truth in the Sciences.* In the appendices to this book, obeying his own strictures, he presented the method of formal mathematical induction, derived the basic principles of analytic geometry from fundamental principles, and made major contributions to the field of optics. The book is a *tour de force* in research methodology.

But the real hero was Galileo Galilei (1564–1642). Unlike Francis Bacon and René Descartes, Galileo was one of the first scientists in the Scientific Revolution to conduct experiments using *sensors* to gather empirical data, such as the thermometer, timing mechanisms, and the telescope. With the telescope Galileo was able to acquire direct observational evidence to support the theories of the Polish astronomer Nicolaus Copernicus, who had hypothesized a century earlier that the Earth revolved around the Sun, not the other way around. The Copernican theory had been repeatedly rejected outright by the Catholic Church as counter to classical teachings and the Holy Scriptures. But now for the first time, the Church could no longer simply attribute the idea to the misguided ravings of heretics; now there was direct evidence that was difficult to ignore. Something had to be done. In 1633, the Office of the Holy Inquisition arrested Galileo and tried him for heresy. Their case was so precarious that they had to forge false evidence and show Galileo the instruments of torture to intimidate him into recanting, which he did.

Galileo, who had martyred himself in defense of the hypothesis of a dead man, was sentenced to spend the remainder of his life under house arrest and forbidden from publishing and interacting with his peers. He died penniless and blind in 1642, still a prisoner in his own house. He was lucky they didn't burn him at the stake. The immediate effect of Galileo's trial and conviction was to put a complete stop to all meaningful scientific inquiry in the Roman Catholic countries around the Mediterranean, and the Scientific Revolution shifted its focus to Northern Europe and nascent America once and for all.

Galileo's real crime was not that he had gathered evidence that obviated the Earth-centered universe; most intellectuals (and ship captains) of the time had already come around to the Sun-centered perspective, which made so much sense. Galileo's real crime was that he had *published* this evidence (*Dialogue Concerning the Two Chief World Systems,* 1632) in direct disobedience of the Pope (Urban VIII, Maffeo Barberini). He did so because he fervently believed that this was the final and essential phase in the process of scientific investigation: validate your results by presenting them to your peers for critical review. This idea was entirely new. Not only had this been largely impossible before the advent of the printing press, but it also requires a degree of intellectual courage and honesty that puts a strain on the most honorable of scientists. It is not easy to present your new ideas to your peers, inviting them to shoot down your methods and results with all the intellectual ammunition they can muster. Moreover, these peers are sometimes your intellectual (and perhaps personal) adversaries who are actively gunning for you, vying for the same meager sources of funding, and busy positioning themselves for maximum leverage with a meager group of patrons. In this respect, things are not much different today.

Galileo put the final piece of the modern Scientific Method in place. Yes, observe and understand. Yes, hypothesize. Yes, run the experiment and measure performance. But when all is said and done, your own conclusions are *not enough*. You must *actively* seek critical reviews of your work from others working in your field. *Your peers* must be the final arbiters, or you must show just cause why their criticism should be ignored. To validate the conclusions of science or engineering tasks, they must survive the scrutiny of your peers. One can restate this requirement as a kind of meta–null-hypothesis: successful **peer review** must reject the assertion that the conclusions of a task are inconclusive or incorrect.

Needless to say, this final phase of the modern Scientific Method has been a thorn in the sides of many scientists and engineers, past and present. It seems to make the whole process vulnerable to political and personal pressures. But with sufficient care and oversight, peer review compares very favorably with the alternative — no peer review — which invites unbridled self-deception.

9.2 The Modern Scientific Method

The historical process just described did not cease with Galileo; that was merely a moment of profound insight, a critical point in the Scientific Revolution. The subsequent influences of many great thinkers over the centuries, including Johannes

Kepler (1571–1630), Joachim Jungius (1587–1657), Johann Baptista van Helmont (1579–1644), Marin Mersenne (1588–1648), David Hume (1711–1776), Johann Wolfgang von Goethe (1749–1832), Thomas Edison (1847–1931), Bertrand Russell (1872–1970), Karl Popper (1902–1994), Peter Medawar (1915–1987), and Thomas Kuhn (1922–) have refined the concept continually as new demands were placed on science and engineering. The modern Scientific Method focuses on the formulation of hypotheses that can be rejected either in closed form or by experimentation, dubbed the *hypothetico-deductive method* by Karl Popper and Peter Medawar. This method argues that no hypothesis can ever be completely proved, but it can be disproved or rejected. Alternatively, often it can be adapted and modified so that it gradually converges more and more closely to the truth. As Peter Medawar asserts: "A scientist is a searcher after truth, but complete certainty is beyond his reach."

This book does not seek to model the way in which successful scientists and engineers think and work. Rather, it attempts to synthesize a hybrid of historical lessons, philosophical principles, and practical constraints into a consistent and comprehensive framework of knowledge and processes that may be applied to a wide variety of research and development tasks. The proposed hybrid is introduced by the following definition of the modern Scientific Method:

> **Definition 9.1 The Scientific Method** comprises four sequential phases — Analysis, Hypothesis, Synthesis, and Validation — which are applied to a task iteratively and recursively to achieve the objective of the task.

Figure 9.1 presents a list of the major steps of each phase of the Scientific Method for both research and development tasks. Each step is a subtask of its parent phase, although its name in the task plan is often replaced by something more relevant and specific to the actual project. For example, for a theoretical research task, the step Conduct Experiments might be more properly called Check Derived Equation. In addition, when task names are used as section headers in a project report, they are often converted from imperative phrases to substantive phrases. For example, the section describing the step Solicit Peer Review might be titled Evaluation by Oversight Committee or, in the case of a doctoral research project, Dissertation Defense.

The next four chapters address the various problems and issues that arise in each phase during the planning and conduct of a research or development task. The remainder of this chapter presents a brief overview of the four phases and their internal steps by way of an introduction to the overall strategy of the modern Scientific Method.

The purpose of the Analysis Phase is to gain a clear and comprehensive understanding of the task at hand, to establish many of the governing propositions that constrain the ways in which the task and its descendant tasks may be accomplished, and ultimately to formulate a single specific and reasonable objective for the task, consistent with the imposed constraints. This process is a kind of

The Scientific Method

The iteration of four recursive phases for
the planning, conduct, and stepwise refinement
of a research or development task

Analysis	Describe Problem
	Set Performance Criteria
	Investigate Related Work
	State Objective

Hypothesis	Specify Solution
	Set Goals
	Define Factors
	Postulate Performance Metrics

Synthesis	Implement Solution
	Design Experiments
	Conduct Experiments
	Reduce Results

Validation	Compute Performance
	Draw Conclusions
	Prepare Documentation
	Solicit Peer Review

Figure 9.1 The four phases of the Scientific Method and their internal steps

"intellectual reconnaissance," during which the task team observes and organizes the details of the problem landscape in preparation for the selection of an effective solution to the problem in the Hypothesis Phase that follows. In many ways, the Analysis Phase functions like a funnel. A description of the problem is poured in the top and is progressively constrained by the imposition of practical requirements and related knowledge, so that a realistic and feasible objective for the task finally emerges at the bottom of the funnel. What begins as a broad perspective on the problem is systematically focused down to a single realistic and highly specific task objective. Chapter 10 (Analysis) discusses the Analysis Phase in detail.

Once the task objective has been firmly established, the Hypothesis Phase begins. The purpose of the Hypothesis Phase is to specify a detailed and comprehensive solution to the problem, to assert what is expected from that solution as a set of goals, and to define what factors will be varied to measure how well the task objective is achieved based on a set of postulated performance metrics. The solution can employ existing methods or new methods, or some combination of the two. Chapter 11 (Hypothesis) discusses the Hypothesis Phase in detail.

In the Synthesis Phase, the implementation of the solution specified in the Hypothesis Phase takes place. Following a rigorous experiment design that imposes the constraints of the governing propositions and factors, the implemented solution is tested through experimentation, and the results are reduced to the form necessary for the computation of the performance metrics. Bear in mind that even highly theoretical research must employ some experimentation, even if only to confirm the validity of a computation or a mathematical result derived in closed form. Chapter 12 (Synthesis) discusses the Synthesis Phase in detail.

In the Validation Phase, the performance is computed from the reduced results of the experiments using the metrics postulated in the Hypothesis Phase. Based on these performance values, the appropriate conclusions are drawn, stating whether or not and to what extent the task objective has been achieved. Finally, complete and comprehensive documentation for the entire task is prepared and submitted for critical peer review. Chapter 13 (Validation) discusses the Validation Phase in detail.

That's all there is to it: analysis; hypothesis; synthesis; and validation, and you're done. However, as you might have guessed, in reality it is never quite that simple. The devil, as they say, is in the details.

9.2.1 ITERATIVE EXECUTION

It would be naive to imagine that a task can be successfully accomplished with just one pass through the four phases of the Scientific Method. Even the simplest tasks often require several attempts to achieve the objective. For example, as you reach over to pick up a glass of lemonade from the table, you notice belatedly that the surface of the glass is covered with a thin film of condensation. Because this will probably make the glass slippery, you change your tactics for taking hold of it. Once the glass is firmly in your grasp, trying to lift it reveals that it is stuck to the table, and you change tactics again. Finally, as you lift the glass you observe that it is very full and reduce the speed of movement to your mouth. Life is full of surprises, and so is R&D.

For this reason the four phases of the Scientific Method are specifically designed to be performed iteratively, using stepwise refinement for any or all steps of the phases until the objective is achieved or the task is abandoned. The decision to return to an earlier step to revise some aspect of the task plan (or to abandon the task) can be made at any step in any phase of the process. The most common decision point is either at the end of the Synthesis Phase, when the results indicate that the task method was not successful, or at the end of the Validation Phase when the conclusions do not survive peer review. What follows are some typical actions that may be taken to attempt to recover from such failures, listed approximately in increasing order of the cost of the remedial action, *i.e.,* the number of steps that must be repeated:

- Return to the second step of the Synthesis Phase to refine the experiment design;
- Return to the second step of the Hypothesis Phase to modify the hypotheses or goals;
- Return to the first step of the Hypothesis Phase to modify the proposed solution; and
- Return to the second step of the Analysis Phase to loosen some performance criteria.

Each iteration incurs an additional cost in both time and money. Thus, each iteration following a failure to achieve the task objective should be preceded by a decision whether to continue to pursue the task objective or to abandon the entire task as a lost cause. If another iteration of the task will incur a large and unexpected cost, management will probably want to be involved in this decision to try to avoid a violation of Gresham's Law, which admonishes against throwing good money after bad.

Although abandoning a task is usually an unattractive and discouraging conclusion, task failure is a very common occurrence; the vast majority of research and development tasks are unsuccessful.[1] It is important, therefore, to find out as quickly as possible whether a task is likely to be successful or not. A good way to do this is to mount a series of preliminary **feasibility pilot** tasks (see Chapter 3).

Feasibility pilots (also called **QD pilots**, for **Q**uick-and-**D**irty) are simplified and rather informal investigations designed to quickly and inexpensively explore the feasibility of a proposed solution or an experiment design. Such pilots are not usually reported in the final project documentation; no one is very interested in a chronicle of the failures that preceded success. However, no matter how quick-and-dirty a QD pilot is, it is extremely important for the task team members to *record the details of every QD pilot in their research notebooks.* It is so very easy to lose track of what has been investigated and what has not, especially when the QD pilot tasks are being performed quickly, one after another, with little formal planning and management oversight. The heady excitement of converging on a promising solution or the frustration of repeated failures in a series of QD pilots can easily distract the researcher from keeping careful records in the interest of "saving time," which often seems to be slipping away uncontrollably. Almost inevitably, this lack of a detailed audit trail for the QD pilots leads to unnecessarily repeated trials, resulting in a lot of frustration and wasted time, simply because the researcher lost track of the thread of the search by trying to keep it all in the head: "What in blazes did I do yesterday that yielded those really good results?!"

---

[1]Of course, such failures are not usually reported in the literature. I have always thought it would be a good idea to publish a *Journal of Unsuccessful Research* (with anonymous authorship, of course) to save researchers from unnecessary expense and anguish by informing them about things that just don't work, *i.e.,* good idea, no cigar.

or "This value didn't work yesterday when I tried it! Now it does! What's going on?" Every one of us has experienced this kind of frustration, and it is a terrible waste of time and energy that can be largely avoided through rigorous and comprehensive record keeping. It is just a matter of discipline. Wax on, wax off.

QD pilots can be used at every step of *every phase of the task,* not just when exploring the feasibility of a proposed solution. For example, during the Related Work step of the Analysis Phase, a quick bibliographical search on the Internet can provide a survey of methods that have been used in the past to solve the research problem, which may profoundly affect the funding and time required to achieve the task objective. Similarly, a series of "dress rehearsals" with a small number of subjects (perhaps colleagues or team members) will help to debug a complex experiment protocol, even though these pilot subjects may not be demographically equivalent to the subjects in the cohort to be used for the "real" experiments. In the following chapters on the four phases of the Scientific Method, there are specific suggestions for where and how a QD pilot might be useful to estimate the feasibility of proposed task components or to identify potential problems and obstacles.

It is extremely important to remember that the results of feasibility pilots are *not* conclusive. Once a feasibility pilot has yielded a potentially useful result, a *formal* task must be rigorously designed and executed to confirm and validate this informal result.

> **TIP 12**
> Feasibility or QD pilots are useful for exploring the feasibility of proposed task components. Every QD pilot, although perhaps only minimally planned and not usually reported in the final project documentation, *must* be fully recorded in the cognizant team members' research notebooks. Once a QD pilot has yielded a potentially useful result, it *must* be confirmed and validated by planning and executing a formal task.

It is important to remember that planning is an integral part of every project, not a separate activity that is simply tolerated as a kind of unfortunate overhead. Very often both managers and project team members are impatient with planning, anxious to "just get on with it." Because these two groups generally agree on this, there is a strong mutual tendency to minimize or even skip planning efforts, resulting in a large increase in the overall task time. By the same token, however, it is clearly important to minimize the *overall* task time to avoid wasting unnecessary time and effort. The ever-present problem, then, is how to minimize the overall task time, while also ensuring that meaningful results are obtained as quickly and efficiently as possible.

What is the appropriate amount of time that should be spent *planning* a task, as opposed to *executing* it? Clearly, the answer to this question depends on a great many factors that are very difficult to codify: the difficulty and scope of the task, the experience of the task team, the resources available for the task, and so forth. Nonetheless, we can attempt to address this question by constructing a *very* simple model for planning and executing a task to get a general idea of the optimal task planning time.

This model rests on two major simplifying presumptive propositions. First, we postulate that the necessary number of iterations for stepwise refinement to complete a task is inversely proportional to the *task planning time,* expressed as percent of the *ideal task time* (*i.e.,* if the task were executed without any errors or delays whatsoever). Under this postulate, as the amount of planning time approaches zero, the number of necessary iterations approaches infinity, *i.e.,* if no time is spent for planning, the task process is a random walk. At the opposite extreme, if an infinite amount of time is spent for planning, then the number of necessary *repeats* of the process approaches zero, *i.e.,* the task is successfully accomplished after one pass (no repeats), but the total task time is still infinite because the planning time is infinite (one just never gets around to executing it). Second, we assume that on the average each iteration repeats only half of the ideal task time. Figure 9.2 presents the behavior of this simple model, which has a saddlepoint where the planning time minimizes the overall task time.

The saddlepoint in the curve in Figure 9.2 suggests that the appropriate amount of planning time is about 70% of the ideal task time. At this point, the

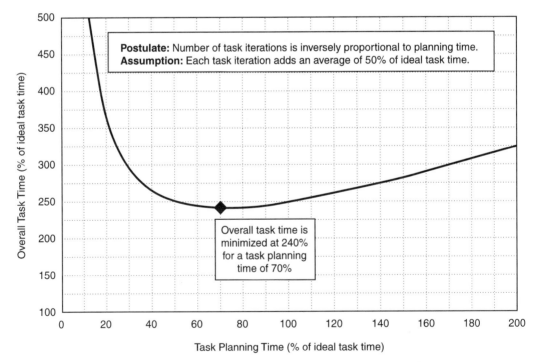

Figure 9.2 A simple model of the relationship between overall task time and planning time

overall task time, including planning time, is about 240% of the ideal task time, meaning that every individual step of the task must be performed an average of 2.4 times. For example, if the expected ideal task time were one person-year, this would suggest that the planning time should be about 8 person-months (70% of one person-year), and that the overall task time, including planning, would be about 2.4 person-years (240% of one person-year).

Having said that, please note that it is not appropriate to use Figure 9.2 to obtain actual numbers for the task time ratios. Instead, the curve should be regarded as a very simple abstraction of the general relationship between task planning time and overall task time. At best, one might consider the following conclusions:

- As the task planning time is *decreased* from the saddlepoint, the overall task time increases very quickly, resulting in excessive and wasteful iteration.
- As the task planning time is *increased* from the saddlepoint, the overall task time increases rather slowly.
- The two preceding observations suggest that saddlepoint in the overall task time should be interpreted as corresponding to a *minimum* planning time, rather than an optimum.

> **TIP 13**
> In general, too little planning time incurs much heavier penalties in overall task time than too much planning time.

A short summary of the concepts presented in this chapter so far is probably useful at this point. Iterative execution of the Scientific Method allows the orderly stepwise refinement of a task plan and its components. Large strategic changes, usually driven by partial or complete failures of one or more task components, must be accompanied by explicit and formal modifications of the task plan, which often require the concurrence of project management. Small tactical adjustments can be made unilaterally "on the fly," which, although perhaps not reported in the final project documentation, must be thoroughly and clearly documented in the research notebooks of the team members. QD pilots are useful for exploring the performance of proposed task components and for identifying potential obstacles and problems in the task components with minimal expense. Finally, planning is an integral part of any task that takes place at the beginning and *throughout* the task, especially when iterative stepwise refinement takes place. In general, insufficient planning causes a much higher risk of increasing overall task time than excessive planning, suggesting that the task personnel should engage in more, rather than less planning.

9.2.2 RECURSIVE EXECUTION

As was explained in Chapter 3 (The Project Hierarchy), every research or development project comprises a hierarchy of tasks. Theoretically, a project hierarchy could consist of a single task occupying a single level. However, just as it is naive to assume that a task can be accomplished without iterative stepwise refinement, it is equally unlikely that a complete project plan would have just a single level in its task hierarchy. Even the simplest project is more complicated than that. In fact, from the structure of the Scientific Method just presented, we now know that every project (or task) can be quite properly decomposed into four sequential groups of tasks, each representing one of the four phases of the Scientific Method: Analysis; Hypothesis; Synthesis; and Validation. Thus, every project can be represented with at least two levels in its hierarchy, as illustrated in Figure 9.3. The tasks are grouped by phase, as indicated by the labels in the shaded areas.

As explained in Chapter 3 (The Project Hierarchy), each of the second-level tasks shown in Figure 9.3 is a project in its own right. Therefore, each of these second-level tasks may in turn be decomposed into four groups of subtasks at the third level, one for each of the four sequential phases of the Scientific Method, as shown in Figure 9.4. In the interest of visual clarity, only the decomposition of the second-level Synthesis Task 1.5 is shown, even though several of the tasks at the second level are actually broken down, as indicated by the vertical ellipses. The fact that there are no ellipses below Tasks 1.3, 1.4, and 1.7 in the second level simply means that no further planning detail was required (or at least specified) for these tasks at lower levels.

And, of course, this recursive process of **hierarchical task decomposition** may be repeated indefinitely. Additional levels in the hierarchy may be added until sufficient layers of detail of the project plan have been fleshed out. Some branches in the project task tree may need more levels than others, depending on how much low-level detail is required for each particular task.

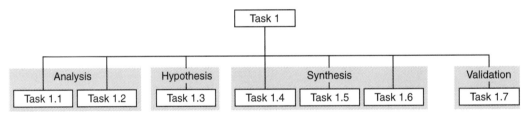

Figure 9.3 Illustration of a two-level project hierarchy

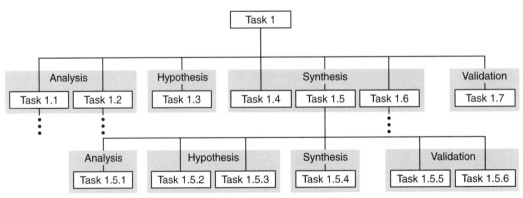

Figure 9.4 Illustration of a three-level project hierarchy (partial)

What does it mean to include several tasks within the same phase, *i.e.,* when several tasks occupy the same shaded area on the project task trees in Figures 9.3 and 9.4? It means that this phase requires more than one descendant task to achieve the objective of the parent task. This is not surprising, because we have already defined four steps, or subtasks, for each of the four phases, as shown in Figure 9.1. For example, the two tasks in the Analysis Phase group in Figure 9.3 might well correspond to the steps Describe Problem and Set Performance Criteria. When these two tasks have been successfully completed, their results contribute to the successful accomplishment of the entire Analysis Phase for parent Task 1. The remaining steps in the Analysis Phase of Task 1 are not broken out, implying (hypothetically) that they are either included in the two descendant tasks that *are* broken out, or are handled directly by the parent Task 1.

Likewise, two of the tasks in the Hypothesis Phase group in Figure 9.4 (Tasks 1.52 and 1.53) might correspond to the steps Specify Solution and Postulate Performance Metrics. Once again, the other two steps in the Hypothesis Phase (Hypothesis and Goals, and Factors) are not broken out, implying that they are either included in the descendant tasks that *are* broken out, or are handled directly by parent Task 1.5.

It is important to emphasize once again that every time a subtask is spawned, it reinvokes the *entire* Scientific Method, from Analysis to Validation. Having said that, in general the deeper the subtask in the project task tree, the briefer its Analysis Phase, because much of the required information has already been included in the Analysis Phase of the higher-level tasks. Thus, the Analysis Phase of a subtask might well be limited to a few sentences or a short paragraph.

The Hypothesis, Synthesis, and Validation Phases of the subtask, however, are often *more* detailed than those of the high-level tasks, because the closer you get to the terminal subtask, the more actual work is being planned and executed. For example, one of Clyde Tombaugh's very-low-level subtasks for his project **Find New Planet** was probably something like **Set Up Telescope Camera**, which almost certainly involved a complex task method that had to be planned in great detail and executed meticulously to ensure efficient, accurate, and consistent use of the equipment at the Lowell Observatory, even though the high-level astrophysical analysis had been accomplished long before.

Even though the task descriptions for the internal steps of a phase are often subsumed by the parent task because their task descriptions are brief and self evident, it is very easy for the number of pages required to *document* a complete project task tree (or project milestone chart) to become very, very large. Fortunately, the very nature of top-down design allows a project task tree to be cut down to an arbitrarily small tree, which represents either a small piece of the project with extremely *high* precision, a large piece of the project with very *low* precision, or anywhere in between. All are useful and accurate representations. Even so, the planning documents can quickly grow into an immense stack of papers that, taken together, provide the necessary detail of the project plan for all its component tasks at many different levels of precision. This is unavoidable, and, in fact, absolutely necessary in the end; after all, it takes an immense stack of planning documents to include all the necessary specifications and blueprints for constructing a hydroelectric power plant or a commercial jet aircraft.

To illustrate the evolution of a project plan based on the Scientific Method supported by its knowledge propositions, excerpts from a case study will be presented from time to time during the next four chapters: **Measure Character-Recognition Performance**. This case study is an abbreviated version of the final report for an actual industrial research project conducted several years ago to measure the ability of inspectors to accurately recognize the numbers and letters on video images of automobile license plates. Only short and often incomplete textual descriptions are included in the excerpts of the case study; ellipses (. . .) at the end of sentences or paragraphs are used to indicate incomplete text, which was much longer in the original project documentation. This compromise, however, does demonstrate that the information presented using the proposed methodology, albeit in abbreviated form, can be largely self explanatory.[2]

The project task tree in Figure 9.5 is a high-level project task tree of the major tasks for the case study. Although Figure 9.5 represents the final version of the project task tree, it evolved gradually from the first version, which was created on the first day of the project. When a research or development project is initiated, it is appropriate to construct an initial project task tree and a milestone chart as soon as possible. Of course, this initial plan can only be a very sketchy and approximate estimate of the eventual task structure and schedule, and the final project task tree and milestone charts will bear little resemblance to it. But all project plans evolve, and it is important to get something down on paper as soon as possible to put the project into some kind of perspective and to begin to assign tasks and resources.

> **TIP 14**
> Because the project task tree and milestone chart present an overview of every step within every phase of the Scientific Method for a project, it is particularly useful as one of the primary planning documents, a current copy of which should always be in the hands of every team member.

---

[2]All of the bibliographical citations in the tables and figures that describe the case study project *Measure Character-Recognition Performance* have been fictionalized to disguise the identity of the company that performed this research project. The important details of the project, however, are factual. Also note that the numbering scheme for the figures and tables within the illustrations for this case study is disjoint from the numbering scheme for the figures and tables of the book itself. The figure and table numbers are prefixed with CS to indicate that they belong to the case study.

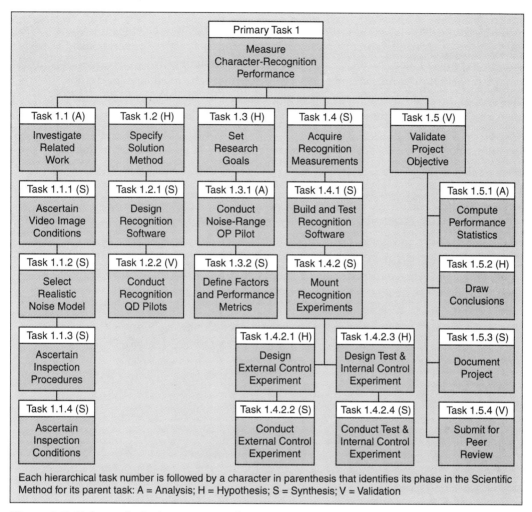

Figure 9.5 Task tree for Project *Measure Character–Recognition Performance*

Not every task that was undertaken in the project is listed explicitly in Figure 9.5. Many phases are not shown, only a few of the steps within each phase are listed at each level, and only three levels in the hierarchy are shown. Clearly, then, this is a high-level description of the project, and the omitted phases and steps are subsumed either by the parent task or by another task at the same level. In point of fact, during the course of the project the members of the project team constructed many additional task trees, which decomposed the tasks listed in Figure 9.5 into hierarchies of subtasks that were important for each member's specific responsibilities.

As mentioned in Section 3.2, the **milestone chart** is another useful way of representing a project plan to include scheduling information. The team for the project *Measure Character-Recognition Performance* constructed milestone charts for the high-level task trees. At the low levels in the task hierarchy, however, milestone charts are often omitted because the scheduling information is too inaccurate and volatile at the corresponding time precision. After all, the times for events on a milestone chart for the homeowner's subtask *Remove Screws* would have to be expressed in units of minutes, which would have been a bit silly. On the other hand, rather ironically, it was during this particular subtask that the entire project failed, partly as a result of poor scheduling!

Even though some subtasks are not explicitly scheduled and/or some subtasks are not explicitly broken out in the task tree, these subtasks *happen* and therefore need to be planned. If little or no effort is applied to planning, then either the planning process has simply been overlooked, or the task members think they can keep all of the necessary information and relationships in their heads and plan on the fly. Either way, this can be a catastrophe waiting to happen.

> **TIP 15**
> For R&D scientists and engineers: Resist the urge to "make it up as you go along." Plan the project thoroughly in advance, but allow the plan to evolve as experience and pilots reveal the flaws and potential improvements in the plan. Be prepared to give management time and cost estimates whose accuracy they can trust. If you discover you will not be able to make a deadline, notify your manager immediately; don't wait until the last moment. Understand that managers see and deal with a larger perspective than you do. Accept that the workplace is not a democracy.

> **TIP 16**
> For R&D managers: Support your scientists and engineers with the time and resources necessary to plan and stepwise refine these plans throughout the lifetimes of projects. Limit the work of each person to one or two concurrent projects. Don't steal time from project scientists and engineers for "fire drills"; hire special staff members for that purpose. Don't micromanage. Don't nickel-and-dime. Try not to burden your scientists and engineers with bureaucratic tasks; that's *your* job. If you have no choice, then explicitly acknowledge that the associated time and costs of the bureaucratic task are not a part of the ongoing project budgets, and extend all deadlines appropriately.

Making intelligent decisions about how much detail is required in a project task plan, how the steps in each of the four phases should be appropriately grouped and decomposed, and how and where the relevant governing propositions for the project should be stated depends on two issues: 1) the size and nature of the specific task and its components; and 2) a solid grasp of the purposes and functions of the four phases of the Scientific Method and their internal steps.

There is no way, of course, that this book can address the first issue, other than by example and analogy. However, the second issue is addressed squarely in the next four chapters, which present a general methodology for the planning and execution of R&D tasks based on the Scientific Method.

> **Exercise 9.1**
> Construct a set of task trees and their associated milestone charts for a project you are currently planning or undertaking.

CHAPTER 10

Analysis

In the seventh month of the human gestation period, all the basic components in the central nervous system of the fetus have finally been assembled and installed, and cognition begins. The fetus begins to gather sensory information from both its external sensors (perception) and internal sensors (proprioception). Although it is pitch black in the womb and the vision system is not stimulated, the fetus becomes aware of external sounds, pressure, tastes, temperature, and things touching its skin, as well as its own and its mother's heartbeat. Voices, though heavily muffled, can be heard through the layers of tissue and muscle that separate the fetus from the outside world. Ironically the mother's voice usually goes unheard, because she cannot bring her mouth close enough to her own abdomen, although there is some sound conduction through her body. The father, siblings, and friends, however, can easily press their face to the mother's abdomen and talk and sing songs to the unborn child. Even very loud sounds like electronically amplified music, thunder, and heavy machinery can probably be heard by the curious infant.

Until it emerges into the outside world, there is little the third-trimester infant can do to respond to these muffled stimuli, other than kick its legs occasionally and move its arms. At this point in its development, the infant must be largely content with **analysis**.

The word "analysis" derives from the Greek word *analusis* (αναλυσης) meaning "a dissolving." This word in turn stems from the Greek word *analuein* (αναλουειν) meaning "to undo," which consists of two parts, *ana* (ανα) meaning "throughout" and *luein* (λουειν) meaning "to loosen." These etymological roots imply that the fundamental process of analysis is to disassemble something into its component parts. And that is the objective of the Analysis Phase:

> **Definition 10.1** The objective of the **Analysis Phase** of the Scientific Method is to gain a thorough understanding of the components of the problem domain, leading to the formulation of a single specific and reasonable task objective.

As shown in Figure 9.1, the process of analysis can be broken down into four steps (subtasks): Describe Problem; Set Performance Criteria; Investigate Related Work; and State Objective. How much flexibility do you have in the order in which these steps are executed? First of all, it seems unlikely that the Analysis Phase would begin with any step other than a description of the problem, or end with any step other than a statement of the objective. However, the order of the two middle steps, Set Performance Criteria and Investigate Related Work, may be switched if appropriate. Ordinarily the criteria for any solution proposed for the task must be well understood before the investigation of previous R&D projects (Related Work) is begun. However, it is also easy to imagine situations that require the results of the investigation of the Related Work to establish appropriate performance criteria. A little bit of common sense will usually reveal the natural and appropriate order of these two subtasks. Note also that the order in which the steps are *executed* could well be different from the order in which they are *reported*. More will be said about reporting and documenting a project in Chapter 13 (Validation). In the general discussion of the purpose and function of the steps of the Analysis Phase that follows, the four steps will be discussed in the same order in which they appear in Figure 9.1, simply by default.

Finally, it is important to remember that the overall goal of the Analysis Phase is to winnow down the problem to a single specific task objective. What begins with a broad and rather abstract perspective at the problem description step is systematically constrained by the imposition of governing propositions, one after the other, until a narrow but highly detailed and reasonable objective can be formulated as the final step of the Analysis Phase.

10.1 Describe Problem

As was pointed out in Section 1.1, the objective of a *research* project is the *acquisition* of new knowledge, while the objective of a *development* project is the *application* of existing knowledge to create a device or an effect. However, when a project is first conceived by a customer or a manager or a scientist or an engineer (or, in fact, by anyone), little thought is given to whether it should be classified as a research or development project. Quite appropriately, the major focus is squarely on a **problem**.

When you visit the doctor, the meeting typically begins with the doctor asking, "What seems to be the problem?" Your response might be "I have a sore throat," or "I'm just not feeling very well." Regardless of the precision of the answer, you have characterized your problem with a *complaint*. On the other hand, it is certainly possible that you will respond to the doctor's question with a *question* of your own: "What is this rash on my arm?" You might even respond with a *requirement:* "Please find out what's causing these terrible headaches!"

All three kinds of answers ("I have a sore throat" or "What is this rash on my arm?" or "Please find out what's causing these terrible headaches!") prompt the doctor to launch a task to find out what is wrong with you, which might be called *Diagnose Illness*. Looking ahead briefly, during the remainder of the Analysis Phase, the doctor will explore the background of your problem by asking you how long you have been symptomatic, what you have been eating, what you have been doing, where you have been, what medications you have been taking, and so forth. Eventually the Analysis Phase will culminate in the formulation of a specific objective: perhaps to conduct a battery of specific tests to determine the probable cause of your illness. Following this Analysis Phase, the Hypothesis Phase will specify each test to be run, and the Synthesis Phase will conduct these tests and gather the results. Finally, in the Validation Phase the doctor will sit down with you and state his conclusions, *e.g.*, "You have strep throat," or "You have a heat rash." This diagnostic task, initiated by a question, a complaint, or a requirement, is a **research task**, because it acquires *knowledge*, which takes the form of the doctor's diagnosis.

At this point, you may well ask the doctor to launch a new task: "Please treat the strep throat," or "Please treat the rash." Each of these new tasks is a **development task** (*Cure Sore Throat* or *Cure Rash*), because it seeks to apply existing knowledge to create an effect, *i.e.*, a cure. On the other hand, you might decide to seek another professional opinion, in which case you would repeat the previous research task to obtain another diagnosis from a different doctor.

Sometimes your visit to the doctor begins differently. When the doctor asks, "What seems to be the problem?", your response is "I would like a flu shot." Instead of a question or a complaint, this is a request or a requirement. The doctor then launches a different task called *Administer Flu Shot*. Looking ahead briefly once again, the doctor completes the Analysis Phase by asking you a few questions about your physical condition and medical history to forestall possible adverse reactions to the flu serum. After formulating the objective to inject you with the current flu serum, the doctor then turns you over to a nurse for the Synthesis Phase, and she administers the injection. In the Validation Phase, the nurse evaluates how the needle penetrated the skin, how well the serum flowed into the muscle, and how much the puncture bled after withdrawing the needle. This task, initiated by a requirement, is a development task, because it *applies* existing knowledge to create an effect (immunization against the flu).

Over the next few months, either you get the flu or you do not. If you do, and you go back to the doctor and complain that you got the flu anyway, he will just shake his head in commiseration and admit that a flu shot is not a 100% guarantee against getting the flu, probably suggesting you try again next year. Nonetheless, the task objective was achieved; you got your flu shot. Things would have been different if you had handed the doctor a different problem: "I must avoid the flu this season with virtually complete certainty." Assuming he would accept this challenging task, his solution method would have undoubtedly been quite different, perhaps suggesting that you go into complete environmental isolation for the entire flu season, such as one might do with a patient who suffers from a severe autoimmune deficiency.

To summarize, the task *Diagnose Illness*, which was initiated by a question, a complaint, or a requirement, is a **research task** because its objective is the acquisition of new knowledge (a diagnosis). The possible subsequent tasks *Cure Sore Throat* or *Cure Rash*, which were initiated by a requirement, were **development tasks** because their objectives are to create an effect (a cure). The task *Administer Flu Shot*, which was initiated by a requirement, is a **development task** because its objective is to create an effect (immunization against the flu). Sometimes it is a fine line that discriminates between research and development tasks, but it is nonetheless useful to try to differentiate between them, so that the remaining methodology of the task may be planned appropriately.

It is often useful to summarize the task problem with a single statement:

Definition 10.2 A **problem statement** is expressed as an interrogative sentence, a declarative sentence, or an imperative sentence that summarizes a question, complaint, or requirement, respectively.

Although the sentence that expresses the problem statement should be as straightforward as possible, it may have one or more conditioning clauses or phrases. If it seems difficult to express the problem with a single question, complaint, or requirement, then you may be attempting to include more than one problem in the task. This is a common error that leads to confusion in the task plan, because the different problems require very different task objectives that interfere with each other, a bad situation that is made even worse when the planning extends to the next phases of the Scientific Method. In fact, confusion in a task can often be directly traced to multiple and conflicting task objectives. Table 10.1 lists some examples of well-formed problem statements.

Therefore, when the task team gets its marching orders, it will either be in the form of a question ("Why doesn't this thingamajig work?"), a requirement ("Develop a software application to do thus-and-so."), or a complaint ("This widget gets too hot."). However, the name of this step in the Analysis Phase is *Describe* Problem, not *State* Problem and it is essential that the summary problem statement be accompanied by a full description of the problem including sufficient background information and motivation to place the task in the proper context. To this end, managers are well advised to explain the full context of the problem and allow the task team to thoroughly investigate the origins and implications of the problem, so that they do not make wrong assumptions and come up with unrealistic or unreasonable solutions. Table 10.2 presents the Problem Description for the case-study project *Measure Character-Recognition Performance*.

There is the story (probably apocryphal, but certainly not unbelievable) of a professor in an introductory physics course who gave his students a short quiz consisting of a single laconic problem statement: *Describe how to use a barometer to measure the height of a building*. Not atypically, the only qualification accompanying the problem statement was: "No questions will be answered during the quiz. State your assumptions and proceed." The story has it that one of the brighter (and, admittedly, smart-alecky) students described *11* different solution methods, *none of which used air pressure*. For example, drop the barometer from the top of the building, time its fall, and apply the Newtonian equations for rectilinear motion. Or, mark off the height of the building using the barometer as a measuring stick and then multiply the result by the length of the barometer. Or, set the barometer upright on the ground at a known distance from the building, find the

Table 10.1 Examples of problem statements

Complaint:	Customers complain that the battery in the portable drill (Part #156A-90) loses its charge very quickly when it is not being used.
Requirement:	Design a more effective and reliable windshield wiper.
Question:	How well can humans recognize the characters on license plates when the surfaces of the plates are severely degraded by dirt and debris?
Complaint:	The user interface for Model 902 is difficult to understand.
Requirement:	Design a speech recognition system for voice control of automobile appliances, such as entertainment and air conditioning equipment.
Question:	Why are red and green used for traffic lights to indicate "stop" and "go," respectively?
Complaint:	My dog has fleas.
Requirement:	Land a team of astronauts on the Moon and bring them safely back.
Question:	Is the sample of vehicle speeds normally distributed?

point where the straight line connecting the top of the barometer and the top of the building intersects the ground, and apply some trigonometry; and so forth. One solution even suggested examining the original blueprints, using the barometer as a paperweight to keep the blueprints from rolling up!

On the other hand, too much guidance can stifle innovation. There is another story (this one true) of a consultant to the Pentagon back in the early 1970s, who had been asked by the brass to help design a large electronic map of the geographical locations of the US tactical and strategic weapons systems. They began their initial briefing of the consultant by telling him that they wanted the continents and political borders outlined in black, using different shaped symbols for each kind of weapons system. The consultant asked, "That's fine. Would you like the display to use color to help identify the different objects and borders?" The officers sighed: "That would be wonderful, professor, but of course we know that's totally impossible. Color display technology for large wall screens just doesn't exist yet." The consultant rocked back in his chair for a moment, and then said, "Gentlemen, why don't you just tell me what you would like, in the best of all possible worlds, and *I'll* tell you whether it's possible or not and what it'll probably cost. That's what you hired me for, isn't it?"

Sometimes the criteria attached to a problem statement are deeply entrenched in the psyche of the research and development community based on historical traditions. For example, some historians claim that the engine was installed at the front of the first automobiles, sometimes on long outriggers, simply because that was where and how the horse was attached to the carriage.

Although the Problem Description must include sufficient background material to put the task in the proper perspective, it should impose as few initial conditions as possible, so that there is plenty of room for innovation. For example, take a look at the fifth example in the list of problem statements in Table 10.1: "Design a speech recognition system for voice control of automobile appliances, such as entertainment and air conditioning equipment." Notice that the problem statement (a requirement) also imposes the *solution* method: speech recognition for voice control. Has someone studied this problem carefully to make sure that this is the best and most reasonable way to solve the problem? What is the problem? Might there be better solutions? Even if some limiting conditions must be imposed, either by management or by the task team itself as a result of their comprehensive analysis, the proper place to specify them is in the next step of the Analysis Phase, Set Performance Criteria, which is discussed next in Section 10.2.

Exercise 10.1
For each task listed in Table 4.1, compose a reasonable problem statement as either a question, a complaint, or a requirement.

Exercise 10.2
Compose a reasonable and accurate primary problem statement for a project you are currently planning or undertaking.

Table 10.2 Problem description for Project *Measure Character-Recognition Performance*

| |

Primary Task 1: Problem Description

As a part of its R&D program to develop innovative highway and traffic surveillance and control systems, Acme Corporation plans to develop an automated vehicle license-plate character-recognition system that is more reliable and at least as accurate as humans are. Currently, license-plate numbers are read manually from images of license plates captured by fixed video cameras overlooking highways and roads. Often the license plates are worn, dented, and spattered with mud and debris that partially obscure the letters and digits. Although it appears that manual inspection can yield high recognition accuracy under these conditions for short times, the task is very labor intensive and incurs high error rates when the inspector gets tired or bored. Clearly, the performance of an automatic character-recognition system, which would avoid these problems, must be at least as good as the *peak performance* of trained human inspectors. Unfortunately, there is very little data available from formal experiments on human character-recognition performance under comparable conditions. . . .

> **Problem Statement:** How well can trained license-plate inspectors recognize partially obscured characters on license plates?

10.2 Set Performance Criteria

Performance criteria are requirements that any proposed solution to the problem must fulfill. Performance Criteria are stipulative knowledge propositions, which may be characterized as facts, accompanied by appropriate explanations and justifications. Furthermore, if in the best of all possible worlds the project personnel would probably rather not impose these criteria, they are biased propositions, *i.e.,* restrictions, limitations, or exclusions.

Performance criteria are very common in the course of our everyday lives: "You may watch television for another half-hour, but only if you have already completed all your homework for tomorrow"; or "Stop on your way home and pick up a quart of eggnog, but be home no later than 7:00 pm"; or "Please buy us two tickets for the opera on Saturday evening, but don't spend more than $45 each."

The eggnog example illustrates how the imposition of *conflicting* performance criteria can lead to task failure: suppose you cannot find any eggnog at any stores before you run out of time. The example about the opera tickets illustrates how *brittle* performance criteria can cause task failure: suppose there are simply no tickets available for the opera on Saturday evening for less than $45. *Excessively precise* criteria can also result in unnecessary or unintended task failure; imagine how annoyed your companion would be if you announced that unfortunately the tickets for the opera were $46, so, of course, you didn't buy them.

Brittle performance criteria are often highly counterproductive or unfair. For example, if the definition of the maximum income level for receiving food stamps is $18,000 for a family of four, then the family earning $18,000 eats adequately, while the family earning $18,001 goes hungry. If at all possible and appropriate, performance criteria should be soft rules, allowing judgment calls based on the final success or failure of the overall project objective: "If your family is hungry, you are entitled to receive food stamps." Such soft rules, however, are difficult to enforce; bureaucrats hate them, because they must make individualized decisions instead of doing things by the numbers. The US presidential election of 2000 illustrated this problem, as officials struggled to decipher each voter's intent on the punched-card ballots in Florida.

Interestingly enough, the Constitution of the United States of America, which was ratified by the citizens of the United States of America at the end of the eighteenth century, is an elegant study in soft rules. The founding fathers purposely included a great deal of fuzzy language, so that the Congress would have some latitude when they passed new laws, while the Supreme Court interpreted whether or not the laws, individually and collectively, exceeded the intentions of the authors of the Constitution. This flexible approach has been one of the main reasons for the astonishing durability and stability of the representative democracy in the United States of America for well over 200 years, the longest uninterrupted constitutional government in the history of human civilization.

Most development projects are tightly constrained by fixed conditions and parameters imposed on them by demands of the marketplace or application domain. For example, if a large aerospace company has decided to develop and manufacture a new passenger aircraft, the marketing department may have ascertained that the new equipment must be capable of carrying as many as 450 passengers (or the equivalent load of freight) with a maximum range of no less than 10,000 kilometers. Presumably, these criteria were the conclusions of a marketing study, conducted as an initial task in the project, which carefully determined the most profitable routes and the competitive equipment of other aircraft manufacturers. Similarly, a microchip manufacturer may conclude that they could earn good profits from a new microprocessor that would run at least three times faster than the best competition, as long as the power requirements and generated heat of this new microprocessor did not prevent its use in compact laptop computers. From another perspective, the United Nations may decide to intervene militarily in a country that is being attacked by a neighboring country whose intention is to annihilate the indigenous population. The UN military commanders may be constrained to methods that guarantee a swift exit strategy for the occupying forces once political and economic stability has been established in the country being rescued. In addition, the political leaders of the military action may insist that all tactics be designed to avoid casualties in the UN military forces. This was the case in the rescue of Kosovo from Serbian oppression in the 1990s.

As mentioned previously, sometimes the imposed performance criteria are so strict that they obviate the achievement of the objective, requiring stepwise iteration of the task to revise and refine the original performance criteria or, in some cases, the complete abandonment of the task short of success. Pilot tasks during the Analysis Phase are often useful to determine the feasibility of a proposed task objective as constrained by the imposed performance criteria. The

R&D engineers designing a new aircraft often mount sophisticated wind-tunnel pilots to measure the flight characteristics of a proposed aircraft envelope, even though such experiments may be conditioned by brittle and approximate scaling assumptions. Microprocessor chip designers will often compute the heat dissipation of a proposed chip design long before a prototype is constructed, even though predicting the thermodynamic behavior of microscopic VLSI technology is rather artful. In the case of military action, pilot tasks take the form of live-fire training exercises under conditions as realistic as possible, even though the actual battle may well be dominated by contingencies that are very difficult to foresee. Similarly, the examination of historical military battles functions as a kind of vicarious pilot in the planning and execution of a military project, although advances in technology often diminish the usefulness of such historical lessons. At the outset of World War II, the battleship was quickly recognized as an unwieldy dinosaur, and air power soon reigned supreme. Today, smart missile systems are the strategic and tactical weapons of choice, and military planners regularly conduct realistic war-gaming exercises to measure their effectiveness under a variety of political and technological performance criteria.

Three of the example running tasks introduced in Table 4.1 are listed as development tasks: *Remove Screws*, *Launch Spacecraft into Orbit*, and *Compile Summaries*. Let us speculate about some of the performance criteria that might be imposed for these tasks.

Perhaps the most obvious performance criterion for the task *Remove Screws* was the requirement that the task be finished, the hinges fixed, and the door re-installed by the early evening before the house guests were due to arrive. In fact, this performance criterion was really imposed in the primary task of the project, namely *Fix Squeaky Hinges*. Nonetheless, a performance criterion for a parent task can and usually does constrain the conduct of the descendant tasks.

The parent project that included the subtask *Launch Spacecraft into Orbit* was *Land Astronauts on the Moon*. When President John F. Kennedy announced the intention of the United States to carry out this high-level project, he made one critical performance criterion very clear: the astronauts must be returned safely to Earth. Without this criterion, the entire project would have been far less expensive and difficult, although it may have been rather hard to find (and ethically highly questionable to solicit) qualified volunteers for a one-way trip.

Taken at face value, the task *Compile Summaries* has no obvious performance criteria imposed on it. However, because this development task is part of the Investigate Related Work step of the Analysis Phase of a doctoral student's research project, there is indeed an important performance criterion. By convention, every doctoral student is required to conduct a successful research project whose results are *original, i.e.,* that contribute new knowledge to the state of the art. To ensure this, the doctoral student must make sure that the literature search on the state of the art is complete and comprehensive, leaving no stone unturned. Only in this way can the reviewing committee be convinced that the work is original. The burden of proof is on the student.

Doctoral students have an additional special constraint imposed on their research: it must not only be *original,* it must also make a *significant* contribution to the state of the art, *i.e.,* the results and conclusions of the research cannot be trivial or inconclusive. Although the decision whether the contribution is *original* can presumably be made objectively, the decision whether it is *significant* must be made subjectively by the reviewing committee.

In contrast with development tasks, other than time and/or cost limitations, pure research tasks normally must meet few performance criteria, if any. In fact, because they seek unbiased knowledge, care must be taken not to impose performance criteria on a research project that could violate the criterion of **objectivity** (see Chapter 5). The fact is, however, that such criteria are by no means unprecedented. For example, in the early seventeenth century, the Roman Catholic Church explicitly stated that any research concerning the behavior of the solar system was acceptable as long as it did not deny that the Earth was the center of the universe, which was an impeachable article of faith. Unwittingly, they had singled out the correct answer as the only forbidden answer. This constraint, coupled with the terrifying penalties for not observing it, snuffed out all scientific research in the predominantly Catholic countries on the Mediterranean for hundreds of years (although the arts flourished!).

As mentioned at the beginning of Chapter 10, the middle two steps of the Analysis Phase, Set Performance Criteria and Investigate Related Work, may be executed and/or reported in either order. If the performance criteria are dictated by management or perhaps by the nature of the problem itself, then usually they will be established first, and then the related work will be investigated subject to these constraints. On the other hand, even if there are no performance criteria imposed by management or by the nature of the problem, the investigation of the related work may well reveal some necessary constraints or limitations that must be recognized and accepted in order to achieve the task objective. In fact, both situations may occur in the same task: management may lay down some necessary criteria, and the investigation of the related work may add additional constraints. In such cases, there is no reason why the two steps cannot be combined in the reports and documentation that are generated for the project.

Table 10.3 presents the Performance Criteria set for the case-study project *Measure Character-Recognition Performance.* Note that this industrial research project *did* impose performance criteria, because there was a specific need expressed by management.

> **Exercise 10.3**
> State the primary performance criteria for a project you are currently planning or undertaking.

10.3 Investigate Related Work

Once the problem has been fully described, including a precise formal problem statement, and all of the *a priori* performance criteria have been imposed, the project team may begin the process of finding out as much as possible about what has been done to address the problem and achieve similar objectives in the past. One of the most important reasons to investigate past and current work related to the task objective is to avoid "reinventing the wheel." A review of the literature may reveal that the task objective has already been achieved, and the results of that past project may simply be acquired and applied to the current project. Without a doubt, if the project is being undertaken in an industrial venue, the project manager will be delighted if one or more of the subtasks can simply apply knowledge, a device, or an effect that is already available, even if it must be purchased. Buying a solution is almost always less expensive than replicating it.

By way of example, in the late 1970s a consultant was hired by a firm that designed, manufactured, and installed passenger elevator systems. The management

Table 10.3 Performance criteria for Project *Measure Character-Recognition Performance*

Primary Task 1: Performance Criteria
Performance Criterion 1: The experiment must use computer-generated images of *serif* and *sans serif* characters which have been obscured by a realistic simulated noise model. (exclusion)
To use video images of real license plates was impractical, because the true characters on the plate cannot be ascertained without examining the actual plates, which were not available. . . .
Performance Criterion 2: The subjects for the recognition experiments must be inspectors who have been trained to read the numbers on video images of license plates.
To avoid learning curves during the recognition experiments, the subjects had to be trained inspectors who were accustomed to the procedures and domain of the job, bringing with them any subtle skills they may have acquired that could assist them in the recognition task. . . .
Performance Criterion 3: The level-of-effort for the research task team must not exceed one person-year over an elapsed time of six months. (limitation)

of the firm told the consultant that a programmer on their engineering staff had recently designed and tested a program that used a new, highly sophisticated scheduling algorithm to control banks of elevators in office buildings. When the programmer had completed the program, it was installed in a large and well-known skyscraper in New York City. The programmer burned the code into read-only memory (ROM) in the computer control system, and then, after determining that it was operating correctly, destroyed all copies of the source code except one, which he hid. He then went to the management of the elevator company and told them they would have to pay him a bonus of $50,000 to get the only existing copy of the source code for the scheduling program.

The consultant was asked by the firm's management if it was possible to reverse-engineer the machine code in the ROM, so that they could avoid paying the blackmail. The consultant assured them that it was certainly technically feasible and then contacted several software firms that might be able to do such a job. The lowest bid for the reverse-engineering task was $90,000 with an elapsed time of about 6 months. Clearly, the extortionist had done his due diligence. The consultant returned to the elevator company management and recommended they simply pay the programmer off and secure the copy of the source code.

Traditionally, investigating the related work is carried out by performing literature searches of the relevant publications, either in print libraries or at sites on the Internet. However, there are many other ways that information about past or current work related to the project can be acquired, some of which are easily accessible but often overlooked, and some of which are not so easily accessed or usually avoided for obvious reasons:

- professional journals
- conference proceedings
- books and monographs
- professional studies and investigations
- Internet reports and databases
- newspaper and magazine reports
- manufacturer's technical specifications
- discussions with colleagues
- reverse engineering
- industrial espionage
- blackmail or theft
- interrogation

Not all of these sources and acquisition methods for information are recommended. Some of them raise clear ethical and legal questions. But make no mistake: all of them are used every day.

Regardless of the source, care must be taken to determine the validity of the information. This is often quite difficult, because the circumstances surrounding the generation of the knowledge by the source may be unclear or suspect, casting doubt on its validity. The report may itself be a secondary source (like an encyclopedia), generally requiring that the primary source be located and digested to establish the validity of the knowledge firsthand. The transfer of information from the original source through a series of secondary sources can cause severe distortion and thus play havoc with the three criteria for valid knowledge: reproducibility, completeness, and objectivity (see Chapter 5).

The sources for all information brought to a project as domain knowledge must be clearly and completely cited. (This requirement alone may deter the use of the shadier methods given in the preceding list.) This is not only necessary for reasons of professional courtesy, but also because it allows the entire epistemological basis of the knowledge domain of the project to be scrutinized and carefully audited as part of the peer review process in the Validation Phase (see Chapter 13).

In contrast with industrial projects, if a doctoral student discovers during the investigation of the related work that the objective of his or her research project has already been achieved by someone else, this is *not* good news, because every doctoral student is required to make an *original* and significant contribution to the state of the art. Likewise, if a search of the related work reveals that the project objective has not yet been achieved, then it behooves the doctoral student to publish the solution method with some preliminary results *as soon as possible* to avoid being scooped by someone else. Once the idea has been protected through publication in the open literature, the doctoral student has established clear evidence that he or she thought of the idea first. Occasionally, a doctoral student goes to defense without publishing in the open literature and without undertaking a thorough search of the related work, only to discover at the last moment that the work is not original. Such a tragedy need never happen.

10.4 State Objective

Once the problem has been clearly described, the performance criteria have been clearly and completely defined, and the related work has been thoroughly investigated, then presumably the task domain has been appropriately and sufficiently focused down and constrained. Most of the governing propositions for the task have been stated and justified (although some must wait to be presented in subsequent phases). At this point, a specific, clear, and detailed task objective may be formulated. As defined in Chapter 4 (The Project Task), the **task objective** is a statement of what the task is intended to achieve, expressed as an infinitive phrase. Throughout the book so far, many examples of task objectives have been presented, so very little further discussion of its form and function is necessary. However, two important final points need to be made.

First, the task objective is the baseline by which the success or failure of the task will be evaluated. Thus, consistent with the definitions of accuracy and error presented in Section 8.1, it may be said that the *accuracy* of the task will be measured by how well a task achieves its stated objective. With this in mind, it behooves the task team to formulate an objective for their task that expresses a *single, highly specific and realistic expectation,* tightly constrained by the specific governing propositions that have been incrementally imposed in the preceding steps of the Analysis Phase (Describe Problem, Set Performance Criteria, and Investigate Related Work). The performance of the task team will be judged by how well it achieves the task objective.

Second, although many other task components may be unilaterally modified by the task team as a part of the process of stepwise refinement to recover from partial or total task failure, to modify the *task objective* often requires the explicit permission of R&D management, especially in development projects whose requirements originate from on high. This is another good reason to spend the necessary time to formulate a clear and specific task objective that is explicitly acknowledged as reasonable and acceptable by both the project team and R&D management.

> **TIP 17**
> The task objective is a *contract* between the task team and management. It often has a political component. Word it carefully.

It would be wrong to conclude that the Analysis Phase is invoked only once in a project at the primary task level. On the contrary, as explained in Section 9.2.2, *all* four phases of the Scientific Method are recursively invoked for *all* the tasks in the project task tree. Thus, whenever a subtask is invoked in the middle of any phase, one drops down one level and re-initiates the Scientific Method beginning with the Analysis Phase. When all phases at this lower level have been completed, one returns to the parent task from which the recursion was initiated.

It is true, however, that the Analysis Phases for levels below the primary task level are usually quite brief, because most of the necessary analysis has already been undertaken at the primary task level. Often the name of the task itself is an adequate statement of the objective of a low-level task. Any additional required information can usually be summarized in a paragraph or two.

Tables 10.4a and b, and 10.5 present an abbreviated version of the Related Work and the Primary Objective for the case-study project *Measure Character-Recognition Performance.* Because this project relied heavily on the results and conclusions of prior research and the field experiences of the trained license-plate inspectors, the Related Work section of the final project report was 15 pages long.

Exercise 10.4
State the complete and detailed primary objective for a project you are currently planning or undertaking.

Table 10.4a Related work for Project *Measure Character-Recognition Performance*

Task 1.1: *Related Work*

Most of the research in optical character recognition has attempted to recognize letters that are embedded in words, using a dictionary to assist the final classification of each letter. Contextual information of this kind greatly enhances the probability of accurate recognition (Jones and Tatali 1994). However, the objective of the higher-level parent project in the R&D department requires the recognition of numbers on license plates. In such cases, there is no contextual information available to help with the recognition task, because the character strings on license plates are randomly generated (subject to a few limitations) (USDOT Publication #124-789A3). Thus, the results and conclusions of past character-recognition studies using text (Jones 1982, Green and Johnson 1990, Brown *et al.* 1979) were not useful for estimating the typical license-plate character-recognition performance of human inspectors. . . .

Task 1.1.1: *Video Image Conditions*

As prescribed by Performance Criterion 1 (Table 10.3), the images of the license plates for this project had to be realistically simulated. Examination of video images of license plates acquired by existing cameras that observe highway traffic (USDOT Publication #106-111T1) revealed that a typical character on a license plate is captured at a vertical resolution of about 9 pixels. The horizontal resolution varies with the width of the character; the letter I is only 1 pixel wide, while the letter W is 8 pixels wide.

There are no uniform international typeface standards or conventions for license plates. Because the Acme Corporation wants to market the proposed automatic license-plate recognition system to many countries that use the Latin alphabet and decimal digits on their license plates (Internal ACME Memorandum AZ119.97 1998), the current project had to measure the recognition performance of human inspectors for both *sans serif* and *serif* fonts of the 26 uppercase Latin characters and 10 decimal digits. Figure CS.1 shows a *sans serif* and a *serif* letter D at the required resolution centered in a 21 × 21 pixel frame without noise. . . .

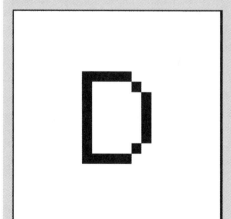 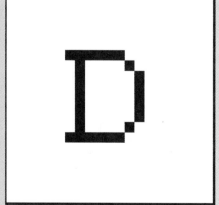

Figure CS.1 Sample simulated video image of a *sans serif* and a *serif* letter D

Table 10.4b Related work (abbreviated) for Project *Measure Character-Recognition Performance*

Task 1.1.2: *Realistic Noise Model*

The open literature reported several research projects that had used noise models to simulate the obscuration of text by dirt, debris, and normal wear-and-tear. The noise model selected for this project was postulated by Jacobi (1993), as illustrated in Figure CS.2. This choice was justified based on. . . .

Postulate 1: A realistic noise model consists of 32 discrete gray levels uniformly distributed over the dynamic range from 0 (white) to ±255 (black), whose probabilities are normally distributed about an average gray level of 0 (white) with a specified standard deviation (**noise level**).

Definition 1: The **noise level** is the standard deviation of the normal distribution of the noise generated by the model specified in Postulate 1 and as shown in Figure CS.2.

Task 1.1.3: *Manual Inspection Procedures*

Based on observations of the license-plate inspectors on the job at the license-plate inspection facility and extensive oral interviews with them, a profile of the data-management and analytical procedures undertaken by them during typical inspection tasks was compiled. . . .

Task 1.1.4: *Manual Inspection Conditions*

Based on an extensive inventory of the layout and environment of the license-plate inspection facility, a complete list of recommended specifications for an accurate model of this facility was compiled, including ambient and workspace lighting conditions, temperature regimes, background acoustic noise levels, inspection duty cycles, *etc.* . . .

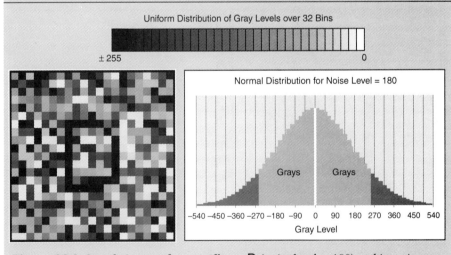

Figure CS.2 Sample image of *sans serif* letter D (noise level = 180) and its noise distribution

Table 10.5 Primary objective for Project *Measure Character-Recognition Performance*

Primary Objective 1:	To estimate the peak character recognition accuracy and response times of trained license-plate inspectors for both *serif* and *sans serif* computer-generated 9-point characters embedded in a 21×21 pixel frame with superimposed noise that realistically simulates the typical range of the actual obscuration of license-plate numbers due to dirt, debris, and normal wear-and-tear. (1 person-year over 6 months)

CHAPTER 11

Hypothesis

When the human (or mammalian) infant emerges from the womb at birth, it is immediately flooded with sensory stimulation. Even though its eyes are still closed, just imagine the sudden torrent of new sounds, smells, tastes, and other sensations that wash over the infant. The doctor gently smacks the infant's bottom, and for the first time air rushes into its lungs. Unfortunately (?) we cannot remember this apocalyptic event. Perhaps we must bury the accompanying sensory explosion in our unconscious mind for the sake of our long-term mental stability.

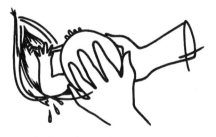

Very little time passes before the infant opens its eyes, and the sensory input data rate jumps by a factor of 100. Curiosity overcomes fear, and the infant, already somewhat practiced in the art of analysis, begins to put things together. Bits of information are combined and organized, guiding and influencing the genetic mechanism that is rapidly wiring up the neural connections in its brain. But, as must always be the case, some pieces of knowledge are missing; the jigsaw puzzle has gaping holes in it, and the curious infant must interpolate and extrapolate as best it can to complete the emerging picture. Even though it is still practically powerless to test them, the child seems content simply to entertain an unending stream of **hypotheses**.

The word "hypothesis" derives from the Greek word *hupotithenai* (υποτιθεναι) meaning "to suppose or to speculate," which consists of two parts, *hupo* (υπο) meaning "beneath" and *tithenai* (τιθεναι) meaning "to place." These etymological roots imply that the fundamental process of hypothesis is to speculate about an underlying cause-effect relationship. And that is the objective of the Hypothesis Phase:

> **Definition 11.1** The objective of the **Hypothesis Phase** is to propose a solution to achieve the task objective, a set of goals and hypotheses for this solution, and the factors and performance metrics for testing the validity of the solution.

By itself, an hypothesis is nothing more than a speculative knowledge proposition, which we have declared *persona non grata* for science and engineering projects in Chapter 6 (Categories and Types of Knowledge). Thus, to elevate such a proposition to legitimacy, we must undertake to establish its validity through a formal process of Synthesis and Validation, which are the final two steps in the Scientific Method. Without such rigor, an hypothesis remains idle speculation.

As shown in Figure 9.1, this process of hypothesis can be conveniently broken down into four sequential steps (subtasks): Specify Solution, Set Goals, Define Factors, and Postulate Performance Metrics. Unlike the flexibility in the order of the steps for the Analysis Phase, these steps are usually undertaken in the stated order.

11.1 Specify Solution

The **solution** used to achieve the task objective is one of the two components of the **task method** defined in Chapter 4 (Section 4.2, Task Method). The other component is the **experiments**, which measure the effectiveness of the solution when it is applied to the task unit. The experiments are designed and conducted in the Synthesis Phase and are discussed in detail in Chapter 12 (Synthesis).

As defined in Section 4.2, the solution consists of a **mechanism** and a **procedure**. For some research and development tasks, the mechanism may already exist and may simply be purchased (or otherwise acquired) and applied to the task unit. For example, if your objective is to measure the length of a table, you will probably elect to use a standard tape measure. There is no need to invent or create a new mechanism, and the procedure is straightforward. On the other hand, when NASA was charged with landing astronauts on the Moon and bringing them safely back to Earth, not only were many new mechanisms required, but the procedures had no precedents and had to be designed from scratch. A third possibility is that the mechanism is easily acquired, but new and special procedures must be devised. This happens when a standard product is being tested in a new and unusual venue, such as finding out whether a particular brand and model of truck can be used reliably in the Antarctic. Finally, when a new drug is being developed by a pharmaceutical company, the mechanism may be an extremely complex biochemical process that has never been attempted before, while the procedure for the task is well understood, having already been used many times in the past with new drugs.

Regardless of whether it is devised especially for the task or whether it already exists, it is this step in the Hypothesis Phase where the solution is specified. For all the components of the solution that are new, the specification must include all the necessary engineering drawings, schematics, recipes, flowcharts, pseudo-code, algorithms, mathematics, *etc.* For those components that already exist, a short description accompanied by appropriate citations, technical reports, and/or manufacturer's specifications may be sufficient.

Sometimes the task team intends to try several different solutions, not simply as a result of step-wise refinement triggered by a failure of the originally proposed solution, but as a *bona fide* part of the project plan. For example, in his original plan for the example running task **Repair Squeaky Hinges**, the homeowner specifies two alternative solutions: if possible, try to **Straighten Hinges**; if that fails, then **Buy New Hinges**. Usually the purpose of trying several solutions is to pick the best one. But for some tasks, more than one of the proposed solutions may be necessary or desirable, each being important in its own right. For example, a word processor allows the user to convert lightface text to boldface using a function key, a drop-down menu selection, or even a spoken command, because different people prefer different solutions. Aircraft designs require several different solutions for the same basic function, because redundancy helps to reduce the risk of a catastrophic accident from critical system failures.

If several different solutions are required for the project, then the solution-specification task should be decomposed at this point into several different subtasks in the project task plan. As shown in Figure 3.2, this was the case for the homeowner's task **Repair Squeaky Hinges**; straightening the hinges and buying new hinges are very different solutions. This was also true for the 11 different solutions that the imaginative physics student proposed for finding the height of a building using a barometer. In such cases, the high-level solution-specification task in the project task tree has a descendant task for each of the proposed solutions.

As suggested in Section 4.2, explaining the art of inventing, or even selecting, an appropriate solution to a problem is well beyond the scope of this book — perhaps beyond the understanding of the wisest philosopher. Problem-solving is an elusive talent, the product of some ineffable synergy of intelligence, education, and experience. Faced with this mystery, the best we can do is offer an example of how the solution, once invented, is included in the project plan. Table 11.1 briefly summarizes the solution specifications for the major tasks in the Hypothesis Phase of the case-study project **Measure Character-Recognition Performance**. (See Figure 9.5)

Table 11.1 Solution specification for Project *Measure Character-Recognition Performance*

Task 1.2: *Solution Specification*

To achieve the primary objective (see Table 10.5), a software application was designed to simulate the video images of license plates inspected by the license-plate inspectors on the job. The simulated images of license plates were presented to a small cohort of trained license-plate inspectors to measure their ability to identify the displayed characters in the presence of simulated, but realistic noise. . . .

Task 1.2.1: *Recognition Software Design*

Each subject in a small cohort of trained license-plate inspectors (*task unit*) was asked to identify each instance in a sequence of randomly selected characters (letters or digits) displayed on a CRT monitor screen capable of supporting at least 256 gray levels (*inducer*). Each 9-point character was centered in a 21 × 21 pixel frame, partially obscured by a noise level specified by Definition 1 (Table 10.4b). The character was surrounded on all sides by a margin of at least 6 pixels with the same noise level as that which obscured the character. Both the character and the noise were generated as specified by Postulate 1 (Table 10.4b). To the right of the character frame was an array of 36 buttons, one for each of 26 uppercase Latin letters {A through Z} and the 10 decimal digits {0 through 9}, which were used by each subject to select the recognized characters using the mouse (*sensor*). (See Figure CS.3 in Table 12.4.) Each generated and selected character was recorded by the computer, along with the time of the selection. . . .

Task 1.2.2: *Recognition QD Pilots*

Periodically during the software design phase, preliminary versions were evaluated with a series of informal feasibility pilots, using members of the project team as subjects. . . .

11.2 Set Goals

Once a solution for a problem has been specified in complete detail, the next step is to set the **goals** for the task, which specify how the response of the task unit will be determined when the solution is applied to it. These goals are accomplished with a set of experiments designed and conducted in the Synthesis Phase. From the perspective of the Scientific Method as a whole, the results of these experiments are the basis of the conclusions drawn in the Validation Phase that characterize how well the solution achieves the task objective.

Stephen R. Covey provides a broad, but useful characterization of a goal on page 137 of his book *The 7 Habits of Highly Effective People:*

> An effective goal focuses primarily on results rather than activity. It identifies where you want to be, and, in the process, helps you determine where you are. It gives you important information on how to get there, and it tells you when you have arrived.

If we translate this characterization into the domain of R&D tasks in science and engineering, the goal of a task specifies a way to determine how well the proposed solution works:

Definition 11.2 Every task has at least one **goal** to determine the response of the task unit to the application of the solution (or part of the solution), expressed as an infinitive phrase.

Goals, like tasks, may be decomposed into a hierarchy of subgoals, in which the descendant goals, taken together, accomplish the parent goal. And in a manner similar to the project task tree, it is often very useful to construct a **goal tree**, which is a graphical representation of the goal hierarchy for a project. Each of the lowest-level (terminal) goals in the goal tree is accomplished with an experiment that is designed and conducted in the Synthesis Phase, as described in Sections 12.2 and 12.3 of the next chapter. Thus, each terminal *goal* that is established here in the Hypothesis Phase spawns a corresponding *task* in the Synthesis Phase. The solution for this task in the Synthesis Phase is the set of experiments needed to acquire the knowledge that accomplishes the corresponding goal in the Hypothesis Phase, *i.e., the knowledge that characterizes the effectiveness of the solution for that task.*

It seems logical that research tasks would set goals, because the objective of a research task is to acquire knowledge, which is the function of a goal. But why does a development task need a goal? After all, a development task creates a device or an effect, not knowledge. The answer is because *all* tasks, including development tasks, must acquire *some* knowledge, even if only to confirm that the solution has accomplished the stated objective. It is hard to believe that someone would fix a broken china saucer, for example, without then checking after the glue had dried to see if the repair was successful. The task unit consists of the two pieces of the broken saucer, the solution is to glue the two pieces together, the goal is to measure the strength of the bond, and the experiment is to apply a reasonable amount of stress to the repaired saucer (once the glue has dried) to make sure it will not break again with normal use. As another example, suppose the task objective is to tie your shoelaces (*task unit*). The proposed solution is to tie a slipknot, and the goal is to test the tightness of the slipknot, both of which are specified in the Hypothesis Phase. You accomplish this goal in the Synthesis Phase by tying the knot and then running an experiment in which you tug on the laces to check the tightness of the knot.

These two development tasks are very modest in their scope, and their goals may seem trivial and obvious. However, just imagine how many goals had to be set for the development task *Launch Spacecraft into Orbit*! The equations governing the trajectory of the spacecraft had to be verified, the conditions of the atmosphere had to be measured, expert opinions about the advisability of launching under a wide variety of weather conditions had to be gathered, all the various internal systems of the spacecraft had to be measured, the actual trajectory of the spacecraft during flight had to be compared with the desired trajectory, the vital signs of the astronaut had to be measured, and the risks and possible consequences of malfunctions and emergencies had to be estimated, just to mention a few. Thousands of people and hundreds of millions of dollars were required to accomplish all the goals. Even though this was a development task whose objective was to create an effect (place the spacecraft in Earth orbit), an enormous amount of knowledge had to be collected using many complex experiments at every step of the development process to makes sure the solution methods wore working properly.

Because each experiment task in the Synthesis Phase is spawned by a goal defined in the Hypothesis Phase, the objective of the experiment task is very similar to, if not the same as the goal that spawned it. That is the reason why Definition 11.2 requires that goals be expressed as infinitive phrases, just like objectives. There are many verbs that may be used to express the infinitive for a goal: to measure, to estimate, to count, to compute, to verify, to deduce, to ascertain, to compare, to decide, *etc.* Note that each of these infinitives implies the acquisition of knowledge, *not* the creation of a device or an effect (*e.g.,* to implement, to purchase, to build, to document, to assemble, to launch, to destroy, *etc.*). If you find yourself selecting an infinitive that specifies the creation of a device or an effect, you are probably specifying a *task,* not a goal. Tasks belong in the project task hierarchy (the task tree), and each of these tasks has its own hierarchy of goals (the goal tree).

For some tasks, the knowledge acquired by the experiments that accomplish its goals is sufficient to satisfy the task objective (or part of the task objective). For example, if the boss wants to know how fast the traffic is moving on two highways, the goal might be to measure the traffic speeds using a set of pneumatic hoses stretched across each highway, and the conclusion of the experiment that accomplishes this goal might simply be the averages of the two samples of traffic speeds: "The average traffic speed on Highway A is 65 mph, and the average traffic speed on Highway B is 41 mph." No further conclusions need be drawn to achieve the primary task objective.

On the other hand, the objective for the similar task *Identify Faster Traffic* is to *decide* which highway has the higher traffic speed. To simply report the descriptive statistics for the two speed samples would not be sufficient to achieve objective of the task. A decision must be made.

A task that requires knowledge, but no decision, to achieve its objective is called an **estimation task**. A task that requires the additional step of making a decision based on some acquired knowledge is called a **decision task**. Goals for decision tasks are often expressed using the infinitive *to decide,* although other verbs that *imply* decisions are equally appropriate. For example, the goal for the task that has the objective of simply estimating the traffic speeds on the two highways might be expressed: *to estimate the average traffic speeds on the two highways, A and B, based on measurements with pneumatic detectors.* This is an estimation task. On the other hand, if the task objective is to identify which of the two highways has the higher traffic speed (as is the case for the task *Identify Faster Traffic*), a decision must be made, and the corresponding goal might be expressed: *to identify (or decide) which highway, A or B, has the higher traffic speed, based on measurements with pneumatic detectors.* This is a decision task.

In addition to the goals that acquire the knowledge required to make the decision, every *decision* task must also assert one or more formal statements about the conclusion of the goal. Each of these assertions is called a **research hypothesis** or simply an **hypothesis**.

> **Definition 11.3** A **research hypothesis** (or simply an **hypothesis**) is a declarative sentence that asserts a desired, expected, or possible conclusion of a goal.

For example, the decision task *Identify Faster Traffic* could assert one of three possible hypotheses:

The average traffic speed on Highway A is higher than on Highway B.
The average traffic speed on Highway A is lower than on Highway B.
The average traffic speed on Highway A is the same as on Highway B.

Any of these statements is an appropriate research hypothesis. However, because the traffic engineer has no desire or expectation that the average speed on one highway is different from the average speed on the other highway, there is no reason for her to assert either of the first two hypotheses. In fact, if she did, it would imply a bias. Therefore, to express the intention of being unbiased and objective, she will probably assert the last hypothesis, which simply states a *possible* conclusion and is a neutral choice.

For the task *Measure Drug Effectiveness*, however, the project team hopes that the drug will turn out to be effective. They would be completely justified, therefore, in asserting a primary research hypothesis that reflects this desire: *Drug XYZ is an effective therapy for disease ABC.*

In fact, regardless of the expectations or preferences in the wording of the research hypothesis asserted in the Hypothesis Phase, the actual *decision* about the hypothesis is made in the Validation Phase, based on a special kind of hypothesis that is *guaranteed* to be completely neutral, called a **null hypothesis**. The formal process of validating a research hypothesis asserted in the Hypothesis Phase by testing the corresponding null hypothesis in the Validation Phase is explained in Chapter 13 (Validation). The purpose of asserting the *research* hypothesis here in the Hypothesis Phase, then, is simply to state the decision required by the task objective and to acknowledge the expectation or preference of the project team, if any.

The goals and hypotheses of the primary task of a project are called the **primary hypotheses** and **primary goals** of the project. Once again, this assignation is relative. In a development project to design a fuel pump for a rocket, for example, the highest-level goals are the primary goals of the project (to measure the fuel-flow rate, to measure the MTBF, *etc.*). However, in the larger view, this project is probably just one of many in a higher-level project that has its own set of primary goals (to measure the range of the rocket, to measure the accuracy of the rocket, *etc.*).

To illustrate the relationships between goals and hypotheses, Table 11.2 revisits eight of the running example tasks listed in Table 4.1, inventing a reasonable (but abbreviated) problem statement, task objective, solution mechanism, representative goal, and, if applicable, research hypothesis for each task. Take a moment to study them and compare how the various statements are composed for different kinds of research and development tasks.

Note that the first task in Table 4.1 (***Remove Screws***) has been replaced by its parent task (***Remove Hinges***) in Table 11.2. The representative goal given for this task, "To count the number of screws still holding the hinge plate in place," must be accompanied by another equally important goal for this task, "To count the number of hinge plates removed," and its corresponding hypothesis, "All four hinge plates have been removed." Both goals must be accomplished and both hypotheses must be validated to satisfy the task objective. As shown in Figure 3.2, the two goals spawn two corresponding subtasks: ***Remove Screws*** (Task 1.2.1) and ***Pry Off Hinges*** (Task 1.2.2). Although both goals apply a solution that uses the same mechanism (largest available screwdriver), the *procedures* for the solution differ: unscrewing *versus* prying.

It is doubtful whether the subgoals and subhypotheses for the task *Identify Faster Traffic* deserve specific mention or explicit decomposition into subtasks. Given the two speed distributions (supplied as part of the governing propositions for this task), the computations of the required descriptive statistics are straightforward and can be easily included at the primary task level.

Exercise 11.1

Create the table entries for Table 11.2 for the other four example tasks in Table 4.1: *Launch Spacecraft into Orbit, Identify Collision Damage, Measure Fossil Age,* and *Evaluate GUI Effectiveness.*

When Andrew Wiles relates his ten-year journey that eventually led to the successful conclusion of his task ***Prove Fermat's Last Conjecture***, he describes an evolving labyrinth of goals spanning a deep and broad hierarchy. As he met dead end after dead end and thought of idea after idea, he constantly revised the hierarchy of goals and refined his task methods, applying both classical and modern tools from many branches of mathematics. Clearly, patience was a necessary resource in the achievement of this ambitious objective. His proof consumes 200 pages of very sophisticated mathematics. Wax on, wax off.

For the task ***Find New Planet***, Clyde Tombaugh hypothesized that photographic evidence would reveal a ninth planet of the Solar System in the predicted orbit. One of the major goals was to use a blink comparator (a brand new device at the time) to compare sequential photographic plates from the telescope camera in search of small changes in position among the myriad of points representing the fixed stars. And there were a *lot* of plates! It took him many months. Once he found such a change in position, he had to test the hypothesis that it was the track of a planet in the predicted orbit (not some wayward comet), which he did by computing the proper sidereal motion that generated this track. One can only imagine the scores of additional subtasks and subgoals that were necessary just to operate the telescope, the camera, and the blink comparator properly and consistently, day after day and night after night. Wax on, wax off.

Table 11.2 Summary of some analysis and hypothesis statements for eight example tasks

	ANALYSIS PHASE			HYPOTHESIS PHASE	
Task Name	Problem Statement	Primary Objective	Solution Mechanism	Representative Goal	Representative Hypothesis
Remove Hinges (Development)	The hinges cannot be fixed until they are removed. (Complaint)	To remove the four hinge plates from the door and door frame	Largest available screwdriver	To count the number of screws still holding the hinge plate in place	All three screws have been removed from the hinge plate.
Identify Faster Traffic (Research)	Which highway has the higher traffic speed? (Question)	To decide if the speed of the traffic on one highway is higher than on the other highway	Standard error of the difference between two means	To estimate the averages and standard deviations of the two speed samples	The averages of the two speed samples are the same.
Prove Fermat's Last Conjecture (Research)	Is Fermat's Last Conjecture valid? (Question)	To mathematically prove Fermat's Last Conjecture in closed form	Classical and modern logic and mathematics	To verify the logic of the proof	Fermat's Last Conjecture is proved, *i.e.*, it is a theorem.
Find New Planet (Research)	Is there a ninth planet in the solar system? (Question)	To find a new planet of the required mass in the predicted solar orbit	Blink comparator to detect object motion between successive telescope plates	To use the blink comparator to detect movement of objects between successive plates	There is no apparent movement of the object.
Measure Drug Effectiveness (Research)	Is drug XYZ an effective therapy for disease ABC? (Question)	To measure the effectiveness of drug XYZ as a therapy for disease ABC	Oral intake of drug XYZ (5 mg per day) and weekly blood tests	To measure the progress of disease ABC in the control and test cohorts	Drug XYZ is an effective therapy for disease ABC.
Compile Summaries (Development)	Gather information about multiple-personality disorder (MPD) (Requirement)	To create a database of summaries of information on MPD published since 1850	Summarization rules and formats for database entries for each category about MPD	To verify that all summaries have been entered in the database in the proper format	All summaries have been entered properly.
Measure Liquid Temperature Variation (Research)	The customer reports that the temperature of the liquid in the vessel is not stable. (Complaint)	To decide if the variation of the temperature of the liquid is greater than specified by the manufacturer	Thermocouple, IR detector, and thermometer using cables that minimize heat transfer	To measure the temperature of the liquid over time using the thermocouple	The temperature of the liquid in the refrigerated vessel is stable over time.
Measure Age of Organic Sample (Research)	What is the age of the organic sample? (Question)	To ascertain the approximate year when the organic sample died	Carbon-14 dating equipment	To measure the emission rate of Carbon-14 atoms	(Not applicable)

The primary hypothesis asserting the effectiveness of drug XYZ in the task *Measure Drug Effectiveness* obviously requires an extensive hierarchy of complex goals to complete the investigation, whose conclusions must satisfy not only the medical community, but also the demanding regulations of the federal government. Note that the primary hypothesis says nothing about the *safety* of the drug, but it is clear that such an investigation would have to be undertaken before the drug could be marketed, should the current task confirm its effectiveness as a therapy for disease ABC. For both projects, the pharmaceutical company is undoubtedly fully prepared with standard solutions and experiments (*i.e.,* task methods) that have been fully tested and approved by cognizant professional and governmental organizations.

The primary goal for the task *Measure Liquid Temperature Variation* probably supports an extensive hierarchy of subgoals for the test equipment that must be designed and constructed. The terminal goals that measure the effectiveness of this equipment will be difficult to accomplish, requiring a great deal of stepwise refinement until a satisfactory overall solution has been implemented. Tasks that involve putting together many pieces of equipment, such as the design of a computer system, usually require many feasibility pilots to ensure their accuracy and reliability. To make matters worse, this task must impose a severe performance criterion: *extraneous* heat loss, which inevitably accompanies the measurement of temperature within an imperfectly insulated system, must be minimized. The engineers must be absolutely sure that all *significant* heat loss occurs in the *vessel,* not in the sensors and channels of the test equipment.

The hierarchy of the major goals and hypotheses for the case-study project *Measure Character-Recognition Performance* is presented as a goal tree in Figure 11.1 and in outline format in Table 11.3. To conserve space, the labels in the goal tree are short abbreviations of the complete goal and hypothesis statements.

Just as the task tree is useful for planning and documenting the hierarchy of *processes* of a project, a goal tree is useful for planning and documenting the hierarchy of *knowledge* acquired for the task. Like the task tree, the goal tree may be expanded to add more detail by decomposing the lowest levels into their component hypotheses and goals, or it may be condensed for management briefings and documentation summaries by revealing only a few high levels. And like the task tree, it may be segregated into small subtrees, each representing a different scope (sub-project) of the primary project. Although the goal tree in Figure 11.1 is a fairly complete representation of the goals of the major tasks of the project, it does not include the goals of the unending sequence of minor tasks that occupied the project team for six months.

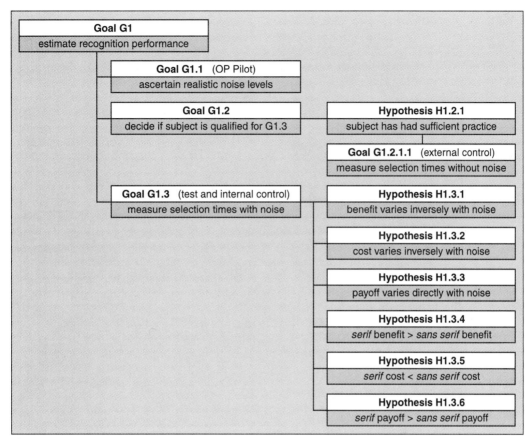

Figure 11.1 Goal tree for Project *Measure Character-Recognition Performance*

Note that a small number of experiment tasks in the Synthesis Phase may be used to accomplish many goals that are set in the Hypothesis Phase. For example, both goals G1.2 and G1.3 and their seven associated hypotheses (H1.2.1 and H1.3.1 through H1.3.6) in Figure 11.1 are all accomplished by the two experiments designed and conducted in Task 1.4.2 in Figure 9.5. This is especially important if goals require the responses of a cohort of living subjects. For reasons of efficiency, a *single, carefully designed experiment* must often suffice to provide the results necessary to accomplish many goals, because such experiments are always very difficult and expensive to plan, design, and run. Moreover, the experiments usually must be conducted with a very small cohort, so repeated trials with naive subjects are out of the question. This issue is discussed in more detail in Chapter 12 (Synthesis).

When the conclusions for goals and hypotheses are drawn in the Validation Phase, these conclusions are used to validate the objectives of the associated tasks. In Table 11.3, Goal G1 is validated on the basis of the conclusions drawn from the results of the subgoals G1.1, G1.2, and G1.3, which include both estimates of performance and decisions about the associated hypotheses (H1.2.1, and H1.3.1 through H1.3.6). Briefly, validation is the process of drawing a meaningful conclusion from the results

Table 11.3 Research goals for Project *Measure Character-Recognition Performance*

Task 1.3: Research Goals

Goal G1: To estimate the ability of trained license-plate inspectors to correctly identify characters on simulated video images of partially obscured license plates

  Goal G1.1: To ascertain dynamic range and increment for the noise-level factor. (OP Pilot Task 1.3.1) Based on their experience, a small number of trained license-plate inspectors were asked to select a reasonable and realistic dynamic range and increment size for the noise-level factor for the formal experiments in Task <None>1.4. . . .

  Goal G1.2: To decide whether each subject is qualified to participate in a test trial

    Hypothesis H1.2.1: The subject has had sufficient practice with the interface.

      Factor: Rejection-confidence threshold

      Goal G1.2.1.1: To measure character-selection times without noise (external control Task 1.4.2.2)

        Factor: Displayed characters = {A...Z} and {0...9} (representation factor)

        Performance Metric: Rejection confidence that absolute correlation of selection times with acquisition order is zero.

  Goal G1.3: To measure character selection times and accuracies with noise (internal control and test Task 1.4.2.4).

    Factors: Displayed characters = {A...Z} and {0...9} for test trials; or no character for internal treatments (representation factor)

      Typeface = *serif* or *sans serif* (n/a for internal control treatments)

      Noise level = 130 to 240 in increments of 10 (from Goal G1.1)

    Performance Metrics: Average selection times $\overline{T}_{control}$ and $\overline{T}_{test}$

      Number of characters selected correctly

      Benefit B = percent of characters identified correctly

      Cost C = Type I error probability (%) that $\overline{T}_{test} < \overline{T}_{control}$

      Payoff $P = B (100 - C)/100$

    Hypothesis H1.3.1: Benefit varies inversely with noise level.

    Hypothesis H1.3.2: Cost varies directly with noise level.

    Hypothesis H1.3.3: Payoff varies inversely with noise level.

    Hypothesis H1.3.4: Benefit for *serif* typeface is higher than for *sans serif* typeface.

    Hypothesis H1.3.5: Cost for *serif* typeface is lower than for *sans serif* typeface.

    Hypothesis H1.3.6: Payoff for *serif* typeface is higher than for *sans serif* typeface.

      Performance Metric: Confidence that the null hypothesis may be rejected, (H1.3.4 thru H1.3.6) based on the standard difference of two proportions

of the task, although not necessarily the *desired* conclusion. The complete process of validation is defined and explained in Chapter 13 (Validation).

Exercise 11.2

For a project you are currently planning or undertaking, define the goals and hypotheses for the major tasks down to a reasonable level of detail. Represent this hierarchy as a goal tree.

11.3 Define Factors

At this point in the application of the Scientific Method to a research or development project, all of the knowledge propositions for the tasks should have been clearly and precisely established, with the possible exception of the methods for reducing results, computing the performance, and drawing conclusions. Many of the parameters and conditions specify constant values or settings that will not be varied during the execution of the task, either because there is no need to vary them to achieve the objective of the task, or because they cannot be varied for some abiding reason (*e.g.,* they represent the effects of an unavoidable natural force). As defined in Chapter 6 (Categories and Types of Knowledge), knowledge propositions that will not or cannot be varied are called **fixed parameters** or **fixed conditions**, or taken together, **governing propositions**.

From those parameters and conditions that *could* be varied as part of the experiment protocol, the investigators select a subset that *will* be varied to observe their effects on the performance of the task unit. These are called **factors**, which were formally defined and described in Definitions 7.8, 7.9, and 7.10. The remaining parameters and conditions are given constant levels or settings and become the governing propositions.

The first step in determining the factors for a task is to make a *complete list of all the parameters and conditions.* In fact, this special list should be started on the first day of the project and updated whenever a new governing proposition is established or a previous one is modified. Then, when the planning reaches the Hypothesis Phase, not only will the list already be largely complete, but the task personnel will have a deep and shared understanding of the items on this list, having participated actively in its preparation for some time. Once they are satisfied that the list is complete, that no stone has been left unturned, then a subset of these parameters and conditions can be moved to the list of factors.

> **TIP 18**
> From the very beginning of a project, start building a list of knowledge propositions. Update it whenever a new proposition is established or a previous proposition is modified.

For a project whose complexity is comparable to that of the case-study project *Measure Character-Recognition Performance*, there might be a few dozen governing propositions, although this is just a guideline; the actual number can vary widely from task to task. For the running example task *Launch Spacecraft into Orbit* there were many *thousands* of governing propositions. No matter how many end up on the final list, there will probably be a lot more than you first imagined. Table 11.4 lists only the *major* domain knowledge propositions (greatly abbreviated) for the case-study project *Measure Character-Recognition Performance*. The code in parenthesis indicates the role of the proposition (See Chapter 6). The last column identifies the propositions that are defined as factors and specifies the number of values assigned to each factor for the experiments.

Table 11.4 Knowledge propositions for Project *Measure Character-Recognition Performance*

Knowledge Propositions (Role Code)	Number of Factor Values
1 The ambient laboratory conditions have no significant effect on recognition performance. (gcsE)	
2 The CRT screen conditions are constant and have no significant effect on performance. (gcsE)	
3 Test characters are uppercase instances of all 26 uppercase Latin letters and 10 decimal digits. (fcp) ⟶	36
4 The typefaces of the characters on license plates are Roman and either *serif* or *sans serif*. (fcp) ⟶	2
5 The design of the user interface has no significant effect on recognition performance. (gcrE)	
6 The colors of the displayed objects have no significant effect on recognition performance. (gcsE)	
7 A realistic noise model consists of 32 discrete gray levels, uniformly distributed. . . . (gcrL)	
8 The noise level is the standard deviation of the normal distribution of the noise generated by the model . . . (gcs)	
9 Simulated video images of license plate characters yield reliable performance estimates. (gcrL)	
10 Noise levels from 130 to 240 gray levels in increments of 10 represent a typical range . . . (fcvL) ⟶	12
11 Recognition performance metric B (benefit) = percent of selections recognized correctly. (gcr)	
12 Recognition performance metric C (cost) = confidence subject has had sufficient practice. (gcr)	
13 Recognition performance metric P (payoff) = BC = product of benefit and cost. (gcr)	
14 The test cohort is an accurate sample of the recognition skills of trained license-plate inspectors. (gps)	
15 There are no statistically significant differences between the pilot cohorts and the test cohorts. (gpcE)	
16 There are no statistically significant differences in the recognition skills of the subjects. (gpcE)	
17 Personal attributes (age, sex, *etc.*) of the subjects have no significant effect on performance. (gpsE)	
18 There are no statistically significant differences in the typographic biases of subjects. (gpcE)	
19 Testing each subject with 6 to 30 characters per trial assures the independence of responses. (gpcE)	

Role Codes

g = governing proposition	+	p = parameter + s = scheme c = condition + r = regime	or c = constant or s = situation	L = limitation +R = restriction
f = factor + p = parameter	or	c = condition + p = policy	or v = variable	E = exclusion
r = range proposition +	r = result or c = conclusion +	(b = benefit and/or c = cost) or		p = payoff

When the project *Measure Character-Recognition Performance* was in the planning stage, many of the project team members wanted to include even more parameters and conditions as factors; after all, they argued, when you have the unusual opportunity to experiment with a specialized and highly trained cohort, why not extract as much information as possible? For example, some team members wanted to experiment with the color of the characters and the background on the CRT display (limited by Proposition 6) to observe the impact of these conditions on recognition performance. Others wanted to see if thicker boldface characters would improve recognition performance (limited by Proposition 4). Some even wanted to find out if recognition performance varied with the age of the subject (excluded by Proposition 17). But then, when the logistic impact of the attendant explosion of the factor space and the small number of available subjects became apparent, the team reluctantly trimmed the choices down to three factors with their stated precisions and fixed the rest, as shown in Table 11.4. It is natural and appropriate for research investigators to be very enthusiastic about the extent of knowledge that can be gleaned from experiments. But ultimately, reason must prevail.

This problem can also occur with experiments that have nothing to do with live subjects. Computer simulations often have a large number of parameters and conditions, many of which are very attractive for investigation as factors. But even computer programs take time to run. Genetic algorithms and high-dimensional clustering methods, for example, often require hours of CPU time to provide the results from one set of factor values. By analogy, if the number of computer runs was the same as the number of factor-value combinations shown in Table 11.4 (864), and the average CPU time for each run was one CPU hour, then the total amount of computer time to run all the experiments would be more than one CPU month! The decision to expend such resources must be made with great care. Mistakes and necessary refinements of the task method that require re-running all the experiments are very common and very expensive!

Once the number of factors and the response load on the subjects have been winnowed down to a manageable number, there is still another important problem to solve: what values should be assigned to the fixed parameters, the fixed conditions, and the dynamic ranges and increments of the factors? It is one thing to announce blithely that 15 of the 19 Propositions in Table 11.4 will be assigned fixed values, and only three will be selected as factors; it is another thing entirely to *select* the values for all these fixed parameters and conditions. For example, even though you may have decided to assume (quite reasonably, perhaps) that the ambient conditions of the laboratory have no effect on recognition performance and should be fixed, you still must decide, in fact, how bright the lights in the laboratory should be. By the same token, if the conditions of the CRT are to be fixed, how bright should the CRT display be, and what color should the characters and the background on the display be? And given that the noise superimposed on the characters and the background is to be varied as a factor, what is a realistic dynamic range and increment for this noise level?

Sometimes the answers to these questions are simply a matter of common sense that can be easily justified and will probably be just as easily accepted during peer review. For project *Measure Character-Recognition Performance*, as an example, although it might have been difficult to justify lavender characters on a red background, the team members for the project readily settled on black characters on a white background, assuming (probably correctly) that such a choice was noncontroversial and the same as most actual license plates. Likewise, it was decided to make the lighting conditions as much like the actual room in which the license-plate inspectors worked every day. This decision triggered a short, but important task in the Analysis Phase of the primary task: *Ascertain Inspection Conditions* (Task 1.1.4 in Figure 9.5).

Some questions, however, could not be answered as easily. For example, what is a realistic dynamic range and increment for the noise-level factor? One way to approach a problem like this is to run what is called an **operating-point pilot**, or **OP pilot** for short. Unlike a QD pilot, whose purpose is to informally explore the feasibility of proposed solutions, the purpose of the OP pilot is to informally search a factor space for suitable values for the governing propositions, and for suitable dynamic ranges and increments for the factors of the formal experiments of the project.

If the value of a knowledge proposition affects performance in a way that does *not* depend on the values assigned to other knowledge propositions, a suitable value for it can be found using a systematic search. For example, to find the dynamic range and increment for the noise-level factor in the case-study project *Measure Character-Recognition Performance*, an OP pilot was run with a small number of subjects who were eligible for the cohort: *Conduct Noise-Range OP Pilot* (Task 1.3.1 in Figure 9.5). As briefly described for Goal G1.1 in Table 11.3, the subjects of this pilot were asked to observe a series of video frames of characters embedded in various levels of noise, and then identify the frames that best represented the maximum and minimum amount of noise typically occurring in video images of real license plates. These recommendations established an appropriate range for the noise-level factor and a reasonable increment between successive noise levels within this range, thus defining the values in the 10th knowledge Proposition in Table 11.4.

Note that the subjects used for this OP pilot were no longer eligible for the formal experiments that followed. In the vocabulary of experimental research, they were no longer "naive" and had been "tainted" because they had acquired specialized knowledge about the task that could bias their responses in the formal experiments. In fact, they were specifically asked not to discuss the OP pilot or the project with anyone until all the formal experiments had been run to completion.

Very often performance is a function of the *joint* values of several or all of the parameters and conditions; the operating point found by changing the value of one parameter while the others are fixed will not be the same as the operating point found by changing the value of a different parameter. By way of analogy, the shortest path up a mountain to its peak is not always straight and depends on your starting point. However, if (and this is a very big *if*) you may reasonably assume that there is only one peak, and that all paths, even if not straight, lead monotonically up to this peak (you should be so lucky!), then you can use the search method known as **gradient ascent**, which guarantees that you will find this single peak (eventually). Even if this assumption is not *entirely* valid, but approximately valid, gradient ascent may still get you there. However, if this assumption is just plain wrong, and there is more than one peak, then all bets are off; it will be very difficult for you to find the highest peak, and (worse yet!) you may never know whether you have or not. This kind of problem is called NP-complete in the jargon of computer science, which in everyday language means that you are in deep bandini. If this occurs, the application of genetic algorithms might be a way to find the peak performance, but even then finding the highest peak is *not* guaranteed, and genetic algorithms are *very, very* CPU intensive. If you need results in a hurry, then all you can do is search in some systematic way (like gradient ascent), keeping track of the best values of the parameters and conditions so far until time runs out, assign these values, and live with them. There is nothing wrong with this approach, because under the circumstances it may be the best you can do. Nonetheless, you are obliged to inform those who read about and review your research that this is what you have done.

> **TIP 19**
> Use OP (operating-point) pilots to search for suitable values for fixed parameters and conditions, and dynamic ranges and increments for factors.

11.4 Postulate Performance Metrics

The subject of performance metrics was introduced in Chapter 6 (Categories and Types of Knowledge) and further explained in Section 7.3 (Range Knowledge).

> **Definition 11.4** A **performance metric** is a postulate that transforms the results of the task into measures of performance for drawing conclusions about the task objective.

Unfortunately, there is little general advice that can be given regarding the selection of appropriate performance metrics, because they are usually very specific to the task at hand. Nevertheless, a few suggestions can be offered.

It important to note that the *final* selection of performance metrics, unlike other knowledge propositions, can be postponed until after the formal experiments have been run. Scrutinizing the experiment results often evokes new and more appropriate ideas for measuring performance. (QD pilots are very useful for exploring these ideas.) For example, if the distribution of a result is multimodal, there is little point in computing the statistical moments, whose values will be more deceptive than informative; different summary metrics are needed. In fact, the distributions themselves, binned into histograms or sorted into Pareto plots, are often the only representations that make sense, because further reductions would obscure important details. Moreover, such distributions can then be used to estimate statistical confidences, which are often the most useful performance metrics for drawing conclusions.

It is also true that selecting the right performance metric can change a failed task into a successful one. This statement is offered with great trepidation; needless to say, the metric should only be changed if the new metric is *more* representative of performance than the one that indicated failure. It is always a great temptation to tinker with metrics until you get the "right" results, but this must never be done at the expense of **objectivity**, one of the three fundamental criteria for the provisional acceptance of knowledge, as stated in Chapter 5 (An Epistemological Journey). This is another reason why peer review is an essential step in the process of validation; your reviewers, hopefully, will not let you get away with inappropriate performance metrics. It is probably best, however, if your own internal value system as a scientist or an engineer proscribes "cooking the books" before you embarrass yourself in front of your peers.

Note that the single performance metric for Hypotheses H1.3.1 through H1.3.6 for the case study project in Table 11.3 is postulated *after* the last hypothesis. As indicated, this simply means that the performance metric applies to all six hypotheses. Mentioning this performance metric just once not only simplifies the project documentation, but makes it more compact.

The reader may have come to the conclusion that a development task *need not* have any factors or performance metrics, because, strictly speaking, its objective may simply be to create a device or effect in a prescribed manner — full stop. This, however, is incorrect. It is impossible to conceive of a development task that would be conducted blindly without measuring anything. In reality, every development task must have a research subtask with at least one goal to decide when the parent task is finished, and another goal to measure how well the parent task objective was achieved.

A good example of this is an industrial manufacturing task, in which sensors are used to monitor the progress and quality of the product as it travels along an assembly line. To accomplish the goals of this manufacturing task, there must be performance metrics that specify how the sensor data will be used to make decisions and evaluate the quality of the products.

As another example, consider the task *Peel a Potato*, which is a development task because it creates an effect, namely, a potato without its peel. One solution is to physically peel the potato with a knife. However, if no goal has been specified for this task, then you are *not allowed* to check the peeled potato from time to time to find the small areas of peel you have missed and to terminate the process when you have removed all the peel, *i.e.,* when you have achieved the objective.

This is one reason why the fields of artificial intelligence and robotics have had so few practical successes over the last 40 years. To make a simple robotic mechanism that can physically wield a knife and accurately follow a set of strict and rigid instructions to carve up a potato is not too difficult. But to create an intelligent machine that uses a video camera to continually guide the process and make sure the peel is completely removed with a minimum loss of potato, requires a level of cognitive sophistication that is beyond the current state-of-the-art in image processing and artificial intelligence. (Don't forget about removing the eyes of the potato.)

This concludes the Hypothesis Phase of the Scientific Method. The research or development project now has a detailed design for the solution and a complete set of goals for each major task. Each goal is accompanied by a set of factors, performance metrics, and, if decisions must be made, a set of research hypotheses. Now we are ready to proceed to the Synthesis Phase to implement the solutions and to design and conduct the experiments to accomplish the goals. The results of these experiments will become the basis for computing the performance measures and drawing the conclusions that determine whether the task objective has been achieved.

Exercise 11.3

List all the major knowledge propositions for a project you are currently planning or undertaking, including the performance metrics. Assign a role code to each. Select those that you would like to be factors. Compute the number of treatments you would have to run to accommodate this factor space. Try not to faint. Make the hard choices, and trim down your set of factors.

   Suggest where an OP pilot might be useful in this project to determine the values to be assigned to one or more of your knowledge propositions.

CHAPTER 12

Synthesis

During the first two years, the child steadily gains enough mobility to test many of its more far-reaching hypotheses and goals: "That dish on the table might have something good to eat in it," or "I would like to hear how that pretty red cup sounds when I bang it on the floor." First crawling and then walking, the child begins to explore the world. The desire to manipulate things is augmented by the need to get somewhere. Walking replaces crawling as the stability problem is gradually solved. Soon the child realizes that being rescued from a fall by doting grownups, while comforting, often abruptly terminates a very important exploratory task, which the grownups, strangely enough, regard as trivial. So the child learns that simply picking oneself up is a better solution, that pieces of furniture provide handy points of purchase, that a railing helps to maintain balance and avoid really bad falls on the stairs, and that running is much faster and much more fun than walking.

 Then, at about age two, it dawns on the child that it just might be able to manipulate *people,* much as it has learned to manipulate its environment. Having long ago discovered it can usually get what it wants by smiling and being nice, the child hypothesizes that it can *avoid* what it does *not* want by being nasty and uncooperative. The word "no" becomes a key inducer in a series of feasibility pilots, supplemented as needed by abundant kicking and screaming. In fact, the child becomes so intent on thoroughly testing its primary hypothesis, that it often **synthesizes** an experiment even when it is tired or bored, just to see if the hypothesis

is valid under the current conditions, however inappropriate. No wonder these are called the Terrible Two's.

The word "synthesis" derives from the Greek word *suntithenai* (συντιθεναι) meaning "to combine separate elements to form a whole," which consists of two parts, *sun* (συν) meaning "together" and *tithenai* (τιθεναι) meaning "to place." These etymological roots imply that the fundamental process of synthesis is to assemble and test something. And that is the objective of the Synthesis Phase:

> **Definition 12.1** The objective of the **Synthesis Phase** of the Scientific Method is to implement the task method (solution and experiments) to accomplish the goals and validate the hypotheses of the task.

As shown in Figure 9.1, the process of synthesis can be conveniently broken down into four sequential steps (subtasks): Implement Solution; Design Experiments; Conduct Experiments; and Reduce Results. Unlike the Analysis Phase, these steps must be undertaken in the stated order. Putting something together is generally an order-dependent process. Having analyzed the problem, specified a solution, and stated the precise goals and hypotheses, we now set out to assemble all the pieces and try it out — the proof, so to speak, being in the pudding.

12.1 Implement Solution

Before the goals and hypotheses for the specified solution can be tested, it must be implemented. Very little specific advice can be offered for implementing solutions; usually the scientists and engineers on the project team already have the necessary skills to achieve their specialized task. However, a few strategic issues are worth mentioning.

There are two different methods for implementing solutions: it can be manufactured, or it can be acquired (or, of course, some combination of the two). Most managers will readily agree that acquiring a solution, if one is available, is generally preferable to manufacturing it, which is usually very expensive. Sometimes, however, acquiring an existing solution requires making troublesome compromises. For example, most existing software packages are usually missing a few of the features that are important for the task, and little can be done to modify them. Likewise, the use of prefabricated buildings for a housing development project may not offer enough variety or options to attract potential homeowners. And everyone knows that TV dinners are just not as tasty as mom's home cooking. Nonetheless, spreadsheet, word-processing, communications, and graphics software packages, which are very common tools for implementing solutions, are enormously versatile. Similarly, no task team would ever dream of manufacturing their own computer systems instead of purchasing one of the many desktop or mainframe machines that are available. Clearly, the task team should strongly consider acquiring an existing off-the-shelf solution, even if only for a piece of the entire solution method.

One example of an acquired solution is a consultant. (A person's knowledge can certainly be the implementation of the solution, or a part of it.) Hiring a consultant is usually much less expensive and much faster than training a member of the task team to do the same job. Moreover, project funds for consultants are generally not subject to overhead, fringe, and indirect costs, making the seemingly exorbitant fees that consultants charge actually quite reasonable.

> **TIP 20**
> If it meets the needs of the task, using an existing solution is almost always more cost-effective than manufacturing it yourself.

Sometimes a component of the solution is a database that must be acquired, which raises a thorny issue: data ownership. For example, a medical image-processing task may require an existing database of MRI images captured by another task in the same or a different project. Such a database is often treated like a love child by its creators, who behave like fiercely loyal and protective parents, loathe to let anyone touch their offspring, let alone traffic with them or question their character or quality. They spent a lot of time and effort building the database, and heaven help anyone who wants to carry it away and do who-knows-what with it!

Some years ago, a research institute undertook a signal-processing research project for an industrial firm that manufactured high-capacity hard disks. As the last step in the manufacturing process, the drives were evaluated for excessive noise by a technician who used a contact microphone to listen to each and every hard disk that came off the assembly line. If the drive motor made too much noise or made strange sounds, the inspector threw it in the discard bin. The problem, however, was that the inspector got bored and tired after just a few hours of doing this, and his error rate went through the roof, indiscriminately throwing out the baby with the bath water. To make matters worse, not surprisingly, the turnover for this job was very high, and it took a long time to train new inspectors. The research institute was given a contract to automate this process by applying state-of-the-art signal-processing technology.

A large database of sound data from the hard drives was sent to the research institute as a basis for the research. This database had been carefully segregated by an expert (one of the inspectors) into two data sets: normal drive sounds and abnormal drive sounds. The project team quickly developed a statistical-learning solution that was trained with the normal sample, and then was used to test the abnormal sample by statistical comparison. Early on, the project team began to suspect that some *abnormal* sound samples had been mistakenly included with *normal* (training) samples, thus tainting the training statistics. To check this hypothesis, they decided to take the questionable samples back to the factory and ask the inspector to reevaluate them.

When they arrived at the inspection facility, they sat down with the inspector, explained their concerns, and asked him to check the samples again. He sat quietly for a moment and then said: "Are you sure these samples were inspected by me?" The project team said yes, they were sure. "Well, in that case," said the inspector, " I don't need to inspect them again. If I evaluated them as normal, then they're normal. I have had a lot of experience in this job, and I know what I'm doing." Full stop. So, the project team took their leave, went back to the institute, and did the best job they could with what they had. (It turned out the project was successful, anyway.)

Sometimes problems with databases arise because the supplier simply wants to be very efficient. For example, suppose a supplier is ready to send you a database of high-resolution images via the Internet. Thinking it makes no difference and without informing you, the supplier decides to save transmission time by compressing the images using lossy methods. When you apply your image-processing algorithms to the images you received, the results seem a little strange and unexpected in very subtle ways. Even though the effects are almost invisible to the eye, you begin to suspect that your sensitive algorithms are reacting to the artifacts and distortion of some compression method. The supplier apologizes profusely for your weeks of wasted effort.

Other times, in a sincere effort to be helpful, the supplier preprocesses the data *just a bit,* such as applying smoothing functions, interpolating missing values, deleting "misleading" data items, *etc.,* all unbeknownst to you. Tampering with data is a serious issue, regardless of the motive.

> **TIP 21**
> Arrange to be given complete, raw, and unpreprocessed data. Expect to deal with political and ownership issues, as well as unsolicited "help."

There are a great many other equally important data management issues, but most of them are concerned with the *creation* and *administration* of data, rather than the *acquisition* of data. These issues are considered in Section 12.2.3, which discusses the management of experiment data.

Although the acquisition of solutions raises a few difficult issues, as every experienced R&D manager knows full well, *manufacturing* a solution is no picnic either. Each field has its own set of special manufacturing problems, and a review of such problems is outside the scope of this author's expertise. However, because the design and implementation of *software* solutions are well within the purview of the author's 30 years of professional experience, a few especially important, and often egregiously ignored, software development guidelines are offered here.

Half of the lines in a computer program should be internal comments and documentation. Documentation is necessary to allow current and future programmers to maintain and modify the programs reliably and quickly. Future programmers are unlikely to include the original programmers, but even if some are still around, they will have all but forgotten what they did and why. Reverse engineering undocumented code is very difficult, error prone, and expensive.

This guideline is likely to evoke groans from programmers. They will maintain that they already know what everything in their modules does (true), nobody else needs to know (false), and it takes too much time and trouble to document so much (irrelevant). Managers are likely to insist on complete and thorough documentation, but unfortunately are often impatient and unable to grant enough time to generate adequate documentation when faced with (often unreasonable) delivery deadlines. This puts the programmer between a rock and a hard place. Managers and engineers are invited to take another look at Tips 15 and 16 at the end of Chapter 9 (Overview).

What about those programs that are written and then run only once to obtain some needed results? If that is ever *explicitly* the case, then the programmer must promise that when the program has finished running the *single* time to get the results needed for the task, *all copies will be immediately destroyed.* But, the programmer would argue hastily, the program just *might* turn out to be useful again sometime in the future, so it is silly to throw it away. Well, now, which is it? A one-shot program or a program worth keeping for possible future use? Ironically, it is precisely in this situation that thorough documentation is *especially important!* The longer the time before the need to resurrect a program arises, the sorrier the programmer will be for not taking the time to document it adequately when it was first written. *Be careful not to confuse expediency with efficiency.*

Because documentation is *essential* for the survival and maintenance of all software, a number of critical documentation guidelines are presented in Table 12.1.

Software development from concept to shrink wrap is accomplished at the rate of one line of code per person–hour. Managers need estimates of the amount of effort that will be required to complete a software development task. Unfortunately, software designers and programmers are often very poor at estimating how much *time* it will take. Therefore, after introducing the team to the project and the required solution, ask the *programmers* to estimate how many *lines of code* there will be in the final product, including internal documentation. This estimate of the number of lines, it turns out, is also a fairly good estimate of the *level-of-effort* for the *entire* project in person-hours. For example, an estimate of about 2000 lines (a rather small application) means that the entire software development project will require about 2000 person-hours or 1 person-year. If this sounds like a lot of time and effort for such a small application, wake up and smell the coffee; software development is very labor intensive and very expensive.

This method of estimation assumes that the objective of the project is a shrink-wrapped software product, ready for delivery to the end-users. Note that most of time for such a project will *not* be spent on programming, but rather on all the other essential tasks: design, testing, marketing, and production. If the software is being developed solely for internal use, then the marketing and production phases are omitted, and the testing requirements may be less stringent. In this case, the typical software development rate increases to about three lines per person-hour.

Use computer scientists or engineers for software design; use programmers for software implementation and coding. Much too often these two jobs are thought to be interchangeable, and both types of people are asked to do both jobs. This is entirely inappropriate, regardless of what the programmers or

Table 12.1 Critical software documentation guidelines

Internal Documentation

---

1   Limit each code module to one or two standard pages for convenient inclusion in reports on facing pages.
2   Introduce each code module with a block of comments that specify:
  • Slate (title, module hierarchy position, programmer, contact info, date, revision number, remarks)
  • Purpose of the module
  • Values of relevant governing propositions
  • Purpose of each major data structure (variable, container, constant, pointer, *etc.*)
  • Intermodule communication parameters
3   Insert a blank line and a row of asterisks before and after each code section to highlight it.
4   Begin each code section with a highlighted comment stating its task function.
5   Clearly indent embedded and nested structures to emphasize the hierarchy of operations.
6   Enter control words in uppercase and boldface (**IF**, **THEN**, **ELSE**, **REPEAT**, **END**, *etc.*).
7   Invent names that are long enough to clearly communicate their task function.
8   Enter shorthand internal documentation statements DURING programming. Periodically clean them up.

External Documentation

---

1   Generate and maintain module hierarchy charts in cooperation with all project designers and programmers.
2   Generate and maintain flow charts or pseudo-code for every code module.
3   Keep accurate and up-to-date maintenance and revision logs.
4   Hire experienced documentation specialists to design user manuals and help files.
5   Maintain internal network sites and Internet web pages (if appropriate) for software documentation and help.

---

designers tell you (often quite indignantly). These two different types of professionals receive (or should receive!) very different kinds of education and training. The requisite skills of a programmer and a computer scientist/engineer are as different from each other as those of an automobile mechanic and an automotive engineer. In general, one would not expect an automotive design engineer to be very good at repairing cars. This does not mean that a programmer cannot *become* a software designer and *vice versa,* but a commensurate investment in additional education and training is required.

   Design software top–down, implement bottom–up, and code inside–out. Most computer scientists and engineers are familiar with the concepts of top-down planning and bottom-up implementation, which are similar to the processes of designing and implementing a hierarchical project plan described in this book. To many of them, however, inside-out coding may be a new concept.

   After the designer (not the programmer) has completed the top-down design, the programmer (not the designer) selects a module at the bottom of the hierarchy as a starting place to begin the coding process. This raises the question: what should be the first program statement that is written down? The conventional wisdom to "start at the beginning and go through until you reach the end" is much too facile for programming tasks. The statement that the programmer thinks is the first one of the module will not retain that distinction for long. This statement will soon be preceded by other statements that must come before it;

indeed, it is very likely that this statement will not even exist *per se* in the final version of the program. This leads to a piece of useful advice: don't even try to think of the *first* statement of the source code for a module; instead, simply think of some statement, *any* statement that must be included in the module. Start there, and build the code around it. This is inside-out coding.[3]

When you have an elusive bug, ask for help. Here is the scenario. A programmer has been struggling with a really tough programming bug for days and is about to commit suicide. As a last resort, he decides to ask one of his colleagues to take a look at his code and see if he can find out what is wrong. He dreads this moment. Not only is it embarrassing, but he also knows it will take hours just to explain what the program is doing and what the problem is. The colleague looks over his shoulder as the programmer begins the long task of explaining the problem. After about 30 seconds, the colleague interrupts and jabs his finger down on a statement in the code listing, saying quietly: "Why is this variable declared as a short integer?" The programmer is stunned for a moment and then slaps his palm to his forehead: "How could I have missed that!?" It's known as the palm-to-the-forehead syndrome.

Don't wait until you are tearing your hair out. Ask for help. A fresh perspective will probably be immune to your logical pathology and help to side step the repetitive and systematic failure in your search for the bug. Try it. You'll be surprised how quickly and consistently well it works.

Sometimes it helps to formalize this method and set up a buddy system in a professional programming shop. Pairs of programmers are assigned as helpers-on-call for each other, each having the acknowledged right to interrupt the other at any time for help with tough programming problems. Note that it is important to work out a sensible and fair way to assign and rotate buddies. No one gets along with everyone forever.

Table 12.2 presents an abbreviated description of the Solution Implementation for the primary task of the case-study project *Measure Character-Recognition Performance*. Note that this step is not explicitly broken out in the task tree in Figure

---

[3]I think it was my friend and colleague, Michael Feldman, Professor of Computer Science at The George Washington University, a talented expert in applications and systems programming, who first coined this term.

9.5, because the project team decided to limit its inclusion to a brief summary in the primary task description.

Remember that the solution implementation for the *primary* task is just one of many for the entire project. *Every* invocation of the Synthesis Phase at *every* level in the project task hierarchy has a solution to be designed and implemented. Although Table 12.2 describes only the solution implementation for the primary task, the programmer who actually built the software decomposed its implementation into a detailed task tree. In the final report for the project, all of the solution implementations that were not self-evident or trivial were fully documented.

Deciding which solution implementation methods deserve more than just a brief description is a matter of judgment, guided by the three criteria for the provisional acceptance of the knowledge as postulated in Chapter 5 (An Epistemological Journey): reproducibility; completeness; and objectivity. If the method can be clearly and unambiguously understood in an abbreviated form, and if this abbreviated form still provides sufficient information for another engineer or scientist to reproduce the task and get the same results, then this brief description is adequate. However, when in doubt, it is better to err on the side of over-explanation.

Exercise 12.1
Describe the solution implementation methods for the significant tasks in a project you are currently planning or undertaking.

Table 12.2 Solution implementation for Project *Measure Character-Recognition Performance*

Primary Task 1: Solution Implementation
The software application for acquiring the character-recognition data during the experiment trials was constructed in Microsoft Excel. The appearance of the graphical user interface is shown in Figure CS.4 (see Table 12.8) in the description of the laboratory design. . . .
A field was provided for the experiment supervisor to enter the subject ID; this field disappeared after the ID was successfully entered. No keyboard was provided; all input from the subject was accomplished with mouse clicks. Care was taken to disable and/or ignore all illegal mouse-clicks by a subject during data acquisition. Only clicks on the START button (which disappears after being clicked) and the character buttons {A . . . Z} and {0 . . . 9} were accepted and processed. . . .
Results for each subject were recorded in a tab-delimited text file that could be read directly by Excel for convenient reduction of the results. The recorded data items included timeStamp, technicianID, trialType, subjectID, typeface, noiseLevel, and recognition data {correctCharacter, responseCharacter, responseTime}. . . .

12.2 Design Experiments

The purpose of this step in the Synthesis Phase is to design a series of experiments whose results will eventually be used to estimate how well the task objective has been achieved.

> **Definition 12.2** An **experiment** acquires data to measure the performance of the solution under controlled conditions in a laboratory.

Before undertaking the task of designing the experiments, it is a good idea to conduct a thorough inventory of the task components that have been specified, generated, manufactured, and acquired for the task so far, reexamining and reconfirming the specification of each one. These task components will profoundly influence the design and conduct of the experiments. To facilitate this inventory, Table 12.3 lists the major task components. As such, it can also function well as a checklist for inventories throughout the planning process. Note that the last item, experiments, which is included in the checklist for the sake of completeness, has not been specified yet. That's the purpose of this section.

The inventory using the checklist in Table 12.3 should be conducted for every task in each level of the project task hierarchy, skipping only those items that have already been specified at higher levels. Table 12.4 instantiates this checklist for the case study project *Measure Character-Recognition Performance*. Although this seems like a lot of work, it is simply the cost of doing business effectively — of getting it right. As described in Section 9.2.1, project planning consumes a large fraction of the total project time, but pays off handsomely in the long run. Also be reassured that planning proceeds a good deal more quickly and smoothly with practice — like all things. Wax on, wax off.

> **TIP 22**
> Before beginning the design of the experiments, carefully inventory all the components of the task.

> **Exercise 12.2**
> Make a comprehensive inventory of all the components for the significant tasks in a project you are currently planning or undertaking.

Table 12.3 Inventory and planning checklist for the significant components of a task

Component	Reference
DOMAIN	
❏ task objective Section 4.1.1	
❏ task unit	Section 4.1.2
❏ task resources	Section 4.1.3

❏ inducers
       required or unwanted
       artificial or natural
       external or internal
       concurrent or past
	Section 4.1.3.1

❏ sensors
       quantitative
       qualitative
	Section 4.1.3.2

❏ system and environment	Definition 4.7
❏ supervisor	Section 4.1.3.3
❏ channels (input, output)	Section 4.1.3.4
❏ domain knowledge	Chapters 6, 7, 11

knowledge propositions (presumptive, stipulative, conclusive)
         conditions (regimes, situations)
         parameters (schemes, constants)
	Sections 6.2–6.4

governing propositions
         neutral propositions
         biased propositions (limitations, restrictions, exclusions)
	Section 7.1

goals and hypotheses	Section 11.2
factors (policies, variables)	Section 7.2
performance metrics	Sections 6.2, 13.1

benefits
         cost
         payoff
	Section 7.3

RANGE	
❏ devices and effects (development tasks)	Section 4.3
❏ range knowledge (research tasks)	Section 7.3
❏ results	Sections 6.4, 12.4
❏ conclusions	Section 6.4
METHOD	
❏ solution	Sections 4.2, 11.1
❏ experiments	Section 12.2
❏ laboratory	Section 12.2.1
❏ block design	Section 12.2.2
❏ data management	Section 12.2.3

Table 12.4 Task components for Project *Measure Character-Recognition Performance*

Summary of Task Components for the Primary Task

Task Objective: To measure the peak performance of trained license-plate inspectors in recognizing computer-generated 9-point *serif* and *sans serif* characters embedded in a 21×21 pixel frame with superimposed noise that realistically models the actual range of obscuration of license-plate numbers due to dirt, debris, and normal wear-and-tear.

Task Unit: Subject in a small cohort of trained license-plate inspectors

Inducers: CRT display capable of supporting at least 256 gray levels

    Test image (See Figure CS.1 in Table 10.4a)

        Noise [RAEC]

        Characters [RAEC]

    Other visible objects [UNEC]

    Background [UAEC]

  Subject chair [UAEC]

  Cognitive distractions [UNIC]

  Sound, light, and air-conditioning sources [UAEC]

  Smells [UNEC]

Sensors: Mouse used by the subject to select the recognized characters from an array of 36 buttons for the 26 uppercase Latin letters {A through Z} and the 10 decimal digits {0 through 9}

System and Environment: See Figure CS.3

Supervisor: Computer (PC) and technician

Channels: Air for sound, smell, and light (input)

  Air for heat (input and output)

  Subject PNS* (input and output)

  Mouse cable (input and output)

  CRT display cable (input and output)

  Power cables (input)

  Chair (input)

Domain Knowledge: See Table 11.4

Range Knowledge:

  Results: timeStamp

    technicianID

    trialType

    subjectID

    typeface

    noiseLevel

    recognition data

      correctCharacter

      responseCharacter

      responseTime

Figure CS.3 System and environment

  Conclusions:

    Recognition performance as a function of noise level and typeface

    Confidence performance with *serif* typeface > performance with *sans serif*

Solution: See Table 11.1

Experiments: See Tables 12.8 through 12.12 and 12.14

*PNS = peripheral nervous system

Before embarking on the subject of the design of experiments, there is an important pedagogical issue to be addressed. Theoreticians may argue that some research tasks simply do not involve experiments, such as those that have purely theoretical or computational goals. This is not the case. Even though the term "experiment" may not seem to be appropriate for such tasks, *every* task includes an experiment, although it may be rather brief and not typically referred to as an experiment. Research tasks whose solutions are mathematical derivations or computations are no exception, although the term *verification* may be an appropriate alternative for the term *experiment* for purely theoretical research tasks. There are two tasks in the list of running examples that illustrate this situation: **Prove Fermat's Last Conjecture** and **Identify Faster Traffic**.

In the theoretical task **Prove Fermat's Last Conjecture**, Andrew Wiles selected the appropriate mathematical tools and logical methods of proof in the step Specify Solution. He then applied these tools and methods in the step Implement Solution, where he derived his impressive 200-page proof of this age-old conjecture. The next two steps in the Synthesis Phase, Design Experiments and Conduct Experiments, simply verified that the proof was sound. The actual process that Wiles undertook to verify his proof was, of course, too highly specialized to be included here as an example. A simpler example, however, will illustrate the nature and necessity of this process.

Suppose the objective of a research task is to find the two-dimensional plane $z = ax + by + c$ (task unit) that satisfies the following conditions: When $y = 2$ and $x = -1$, then $z = 3$; when $y = -1$ and $x = 4$, then $z = -2$; and when $y = 0$ and $x = 2$, then $z = 1$. One solution is to solve three linear equations simultaneously using the Gaussian elimination method, and the goal is to find the coefficients a, b, and c for the given conditions by applying this solution. (Incidentally, if your algebra is a bit rusty, you might try a different solution, such as a graphical representation, or even perhaps ask your friendly neighborhood mathematician for the answer.) The solution is implemented in Table 12.5.

The result for the last step in Table 12.5 gives the required equation for the desired two-dimensional plane. The Gaussian-elimination solution method has met the goal of the task.

Or has it? Most certainly we have implemented a classical solution method and followed a clear procedure to apply it and to accomplish the goal. But did we get the correct answer? How do we know? Should we just trust the final result and go right ahead and build the machine that uses this result? Or could we have made an error? Is it possible that we made a mistake in the algebra? (Such errors are certainly not unprecedented!) Perhaps we should seek some *evidence* that the final result in step #10 in Table 12.5 is correct (before we launch the rocket).

That is exactly the purpose of designing and conducting an *experiment*: to gather formal evidence that the goal has been met or that the hypothesis may be decided. Only when we are sufficiently certain of this can we proceed to use the results to draw appropriate and valid conclusions about the task objective.

The experiment to check the final result of the algebra problem in Table 12.5 is very simple and brief: substitute any set of the constraints for x, y, and z

Table 12.5 Simultaneous solution of a set of three linear equations

PROBLEM: Find the plane $z = ax + by + c$ for the constraints:		When $y = 2$ and $x = -1$, then $z = 3$
		When $y = -1$ and $x = 4$, then $z = -2$
SOLUTION: Gaussian elimination		When $y = 0$ and $x = 2$, then $z = 1$

Step	Operation	Result
#1	substitute condition values into plane equation	$3 = a(-1) + b(2) + c = -a + 2b + c$
#2	substitute condition values into plane equation	$-2 = a(4) + b(-1) + c = 4a - b + c$
#3	substitute condition values into plane equation	$1 = a(2) + b(0) + c = 2a + c$
#4	multiply result #2 by 2	$-4 = 8a - 2b + 2c$
#5	add result #4 and result #1 (eliminate b)	$-1 = 7a + 3c$
#6	multiply result #3 by 3 (eliminate b)	$3 = 6a + 3c$
#7	subtract result #6 from result #5 (eliminate c)	$a = -4$
#8	substitute result #7 into result #3	$c = 9$
#9	substitute results #7 and #8 into result #2	$b = 5$
#10	substitute results #7, #8, and #9 into the prototype equation	$z = -4x + 5y + 9$

given at the top of Table 12.5 into the final equation in step #10 and see if the goal has been met. For example:

$$z = (-4)(-1) + (5)(2) + 9 = 23 \overset{?}{=} 3$$

Oops! That's not correct. For this set of values for x and y, z must have the value 3, not 23. The experiment has revealed that something is wrong with the implementation of the solution. And sure enough, a review of the algebraic operations in Table 12.5 reveals a sign error in step #9. After correcting that error, the value of the coefficient b is determined to be -5, not $+5$. Of course, having found this error and modified the derivation, we are obliged to run the experiment again to be sure that we have indeed corrected the error, *and* that this is the only error:

$$z = (-4)(-1) + (-5)(2) + 9 = 3 \overset{?}{=} 3$$

That's correct. To complete the verification, it is important to check the results for the other two sets of constraints for x, y, and z in Table 12.5. This is left as an exercise for the reader.

The traffic engineer for the task *Identify Faster Traffic* has an even briefer and simpler experiment to design and conduct. We can reasonably suppose that her task goal is to use Excel to compute the averages and standard deviations of the two traffic speed samples. There is very little reason to doubt the accuracy and reliability of these two Excel functions, which have been tested by millions of users over many years. However, it would certainly behoove her to make certain that she has written the Excel functions correctly (that they are referencing the correct data ranges, for instance). She could do this several ways, one of which would be to add up the speeds and the squares of the speeds for each sample and use the well-known formulas in statistics books to compute the first and second moments from scratch. If the answers are the same as from the Excel functions, then this brief experiment has provided some credible evidence that the solution was implemented correctly. On the other hand, depending on how much is at stake, an adequate experiment might simply be to carefully recheck the formulas on the Excel worksheet.

The primary purpose of these examples is to demonstrate that even research tasks relying solely on computation and derivation must include an experiment, however brief, to gather evidence that the solution has been implemented correctly. If the word "experiment" is bothersome for such tasks, it is completely acceptable and appropriate to refer to these two steps in the Synthesis Phase as "Design Verification" and "Conduct Verification." A rose by any other name. . . .

Many tasks, however, require complex and rigorous experiments to acquire the results necessary to decide whether or not the solution is effective. Just imagine the enormous complexity of the experiments required for some of our running example tasks, especially *Launch Spacecraft into Orbit*, *Measure Drug Effectiveness*, and *Evaluate GUI Effectiveness*, as well as the case-study project *Measure Character-Recognition Performance*. The experiments for such ambitious and challenging tasks require meticulous planning, including: 1) a comprehensive specification for the laboratory; 2) detailed designs of the experiment blocks and their protocols; and 3) clear statements of the regimes for acquiring and managing the data. These three planning tasks interact heavily with each other, and even though they are discussed here in a particular order, they must be planned in parallel with a great deal of stepwise refinement.

12.2.1 EXPERIMENT LABORATORY

According to Definition 4.10, the laboratory for a task is the set of physical regions containing the task unit and all the task resources required during the experiments. This definition certainly does include the conventional image of the laboratory: a large facility filled with electronic, mechanical, chemical, medical, or computer equipment. However, this stereotype turns out to be the exception rather than the rule. In fact, only three of our running example tasks would reasonably be able to conduct their experiments in such a facility. The experiment for the task *Evaluate GUI Effectiveness* probably takes place in a single room outfitted with several computer work stations in carrels; the experiment for the task *Measure Age of Organic Sample* must be conducted in a specialized facility equipped for carbon-14 dating; and the experiments for the task *Measure Drug Effectiveness* are conducted in the standard medical testing facilities of the pharmaceutical company, which have been used many times for testing new drugs. The experiments for the rest of our running example tasks, however, must be conducted in the field, often in several different locations, each playing an essential role in the experiment.

The laboratories for another three of our example running tasks, *Identify Faster Traffic*, *Prove Fermat's Last Conjecture*, and *Compile Summaries*, require only the investigator, a personal computer, a desk, and perhaps pencil and paper. The major effort takes place in the mind of the investigator. Unfortunately, human minds are easily distracted, and simply ignoring or resisting such distractions is almost impossible. Although, many scientists and engineers have trained themselves to focus their intellectual skills tightly on the task at hand, emotions are largely beyond intellectual control in a sane person. Formal experiments have demonstrated that even though a person may feel emotionally in control, deep inside there can be a seething cauldron, and people who insist they can divorce themselves from their own emotions by force of will are either psychopaths or are in denial. The unconscious rules the mind, and there is precious little one can do about it.

Therefore, if an experiment involving humans requires that their emotional reactions be under control, the only reasonable approach is to remove all emotional triggers in the laboratory. Having said that, it may be very difficult to identify these triggers without running sophisticated pilots designed by psychologists who specialize in measuring emotional reactions, which is very expensive and time consuming. Therefore, it makes sense simply to remove *all* unneeded items from the laboratory, including furniture, decorative items, equipment, *etc.*

> **TIP 23**
> Make sure every item in the laboratory has a specific purpose in the experiments. If an item is not on the inventory of required task components, remove it. If it cannot be removed, deal with it proactively. (See Table 12.3.)

The experiments for the remainder of our running example tasks require special or unusual laboratories. At first glance, the laboratory for the task *Find New Planet* appears to be located in an existing specialized facility, the Lowell Observatory in Flagstaff, Arizona. However, the task unit (the predicted region in space), which is in the laboratory by definition, is certainly not inside the observatory. It is connected to the telescope by a narrow conical channel six billion kilometers long that collects the photons of light originating from the predicted position of the new planet, as well as far beyond it. In addition, the gravitational sources that were used to compute the predicted orbit of the new planet are also in the laboratory as the required inducers. Therefore, the laboratory actually comprises the observatory, the field-of-view of its telescope out to its maximum range (many light-years), and at least the Sun and the gas giants. Clyde Tombaugh's laboratory was truly immense.

Fortunately, Clyde Tombaugh's laboratory had been largely prepared for him in advance, both by the project that had built the Lowell Observatory and by Mother Nature. In contrast, a large part of the laboratory for the task *Launch Spacecraft into Orbit* had to be built from scratch at a cost of hundreds of millions of dollars by a huge team of NASA engineers and contractors (although Mother Nature played a big role as an inducer in the laboratory also). The three major Earth-bound facilities were the launch pad at Cape Kennedy, the Manned Space Flight Center in Houston, Texas, and the set of orbital tracking stations around the world.

The laboratories for the tasks *Identify Collision Damage*, *Measure Liquid Temperature Variation*, and *Measure Fossil Age* must all be set up in the field. This usually adds a major degree of logistical difficulty to the project, as well as making it much harder to inventory and measure the effects of unwanted inducers and to keep the experiment in control.

For example, the insurance adjuster for the task *Identify Collision Damage* must inspect damaged cars in a wide variety of locations, trying to sort out the facts from the biases and distortions posed by distraught, frustrated, and sometimes dishonest claimants. Undoubtedly, he has had to invent some very clever solution methods and protocols over the years to separate signal from noise on these channels.

The engineer performing the task *Measure Liquid Temperature Variation* must deal with a double whammy: bias and noise. Because the client believes strongly that the liquid temperature is varying excessively and may resist any other conclusion, he has asked his own representative to be present during the task to "observe and advise." For political reasons, the engineer cannot simply ask the representative to leave the laboratory during the experiments. To make matters worse, the task requires the use of some special sensors, whose heat conduction must be kept to a minimum to keep the experiment in control. This creates a strong potential for experimental error leading to false conclusions that would almost certainly *confirm* the client's claim. Thus, it is in the client's interest for the results to be in error — the classic self-fulfilling prophecy.

The laboratory for the task *Measure Fossil Age* is located on the face of a cliff. The geologist must acquire, install, and use some very complicated gear to climb down the cliff face and search for each geological layer to measure its thickness and constitution. Despite the warnings this book has repeatedly issued about bias that is introduced whenever the investigator assumes the role of one or more of the task resources, for this task the warning must be ignored on two counts. First, only the geologist can examine the cliff face with sufficient expertise to identify the boundaries and constitution of the strata; she *must* be the sensor for the task. Second, because she will actually climb down the cliff face herself, she is also the experiment supervisor, implementing her own protocol. If she were any more deeply involved with the experiment itself, she could probably just skip the experiments and report her hypotheses as conclusions. So, to prevent both bias and the *perception* of bias, she would be well advised to take many photographs during the procedure to help establish the validity of her methods and conclusions during peer review. Of course, taking photographs, measuring distances, and estimating geological conditions while hanging at the end of a rope certainly does add a good bit of complexity to the laboratory and the experiment protocol. Certainly, a series of feasibility pilots is in order.

This raises an important issue about experiment laboratories — safety. Although safety concerns vary widely and are often specific to the task (*e.g.,* chemical, biological, and radiation hazards, and the corresponding safety measures), a few general safety tips are offered in Table 12.6.

It is important to remember that, although human subjects and task personnel may willingly sign release forms even when participating in dangerous experi-

Table 12.6 Safety tips for the design and use of experiment laboratories

! Do not allow anyone with any of the following items near machines with exposed moving parts: long sleeves or hair; neckties or scarves*; watches and jewelry on the hands, neck, or limbs.
! Do not allow anyone to wear metal hand jewelry when working on high-voltage equipment.
! Limit contact with electrical equipment to one hand; touch only your own body with the other hand.
! Require hard hats when work is performed near apparatus that is suspended overhead.
! Require safety glasses in labs where chemical or biological agents could be released into the air.
! Allow no exceptions for safety rules, regardless of the rank or status of the person in the lab.
! With dangerous equipment use panic switches controlled by technicians who have no other lab duty.
! During dangerous experiments, limit the number of people in the lab to an absolute minimum.
! Make video tapes (with a sound track) of all dangerous or ethically controversial experiments.
! Consult your legal staff about all experiments involving humans or animals. The laws are very strict.

*Many years ago, I had the distinct displeasure of watching (helplessly) as an R&D manager caught his necktie in a rotating tool on a lathe. He spent months in the hospital and was disfigured for life. He is lucky to be alive.

ments, in many countries (including the United States) no contract may require a person to surrender his or her civil rights, and one that attempts to do so is not legally binding. If someone is injured during the task, no matter what release forms they have signed, you and/or your organization are legally responsible and may be sued for damages. Your country probably has explicit laws about the conduct of laboratory experiments using humans and animals, and the vivisection laws are often very strict and strictly enforced. These laws apply to pilot tasks as well as formal experiments. Almost certainly, your employer also has explicit policies about such experiments. Study the country's laws and your organization's rules carefully. Be sure to check with your organization's legal staff before conducting any experiments involving humans or animals, no matter how innocuous the protocols and laboratory conditions may seem. In some instances, it may be practical to ask for forgiveness later instead of permission beforehand, but this is *not* such an instance.

Security is also often a problem with laboratories.[3] Because data security is such an important topic, it is treated separately in Section 12.2.3. However, the physical security of the laboratory itself is also often at risk, not so much from intentional breaches, but from thoughtless, careless, or accidental actions, often by those not directly involved with the experiments or the task. At the very minimum, if the laboratory will be used to conduct experiment trials for several days, those who are not task team members must be informed that they will be denied access to the laboratory unless given explicit permission by the task team.

An obvious candidate for exclusion from the laboratory is the maintenance staff of the building where the laboratory is located. Dusting, vacuuming, and cleaning may cause equipment to be moved out of critical positions, power plugs to be pulled out, computers to be accidentally turned on or off, and so forth. Even worse, because these actions will most likely take place during periods when the laboratory is untended, the task team may never even know that these events occurred. For example, after washing the floor, the maintenance worker thoughtfully moves the equipment back in almost its original position, but not quite. When finished vacuuming, the plug that he pulled to plug in his vacuum cleaner is conscientiously plugged back in, cycling some equipment and putting it in an entirely different state. After accidentally turning a computer off while dusting, the worker dutifully turns it back on, causing data loss (bad) or partial corruption that goes unnoticed (worse).

Another threat to the security of the laboratory comes from those who understand and appreciate the rules, but decide that they may ignore them safely. For example, one evening the CEO decides at the last minute to take a distinguished colleague and her family on a personal tour of the building, including a visit to the laboratory, despite the prominent Keep Out signs. While the adults look around with great care and respect, out of their sight one of the small children (who has just been told sternly not to touch anything) sees a red light and cycles the switch below it. Much to his disappointment, nothing happens. Nothing he can see or hear, that is.

But infractions can occur within the ranks as well. During a quiet moment after the experiment trials for the day are concluded, one member of the task team decides to check his e-mail using the supervisor computer. Unbeknownst to him, a virus that has made it through the company firewall enters the machine and discretely changes all the 3's to 4's in every text file whose name begins with the letter R, including the tab-delimited text file called "Results."

Regardless of the perpetrator, the next day the task team goes right on with the experiments in the laboratory. If the gods are merciful, the team will immediately detect the fact that something is wrong. If the gods are only moderately vengeful, the experiments will all be completed and the cohort sent home before any anomalies are detected in the results. If the gods are truly incensed, however, the damage will never be discovered by the task team, a new product will be developed and manufactured based on their research results and conclusions, and several years later some property will be destroyed or some people injured or killed. That this may have been the root cause of some of the otherwise-unexplained technological disasters over the years (airplane crashes, train collisions, pipeline explosions, nuclear accidents) is not beyond the pale.

Of course, alerting visitors and staff by posting clear signs and issuing internal memos is a good idea, but it may not be enough. The hypothetical infractions just described were committed by people who had decided that the Keep Out sign was not intended for them. It happens all the time. Maintenance workers must clean their assigned rooms; you cannot ask the CEO to surrender his key to the laboratory; and task team members, after all, know exactly what they are doing, right? To protect against this kind of violation, it may be necessary to continuously monitor the laboratory environment during periods of inactivity, so that these records can be reviewed before any new trials are conducted, or if that is not practical, at least when something untoward is noted. Table 12.7 lists some options for continuous monitoring of the laboratory.

Table 12.7 Options for continuous monitoring of the laboratory

• Video cameras (visible and IR)	• Power line activity	• Temperature
• Sound and vibration	• Electromagnetic radiation	• Humidity
• Computer keyboard and mouse	• Nuclear radiation	• Smoke and gas

Other precautions can be taken to prevent unintended alterations to the laboratory. If pieces of equipment, such as chairs, tables, and computer displays, must not be moved out of their proper positions for the duration of the experiments, bolt them to the table or floor (or at least mark and document their positions). If power to some pieces of equipment must not be interrupted, then wire their power cords directly to the power line or to uninterruptable power supplies. Cover, block, or disable switches whose settings must not be changed. In summary, minimize the ability to alter the fixed parameters and conditions of critical laboratory components.

Similar precautions can also be applied to software. Make sure that only the software that is required for the conduct of the experiments is mounted on the supervisor computers. Do not connect the supervisor computers to internal or external networks unless such access is required for the experiments. (Such connections are important channels in the task domain and must be included in the inventory described in Section 12.2 and Table 12.3.) Disable the effects of all unused keys on the keyboard. If preprogrammed spreadsheets are used during the experiment, lock all the fields that are not used for direct input.

Finally, during the design of the laboratory, make sure you have a complete list of all the channels that connect the laboratory to the rest of the environment, the system to the laboratory, and the system to the rest of the environment. Don't forget to include the possibility of human beings whose entry, presence, and effects are unwanted and unexpected. Consider each of these channels carefully, examining how each could affect the conduct and results of the experiments. Every one of them is a potential source of error that can alter the task conditions and throw the experiments out of control.

> **TIP 24**
> Deny access to the laboratory for everyone but essential personnel, post and publish this restriction, and strictly enforce it. Disable the adjustment of all fixed parameters and conditions. If necessary, monitor the laboratory continuously during untended periods. (See Table 12.7.)

12.2.2 BLOCK DESIGN

The design of the laboratory must be closely coordinated with the design of the *processes* that take place in the laboratory during the experiments. So far, we have alluded informally to these processes as treatments and trials and blocks that are assigned to the task unit, which was referred to as an object, a concept, a sample, or a cohort of subjects. Hopefully, up to this point these allusions were robust enough to foster an intuitive understanding of the experiment processes. But now we need to sharpen these terms a bit with some formal definitions.

Table 12.8 Laboratory design for Project *Measure Character-Recognition Performance*

Task 1.4.2 *(see Figure 9.5): Recognition Experiments (Laboratory Design)*

The data collection facility for the experiments was a small room (3 by 4 meters) with no windows and blank walls painted a light beige. The room was illuminated by a diffuse 200 watt overhead light whose intensity was set at 80% as a result of the pilot tasks. The temperature of the room was not directly controlled, but varied little from an average temperature of 20°C. The room was furnished with a single table (0.75 meters wide, 1.0 meter long, and 0.72 meters high) and a 5-leg swivel chair on casters with an adjustable back angle and seat height. Data was collected using a Simsong Model 65J467 15-inch CRT display that was bolted to the table, and an Ace Model 56D mouse. A sample of the CRT display objects is shown in Figure CS.4 below. Before each trial began, the character frame was empty (white) except for the START button and the Subject ID field. Both objects disappeared after the Subject ID was entered and the START button was clicked. One second later the question box appeared at the top of the display, and the first frame was displayed. Thereafter, whenever the subject clicked a character button, the question box disappeared, and the subject's response was recorded. One second later, the question box reappeared, and a new frame was displayed. When the trial was over, the words THANK YOU appeared in the middle of the frame. . . .

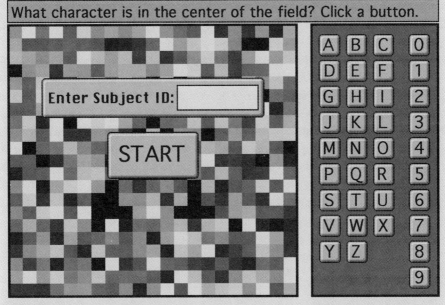

Figure CS.4 Sample CRT display (actual size was 16 cm wide by 11 cm high)

The technician monitored the activity of the subject during the experiment trials via the supervisor computer in an adjoining room. The inducer CRT display was driven by a graphics card in the supervisor computer. The cables connecting the display and the mouse to the supervisor computer passed through the wall separating the two rooms. . . .

Definition 12.3 A **cohort** is a task component consisting of a set of living plants or animals, all of which are assumed to be equivalent for the requirements of the task objective; a member of a cohort is usually called a **subject**.

Definition 12.4 A **sample** is a task component consisting of a set of nonliving objects or concepts, all of which are assumed to be equivalent for the requirements of the task objective; a member of a sample may be called by several names: **instance**; **response**; **data point**; **specimen**; **item**; or **case**.

The difference between a cohort and a sample is based on the difference between living and non-living things. Philosophers and scientists have tried through the ages to formulate a rigorous and universal definition of the word *living* without much success. Fortunately, the distinction necessary for the design and conduct of experiments is usually intuitively clear: a nail is an instance in a sample of hardware; a live frog is a subject in a cohort of live amphibians; a prime number is a data point in a sample of integers; and a living child is a subject in a cohort of living humans. Occasionally, however, the distinction blurs. For example, is a virus a subject in a cohort or a specimen in a sample? Is the artificially intelligent entity HAL in Stanley Kubrick's film *2001: A Space Odyssey* a subject, an instance, or a specimen? Fortunately, it doesn't really make much difference. These are just words, and although a reasonable attempt should be made to use them properly, if you end up calling a dishwasher a subject in a cohort of kitchen appliances, or a man an instance in a sample of men, it might sound strange, but everyone will know what you mean.

A few more terms need to be defined before we can begin the process of designing the experiment block. You will recall that the factors (variables and policies) associated with a goal or hypothesis of a task are those parameters and conditions whose levels and settings are varied during the experiments to observe their effects on the performance of the task unit as it is being altered by the inducers and measured by the sensors.[4] Each factor is assigned a set of values in a governing proposition. The total number of combinations of values that may be assigned to all the factors is the product of the number of values assigned to each one of them (minus any specific combinations that are specifically ruled out). This number is called the **extent** of the factor space. For example, in Table 11.4, which lists the major knowledge propositions for the case-study project *Measure Character-Recognition Performance*, three factors were defined: the policy *character* was assigned 36 settings in Proposition 3; the policy *typeface* was assigned 2 settings in Proposition 4; and the variable *noise level* was assigned 12 levels in Proposition 10. All combinations are possible, so the extent of this factor space is 864,

---

[4]This sentence is a dense summary of a lot of fundamental concepts from previous chapters. Please make sure you thoroughly understand what this sentence means before continuing on. If some of the concepts or words are a bit fuzzy, it may be worthwhile to take a few minutes to review the material in Sections 7.2 and 11.3.

which is the product of the cardinalities of the three sets: (36)(2)(12). Thus, there are 864 possible unique combinations of the values of the factors. Each one of these combinations is called a **treatment**.

> **Definition 12.5** An experiment **treatment** is a combination of one level or setting for every factor.

All 864 possible treatments for the project *Measure Character-Recognition Performance* can be defined meta-syntactically in BNF as follows:

    <treatment> ::= <character> <typeface> <noise level>
    <character> ::= A | B | C . . . | X | Y | Z | 0 | 1 | 2 . . . 7 | 8 | 9
    <typeface> ::= *serif* | *sans serif*
    <noise level> ::= 130 | 140 | 150 . . . 220 | 230 | 240

If the task unit is a single object or concept, then by definition all possible treatments must be assigned to it during the experiment. (Why else would all the combinations have been included in the set of treatments in the first place?) On the other hand, if the task unit is a sample or a cohort, then the treatments may be distributed in some fashion among the members of the task unit. In this case, the question is: how should the treatments be distributed? Before we can answer this, we need two more definitions to improve the semantic precision of the question.

> **Definition 12.6** An experiment **trial** is a complete set of treatments applied to a member of the task unit during the experiment.

> **Definition 12.7** An experiment **block** is a set of trials that provides a cover of the factor space that is appropriate and adequate for achieving the task objective.

Armed with these definitions, we can now rephrase the previous question more precisely: what is an appropriate set of trials that provides an appropriate cover of the factor space for the experiment? Or more simply: what is an appropriate block design for the experiment? Although every block design is tailored to the specific associated experiment, there are three basic strategies for experiment block designs: enumerated, systematic, and randomized.

An **enumerated block design** simply assigns every possible treatment to every member of the task unit. If the task unit is a single object, such an exhaustive strategy is necessary. However, for experiments with a cohort or sample, an enumerated design is often impractical, because the total number of trials required is the extent of the factor space multiplied by the size of the sample or cohort, which can easily get completely out of hand. On the other hand, experiments that are run on computers and automated instruments can often process a fully enumerated design in a reasonable amount of time, even with many factors and large samples.

A **systematic block design** uses a deterministic algorithm to assign different treatments to different subsets of the task unit in a systematic way, eventually covering the entire factor space. For example, as items (members of the task unit) come off an assembly line, treatment #1 is assigned to items 1, 11, 21, 31, 41, . . . ; treatment #2 is assigned to items 2, 12, 22, 32, 42, . . . ; and so forth for all 10 treatments. However, there are inherent dangers in this strategy. If a defect in the items just happens to occur only in every 5th item, then the subset of items assigned to treatment #5 will respond completely differently than any other subset of items would have. And because this response will be very consistent throughout the trial, it may be mistakenly concluded that this represents the normal response of *all* items to treatment #5. Likewise, if we assign a number of different treatments devised for a US marketing survey to every 100th telephone number, the treatment assigned to the sequence starting with 555-0100 (as opposed to, say, the sequence starting with 555-0167) will access telephones belonging primarily to businesses. The danger of causing an unintended resonance between a treatment and a periodic artifact in the task unit is exactly why soldiers break step when they cross a bridge. If they continued to march in lock-step, their synchronized tramping might initiate a sympathetic vibration of the bridge structure that could escalate out of control and collapse the bridge. Systematic block designs are very sensitive to periodic biases. Because of the possibility of encountering such a bias, the use of systematic block designs should be avoided unless a lack of bias in the task unit can be assumed with high confidence.

A **randomized block design** is similar to a systematic block design, except that the treatments for each member of the task unit are sequenced randomly (or *vice versa*). This substantially reduces the risk of severe distortion of the results due to a systematic bias. When a company of marching soldiers changes from synchronized steps to unsynchronized steps, this is analogous to switching from a systematic block design to a randomized block design.

The literature on experiment design also often mentions several other block design strategies, such as cluster (or area) and stratified designs. Strictly speaking, however, these are simply different strategies for extracting samples from a population by segregating the members into subsets that share a common value of some attribute (*e.g.,* age, height, weight, *etc.*). Once this is done, the attribute may simply be treated as another factor for the experiment. For example, suppose a cohort of adults was asked to fill out a questionnaire as part of an experiment. After running the experiment, the project team decided they wanted to measure the performance of men and women separately. If the gender of each subject had been captured on the questionnaire, and a statistically sufficient number of subjects of both genders had been included in the cohort, then it was simply a matter of segregating the responses of the men and the women before reducing the results. If not, the experiment would have to be run again with a new block design that incorporates the new factor.

Now that we have most of the necessary definitions out of the way, we can return to the subject at hand, designing an experiment block for a task. Once again, we shall use our case-study project *Measure Character-Recognition Performance* to illustrate the basic principles and processes.

An enumerated block design for the case-study experiment would have been impractical, requiring very lengthy trials. For example, if every subject in the cohort had been tested with every possible treatment, each would been required to respond 864 times (the extent of the factor space). If each response required five seconds on the average, the subject would have been parked in front of the CRT making selections from an unrelenting stream of characters buried in various degrees of noise for well over an hour! And this does not include the time required for any control treatments. Even if the subjects had been willing to participate in such a lengthy trial, it is very likely that their performance would have soon deteriorated due to fatigue, loss of focus, and just plain boredom, which is *inconsistent* with the objective of the project, *i.e.,* to estimate the subjects' *peak* performance (See Table 10.5).

There is another subtler disadvantage to enumerated block designs for experiments with human subjects. Humans are smart, and during the course of a fully enumerated trial for the case-study project the subject would probably have quickly guessed that the factor space included all 10 decimal digits and 26 Latin characters, even though they were randomly sequenced. This realization might have enhanced the subject's recognition of a partially obscured letter through the process of elimination. Consciously or unconsciously, the subject simply needed to keep track (at least approximately) of the frequencies of the characters displayed so far. Given that the distribution of frequencies of characters was uniform, as the end of the trial approached, infrequent characters would become better guesses than frequent characters: "This character looks like either a B or a D, but I've already seen a lot more D's, so I'll guess this is a B." This is analogous to counting cards while playing blackjack at a gambling casino. It is difficult and the advantage is small, but with sufficient accuracy (and plenty of capital) it is possible to beat the house. For R&D experiments in which responses to the treatments are expected to be independent of each other, designs that allow educated guessing by the subjects must be avoided.

Fortunately, it was not necessary to assign every treatment to every subject in the case-study experiment, because Proposition 16 in Table 11.4 explicitly assumes that all the subjects in the cohort had the same recognition skills. Therefore, the task team was free to use a block design that distributed the 864 possible treatments equally among all the subjects in the cohort, significantly reducing the time each subject spent in front of the CRT.

Although a systematic block design could accomplish this distribution, it was not considered for the case study, because the subjects may have eventually been able to learn the inherent periodicities and patterns of the planned sequences, enabling them to exhibit improved performance for the wrong reason. Instead, the project team used a randomized block design, in which computer-generated sequences of pseudo-random numbers assigned treatments to subjects and made other arbitrary decisions without bias.

Nonetheless, the project team still needed to decide *how many and which* sets of treatments would be randomly assigned to the subjects. They had to consider each factor, one at a time, and define the reasonable subsets of its values. Combinations of these subsets for each factor could then be combined into a set of trials that provided a fair and appropriate cover of the entire factor space, *i.e.,* the experiment block design.

One factor for the case-study experiment was the test character, defined in Proposition 3 in Table 11.4 as a policy with 36 settings. However, the task objective did not require estimates of recognition performance as a function of character, only of typeface and noise level (See Table 10.5). From this it may be inferred that estimates of recognition performance as a function of typeface and noise level were required only over the *entire set* of characters, not for each character. This inference is neither spurious nor dangerous. Although a task objective cannot be expected to explicitly identify every factor for which performance is *not* to be measured (just think how many there could be!), it *must* explicitly identify every one for which it *is* to be measured. Thus, the project team was free to assign any number of randomly selected characters to each trial (subject), as long as: 1) every trial was randomly assigned the same number of characters; and 2) the entire set of 36 characters was applied a whole number of times in the experiment. These criteria assured that all subjects worked equally hard (*i.e.,* each trial had the same number of treatments) and that character biases were avoided (*e.g.,* more O's than A's).

A factor that is not intended as a basis for measuring performance, but is necessary simply for assigning the values of a parameter or a condition uniformly across all treatments, is called a **representation factor**. To avoid possible confusion when describing an experiment block design, the *other* kind of factor defined in Section 11.3 *is* used as a basis for performance estimates and may be referred to as a **performance factor**.

In the case-study project experiment, *character* is a representation factor. Because the task objective makes it clear that separate estimates of performance were required for each of the two settings, *typeface* is a performance factor. Proposition 4 in Table 11.4 defines two settings for the typeface factor: *serif* and *sans serif*. Given the assumption that there was no significant difference among the recognition skills of the subjects, the project team was free: 1) to assign each of the two typefaces to half the subjects; or 2) to assign both typefaces to all the subjects. However, the team was hesitant to assign both typefaces to each subject, because changing the typeface back and forth during the trials might not yield peak performance. Moreover, the conditions would not be realistic, because the typeface of the license-plate characters under real conditions does not change. Thus, it was decided to assign each subject only one of the two typefaces. A byproduct of this decision required an even number of subjects in the cohort, so that each typeface would be tested the same number of times by the same number of subjects.

Proposition 10 in Table 11.4 defines 12 levels for the noise-level factor, from 130 to 240 in increments of 10, which were determined by the OP pilot in Task 1.3.1. Once again, the task objective explicitly required that performance be estimated for each noise level, making it a performance factor. And once again, the project team was free to distribute subsets of this set of levels fairly and equally among the subjects. However, for this factor it was decided that the sets of noise levels that were assigned to the subjects must obey three constraints: 1) the intervals between the noise levels in each subset must be constant and the same for all subsets; 2) the subsets must be disjoint; and 3) the average of the levels in every subset must be 185±5. These constraints were imposed to make sure that the subsets were all similar to each other, that each noise level was represented only once in all the subsets, and that each subset presented a consistent range of recognition difficulty from easy to difficult. As it turns out, only two sets of noise-level subsets satisfy all three of these constraints:

{130, 150, 170, 190, 210, 230} and {140, 160, 180, 200, 220, 240}

or

{130, 140, 150, 160, 170, 180, 190, 200, 210, 220, 230, 240}

The project team now had one last decision to make before the experiment block design could be finalized: the size of the cohort (number of subjects) necessary to ensure that the results and conclusions would be statistically significant. The project team had access to 14 trained license-plate inspectors to participate as subjects in the experiment. However, as described in Table 11.5, a few of them had to be used for the OP pilot (Goal G1.1 and Task 1.3.1) to determine the dynamic range and increment size for the noise-level factor, as postulated in Proposition 10 of Table 11.4.

Subject to this constraint, varying the size of the cohort and the specific factor subsets distributed to each subject allowed the project team to compute how many responses would be generated for each combination of performance-factor values. Statisticians typically require at least 30 responses for a statistically sufficient sample of results. Thus, the number of subjects in the cohort had to be large enough to provide at least 30 responses over the entire experiment for each combination of a typeface and noise-level factor value, the two performance factors. Because the test character was a representation factor, it was only necessary that each set of 36 characters be presented a *whole number of times* to each subject during each treatment (combination of factor values) and over the entire block to avoid character biases.

It may have occurred to you that the size of the cohort is also a *representation* factor of the experiment, because it is not used as a basis for measuring performance, but simply to provide an adequate cover of the performance-factor space and a statistically sufficient sample of results. As stated in Proposition 16 in

Table 11.4, it was assumed that performance would not vary significantly from subject to subject. The project team was only interested in the statistics of the ensemble. The major purpose of representation factors, then, is to provide a **sample** of results from which descriptive statistics of performance can be computed, such as averages and standard deviations.

An experiment trial that includes a representation factor is called a **Monte Carlo** trial, named after the capital of Monaco, which is famous for its gambling casinos. A Monte Carlo trial attempts to measure the effects of a random variable or a stochastic process that is out of the researcher's control, but is assumed to be fair and unbiased, like a roulette wheel in a (legitimate) casino. Of course, if the performance is known or assumed to be constant as a function of the representation factor, then it is not necessary to Monte Carlo[5] the experiment; there is no inherent variance, and one measurement is sufficient to characterize performance.

At this point, the case-study project team had completed the definition of all the feasible subsets of factor values for distribution among the subjects. In addition, the criterion for assuring a sufficiently large cohort or sample for statistical purposes had been set. It now remained to decide which of these subsets would actually be assigned to the subjects during the experiment, as well as the size of the cohort or sample. Table 12.9 summarizes the feasible factor subsets and the cohort-size criteria for the case-study experiment.

Probably the most efficient (and most modern!) way to search for the best combination of factor-value subsets is to construct a spreadsheet, so that the project team can quickly and easily compare the different alternatives. Table 12.10 is an example of the spreadsheet used by the case-study project team to explore and compare feasible and reasonable randomized block designs. The team members simply plugged in different values for the three input rows (a, b, d), and the formulas for the other rows in the spreadsheet enforced the constraints listed in Table 12.9. Unfortunately, this *process* of investigating the various alternatives cannot be dynamically presented in a book (yet). However, because this spreadsheet can be built very easily and quickly, the reader is encouraged to do so and to experiment with the various alternatives.

To find an acceptable design using the spreadsheet shown in Table 12.10, the size of the cohort (a) was input as a number less than 14, and then the values of the other inputs (b and d) were varied within the bounds prescribed in Table 12.9. If the number of responses for each treatment (f and g) was at least 30, and if the 36-character set was repeated a whole number of times for each treatment (f and g), for each subject (h), and for the entire block (e), then the block design was acceptable.

---

[5]Some scientists and engineers object to using the term Monte Carlo as a transitive verb, *e.g.,* to Monte Carlo a treatment, only condoning its use as a noun (acting as an adjective), *e.g.,* to run a Monte Carlo trial. They claim that it began its life as a noun and should remain so. I don't agree with this logic; language is a work in progress. However, to those whom this offends, I apologize for shanghaiing the word.

Table 12.9 Block design criteria for Project *Measure Character-Recognition Performance*

Factor	Feasible Factor–Value Subsets
character	all 36 characters, applied a whole number of times to each subject, treatment, and block
typeface	*serif* or *sans serif*
noise level	{130, 150, 170, 190, 210, 230} and/or {140, 160, 180, 200, 220, 240}
cohort size	smaller than 14 and yields at least 30 responses for each treatment

After careful consideration of the alternative block designs, the project team decided to use six of the inspectors for the OP pilot task to determine the useful noise-level factor values, which was conducted several weeks before the recognition experiment. The remaining eight inspectors were assigned to the test cohort, using the block design in column B (shown in boldface). This gave the project team some flexibility. If one or two of the inspectors failed to show up for the experiments at the last moment (which was certainly possible), they still had one reasonable and useful block design for a smaller cohort of 6 subjects (column A).

By this point in the book it should come as no surprise that an experiment may be quite properly described as a research task unto itself, whose objective is to acquire results for the parent task. The conclusions of this special kind of research task are deemed valid if and only if a critical peer review bestows approval on the design and execution of the experiments.

Unlike a general R&D task, an experiment may be categorized as a **pilot task**, a **control task**, or a **test task**. Pilot tasks have already been described in detail, and because they are intentionally conducted in an informal manner with few formal rules or procedures, we can leave it at that. However, control tasks and test tasks require considerable formal planning and execution. The experiment

Table 12.10 Sample test block designs for Project *Measure Character-Recognition Performance*

	Experiment Conditions		RANDOMIZED BLOCK DESIGNS				Formula
			A	*B*	*C*	*D*	
	cohort size (number of subjects)		6	**8**	12	12	a = input (less than 14)
factors	character: performance-factor combinations		12	**18**	12	6	b = input
	typeface: settings per subject (trial)		1	**1**	1	1	c = 1 (constant per subject)
	noise level: levels per subject (trial)		12	**6**	6	12	d = input (6 or 12)
	total test responses by the cohort		864	**864**	864	864	e = abcd
	test responses per typeface for each noise level		36	**72**	72	36	f = e/d/2
	test responses per noise level for each typeface		72	**72**	72	72	g = e/c/12
	test responses per subject (trial)		144	**108**	72	72	h = bcd
	time per trial for test treatments*		12.0	**9.0**	6.0	6.0	i = 5h/60

*(in minutes): Assumes a response requires an average of 4 seconds followed by a 1-second programmed pause.

for a control task is composed entirely of **control trials**, while the experiment for a test task may be a mixture of test trials and control trials. Test trials, which measure the performance of the task unit under the influence of the required inducers, have already been discussed at some length. Control trials, however, were discussed only briefly in Section 4.2, when the task domain and method were defined and described. Although this brief introduction and the frequent examples thereafter have presumably given the appropriate flavor to the concept of a control trial, its enormous importance for research and development projects requires that we now formalize the concepts of the control trial and its underlying functions.[6]

> **Definition 12.8** A **control trial** measures the performance of one set of task components in the absence of another set of task components to isolate the effects of the included components on performance.

By way of a brief review, a control task always serves a higher-level project and has one of two possible objectives: 1) to identify possible **biases** in the processes of the project tasks, so that the biases can be removed, or their effects minimized or somehow taken into account, a process called **bias negotiation**; or 2) to establish performance **baselines** for the project tasks on which estimates of error or accuracy are based. Some projects require both kinds of control tasks. Some require no control tasks, which implies that all the required baselines and significant biases for the project, if any, are already known and have been fully described and taken into account in governing propositions. When all significant biases have been negotiated, the affected tasks in the project are said to be *in control*.

Only three of the running example tasks introduced in Chapter 4 probably would *not* require control tasks: *Remove Screws*; *Prove Fermat's Last Conjecture*; and *Identify Faster Traffic*. It is likely that all the rest of the tasks would need to establish some performance baselines and/or identify and negotiate some significant biases.

Using a control task to establish a baseline for estimating error and accuracy has already been introduced in Chapter 8 (Limits of Knowledge). Without a baseline, it is impossible to test the hypothesis of a decision task that suggests that the

---

[6]It may have already occurred to you several times that it would have been far more efficient to introduce such formalities nearer the beginning of the book, where the subject was first introduced. Much as I would have liked to have done just that, it turns out that it is impossible to construct a linear exposition of the profoundly nonlinear process of the Scientific Method. As one of my graduate students once remarked in the middle of an intense and convoluted discussion of the Scientific Method: "Peter, wouldn't it be better if you just told us about the entire process of the Scientific Method all at once?" This comment evoked a hearty round of laughter, and his suggestion became affectionately known thereafter as the Pong Option, named for the graduate student. However, although he was joking, his point was and still is frustratingly compelling. How much easier life and R&D processes would be if they were neatly linear like a Charles Dickens novel. But it isn't, and they aren't, and there it is. Wax on, wax off.

task unit will behave in a certain way under induction. For all we know, the task unit might behave the same way in both circumstances, with and without induction. For example, to test the hypothesis that using a new hand lotion is better than using no hand lotion at all requires a control task to measure how well the task unit (the hands of the subjects in the control cohort) performs without any hand lotion, *i.e.,* the performance of the task unit in the absence of the required inducer. Then, when we measure the performance of the task unit after applying the new hand lotion, we will be able to compare the two performance estimates, control and test, and decide statistically whether the hand lotion is effective.

 Note that if conclusive evidence that describes the performance of the task unit in the absence of induction already exists, then mounting a special control task may not be necessary. For example, the statistical risk of getting lung cancer among nonsmokers is already well known and could be used as the baseline for a research project to decide whether smoking increases the risk. There is, however, a danger in using preexisting baselines. Any such research task must explicitly assume that its task unit is sufficiently similar to the task unit used by the baseline project. If it is not, then any increased rate of lung cancer in the new cohort may be due to some attribute that was absent or different in the baseline cohort, such as their lifestyles, diets, medical histories, *etc.* If the conclusions of previously published research projects are used as a baseline for the current project, the cited research must be thoroughly analyzed and meticulously reported in the Related Work step of the Analysis Phase, accompanied by an explicit assumption that the conclusions of the cited research are valid in the context of the current task. The validity of this assumption must be carefully scrutinized during peer review. On the other hand, it is quite acceptable to use preexisting baselines in *feasibility pilots,* remembering that their conclusions must always be confirmed by formal tasks.

Not all baselines are for control purposes. If we want to compare the performance of *our* widget with the performance of *their* widget, measuring the performance of their widget as a baseline is *not* a control task. It does not measure the task unit in the absence of the required inducers. Both widgets will be subjected to the same battery of tests, and each widget is simply one of two settings of a policy in our experiment factor space: *our* widget or *their* widget.

Control tasks are also used to identify and measure biases in task components other than the task unit. As has been discussed, a bias is a consistent tendency to behave in an inconsistent manner under certain conditions. A spring loses its memory (*i.e.,* stops springing back) when its elastic limit is exceeded. An airplane wing has a tendency to stall (*i.e.,* stop providing lift) when the air passing around it drops below a certain speed. A mathematical quotient becomes undefined when the divisor is set to zero. These three examples obey known laws that accurately describe the biases. But all too often, the biases only occur in specific instances and under certain circumstances, often unsuspected and sometimes quite strange. The precision of an element in the CCD array of a video camera (number of gray levels) is cut in half whenever a neighboring element is highly stimulated. A satellite wobbles in orbit as it passes over a gravitational anomaly in the Earth. A male gorilla refuses a proffered banana whenever a certain female gorilla is nearby. An interviewer's voice quavers when asking a subject a question about sex. A witness unconsciously skirts the truth when it is too embarrassing or incriminating.

Random-number generators are sometimes unsuspected culprits. Experimenters using a randomized block design must be very careful to make sure that the pseudo-random number generator is not biased, *i.e.,* that the sequence of generated numbers meets all the requirements for randomness. It must be free of signals that could influence the results in ways that have nothing to do with the action of the inducers or sensors. Tasks are particularly sensitive to such biases when the objective is to measure very small changes in performance as the different treatments are applied. For example, suppose a social scientist hypothesizes that there is a small but significant correlation between violent behavior and reading ability in teenagers. If the random number generator that selects and assigns the treatments is biased even slightly, the small expected effect could be easily overwhelmed by the exaggerated variance of the random sequence caused by the bias. Likewise, if a meteorologist is Monte Carloing an experiment to simulate weather patterns, which are in and of themselves noisy and chaotic processes, the model could predict completely spurious (but eminently reasonable) weather patterns, as an unknown bias in the random number generator is propagated through the time and space of the prediction regime. There are many well-respected methods for testing the efficacy and fairness of pseudo-random sequences, which can be found in most books on simulation methods and many books on statistics.[7] Negotiating such a bias usually involves searching the range of the random number generator for a fair sequence of sufficient length or selecting a better algorithm for generating pseudo-random sequences.

But perhaps the most insidious biases, and the most difficult to negotiate, are those that occur in humans and other intelligent animals who are involved in *any role* in a task: subjects; sensors; inducers; supervisors; and channels. Often these participants have an unconscious desire to help or hinder the achievement of the task objective, and, in so doing, distort the results and perhaps invalidate the conclusions. For example, the technician who supervises a cohort during an experiment may unconsciously assist the subjects with their task. Of course, he is trying to be fair and neutral during the trial, but after all, he is human, and humans have strong unconscious motivations. So a cough here, a sigh there, an uplifted eyebrow here, or a gesture there, may well provide cues that affect the performance of the subject. And this applies to nonhuman as well as human cohorts. Anyone who has a pet knows how sensitive animals are to very subtle signals.

In 1983 a doctoral student conducted a research project whose objective was to train rats to react in a predictable and systematic way to the odor of small concentrations of plastic explosives in the air. He had hypothesized that signals in the brains of trained rats could be used to detect bombs in suitcases being loaded

---

[7]Note that humans cannot generate random sequences manually. The biases in such sequences are like sirens. The readers are invited to try this as an exercise, using standard statistical methods to test the hypothesis of randomness.

onto airplanes. When he stood for defense, the student showed a short video tape of the experiments in the laboratory he had devised. On the left of the TV screen was a transparent plastic chamber containing the rat, which had been fully trained before the video taping session began (training took many days). In the middle of the screen was a computer display that showed the current status of the experiment, while the computer supervised the experiment by administering a random sequence of treatments. For half the treatments, a small amount of gas containing the explosive mixture (the required inducer) was admitted to the chamber, while for the other half of the treatments, a small amount of ordinary air was admitted. Everything was handled automatically by the computer. The treatments with the explosive gas were the test treatments. The treatments with ordinary air were control treatments, whose purpose was to establish the baseline performance in the absence of the required inducer. A neutral substance or effect that is used for control purposes is called a **placebo**.

On the right of the TV screen stood the student, who was providing a running narration of the experiment: "Now, the computer is admitting air — now, air again — now, it's admitting the explosive gas — air — gas — gas — gas — air — gas. . . ." And so forth. His voice droned on for several minutes in the quiet laboratory while the rat scrabbled around in its cage and the computer methodically conducted the experiment. Everything was under control. Or was it?

When the tape finished, one of the professors caught the student's attention. "Jim," he asked quietly, "why did you talk during the experiment?" The student replied, "I wanted to document the tape so that the audience would know what was going on." The professor looked out the window for a moment and then turned back, "Did it occur to you that you were also telling the *rat* what was going on? For every single treatment you announced whether it used the explosive gas or the placebo." The student was bewildered: "But, Professor, the rat can't understand English." "It didn't have to understand English, Jim," the professor responded. "All it had to do was recognize the difference between two very different sounds —'gas' and 'air'— which were *perfectly* correlated with the *exact* conditions it had been specifically trained to differentiate between. Rats, as you have clearly demonstrated, are very smart animals. All it needed to understand was a single bit of information." The student was dumbfounded. "Do you really think the rat was responding to the difference between the two sounds?" he asked. "I have no idea, Jim," said the professor, "but it was a risk you didn't have to take."

Fortunately, only a few of the trials had been recorded on video tape for documentation purposes, and the student had not narrated the other trials. Nonetheless, an analysis of the narrated trials revealed a slight but statistically significant *decrease* in performance from the other trials. Perhaps the rat learned the wrong lesson. Perhaps it was just confused. It doesn't matter. The conditions of the experiment had changed, creating a bias that affected the results. The student had learned a valuable lesson, but it came close to being a very expensive one.

In the mid 1970s, a well-respected vehicle-safety research institute in the United States was funded to conduct a research project whose objective was to measure the ability of drivers to stop their cars safely if the power brakes failed. In those days, power brakes simply amplified the force that a driver applied to the brake. If the force amplifier failed, the brakes would still work, but the driver had to apply a much greater force to achieve the same amount of braking.

The laboratory for the project was located on a special 360-foot test track. At the far end of the track was a brightly colored barrier made of very light materials, so that if a car hit it, it would be easily knocked out of the way, causing no damage to the car and no physical injury to the driver. Each driver in the cohort was asked to accelerate the car from a standstill at the near end of the track, to achieve and hold a speed of 50 mph until reaching the 180-foot mark, which was marked with a yellow line across the track, and then to apply the brakes and bring the car to a halt as close to the barrier as possible without hitting it. Each driver was told that this trial would be repeated 5 times. He was told that this was a test of the driver's braking ability, not the power brakes. The cohort consisted of 35 experienced (but not professional) drivers.

What really happened during the trial (the protocol) was quite different. On the fourth trial, after the driver had started braking, at a distance of 120 feet from the barrier, the force amplifier of the power brakes was disabled, so that suddenly the brakes became very hard to apply. The behavior of the driver and the applied brake pressures were recorded. The fifth trial was a ruse and was never conducted. Every driver was able to stop the car short of the barrier, although there were some close calls. It was concluded that the power brakes were safe under these conditions.

Although greatly abbreviated, this is all the information that was presented in the technical report of the project. There was no independent peer review of the research. Several years later, however, during an informal interview with one of the engineers who had been on the project team, a new fact about the protocol was revealed. Here is a short portion of that interview.

ENGINEER:	After the driver had passed the yellow line on the fourth trial, he applied the brakes. A second or two later as we passed the 120-foot flag, I disabled the force amplifier. Then the driver —
INTERVIEWER:	(interrupting) Excuse me, did you say 'as <u>we</u> passed the yellow flag'? Where were you?
ENGINEER:	I was in the car sitting on the passenger side.
INTERVIEWER:	Why were you in the car?
ENGINEER:	I had to be there to flip the switch that disabled the power brakes.

INTERVIEWER: Could the driver see the switch?
ENGINEER: No, I hid it from his view beside my leg.

And kept a poker face the whole time, no doubt. The potential of this bias to profoundly affect the behavior of the subject completely invalidates the conclusions of the research. That does not mean that the conclusions are wrong. It simply means that there is no way to know what they mean, because there is no way to know whether or not the driver got wind of the impending emergency. However, unless the engineer was a consummate trained actor, during the fourth trial (and perhaps even earlier) he probably gave the driver a sequence of subtle but ever-intensifying nonverbal warnings that something very unusual was about to happen, invoking a nagging feeling of apprehension in the driver that grew stronger and stronger as the yellow flag approached. Had a dog been in the car with them, with its extremely sensitive and precise sense of smell, it would have probably started whining or growling as the critical moment approached. In fact, in another conversation sometime later the engineer reported that several subjects admitted they "knew something was up" in their exit interviews. These interviews were conducted informally and were not included in the project report.

It would have been quite easy and inexpensive to disable the force amplifier by remote control from the sidelines, where a technician could have stood and observed the entire trial with minimal risk of introducing a bias. It would have been even better (although considerably more expensive) if the switch had been tripped automatically by a computer. But it was not stupidity that overlooked this bias, simply lack of experience. The engineer had never designed or conducted experiments that used cohorts of humans or other intelligent animals. His experience was entirely with inanimate task units, like pistons and fuel pumps. After the interview (and with almost 30 years of perspective and additional experience), he readily admitted that the experiment had been biased and the conclusions were meaningless.

Suppose you were asked to investigate a possible distortion in the lens of an ordinary camera. You would never dream of diagnosing the problem by using the faulty camera itself to take a picture of its own lens by using a mirror. The situation when the human mind is used as a diagnostic tool is much the same, exacerbated by our often confident feeling that we are completely objective and in control of our minds. Feeling strongly about something does not make it true. It is important to accept the fact that the human brain is an immensely complicated biochemical computer that intercedes itself between the lens of the eye and the conscious decision-making apparatus in the frontal lobes. Unconscious self-deception is not only possible, but highly likely under the right circumstances.[8]

---

[8] I have met more than a few engineers and scientists who scoff at the idea of the power of subconscious cognitive or emotional biases, calling them just so much psychobabble. They claim that they can easily overcome such biases in themselves and see right through them in other people. One engineer even protested: "I don't worry about such things. When I have to be, I am quite aware of my unconscious thoughts and motivations." Right.

With the proliferation of consumer products for the home and car, the personal computer, and the Internet over the past 20 years, the major industrial R&D emphasis has slowly but inexorably shifted away from hardware toward software, including the design of intelligent human-machine interfaces (HMI). Unfortunately, the engineers and scientists in industry are often not properly prepared to design and conduct experiments with human subjects. There is, however, a very good source of expertise available to them in this critical area. Most of the strategies for identifying and negotiating biases in experiments with humans have been worked out long ago by researchers in psychology and more recently in cognitive science, cybernetics, and management science. Almost every experiment that has been conducted by researchers in these fields has involved humans in some role. Necessity being the mother of invention, over the years they have had to develop very clever ways to identify and negotiate cognitive and emotional biases. Hiring a psychologist with this kind of training and experience is an excellent idea for companies that are expecting to mount an increasing number of experiments with human subjects.

As an example, psychologists who have specialized in language and communication (such as those hired by advertising agencies) are particularly well-suited to advise on the design of experiments that use questionnaires, such as for the example task *Evaluate GUI Effectiveness* that was defined in Chapter 4. In projects like this, the questionnaire is often a qualitative sensor, asking multiple-choice questions that solicit the opinions of the subjects. In other projects, a questionnaire may be intended to measure how well a subject has learned some factual material, like a test in an academic course. It is easy to compose questions that, accidentally or intentionally, trigger irrelevant biases and prejudices in the subjects. In fact, it is hard to avoid it. Advertising agencies write copy to do that intentionally, but a questionnaire used as a sensor for a research project must be as free of bias as possible.

What risk does a bias introduce? Suppose the subject does not know the answer to a question or has no opinion, but is not given the choice of answering *No Opinion* or *Don't Know*. This is called a **forced-choice** questionnaire. In this case, subjects who don't know the answer or have no opinion, must guess. If there is a bias in the subjects that resonates with a semantic bias in the wording of the question or the answers, the subjects may unconsciously select one answer significantly more often than any other. On the other hand, if the subjects *do* know the answer, there is a chance that this answer is the same one that would have been consistently selected if they were guessing. If so, when the distributions of the frequencies for the answers are analyzed after the experiment is over, it will be difficult or impossible to tell if a frequent response was a biased guess or a consistently conscious and unbiased response. The same problem can occur if the bias causes the subjects to unconsciously *avoid* a particular answer. Note that even if a subject is given the option of answering *No Opinion* or *Don't Know* on a questionnaire, a strong bias may still prevent him from choosing this answer when he has no opinion or doesn't know the answer; there is no guarantee that he will guess randomly.

To identify possible biases in questionnaires, the project team must design and conduct a control task. In accordance with Definition 12.8, this control task should measure the performance of the questionnaire (the sensor) in the absence of the task unit and the required inducer. For the example task *Evaluate GUI Effectiveness*, for instance, the task unit is the GUI, which the test cohort (the inducer) is asked to modify and evaluate. To measure the biases, usually long before the test trials are conducted, the test cohort performs an **external control task** consisting of a set of **external control trials** during which each subject responds to all the questions on the questionnaire *without any knowledge of or experience with the GUI*. How can subjects answer questions whose meaning they do not (yet) understand? The answer is, they have to *guess*. That's the whole point. During the control trials, you do not want them to make *knowledgeable* decisions. Instead, you want them to guess, and in doing so, reveal any related biases or prejudices. In fact, the instructions for the control task should specifically encourage the subjects to guess as randomly as possible.

We can simulate such a control experiment right here and now by presenting a couple of hypothetical questions from the questionnaire for the example task *Evaluate GUI Effectiveness*. Given that you have had no experience with the GUI other than the brief descriptions presented periodically in the book, try to answer the following questions as randomly as possible:

What color should the *MALE* button be?
 ❏ pink   ❏ red   ❏ green   ❏ blue
Should the Scroll Bar be on the left or the right side of the window?
 ❏ left   ❏ right
How wide should the very small arrows be?
 ❏ very narrow   ❏ normal size   ❏ very wide

Do you think your choices were unbiased? Did you feel any tendency, however slight, to choose one answer over another? Did you resist it? Do you think resisting your apparent bias helped or hindered your effort to make an unbiased choice? Tricky, isn't it?

When all the results of the external control trials have been acquired, the distribution of the frequencies of the answers for each question may be reduced using the chi-square test to estimate the statistical confidence that the answer is biased, *i.e.,* that one or more of the answers were selected significantly more frequently than the others. If there is no bias (which is the null hypothesis), then the distribution of the counts for the answers will be uniformly distributed. Thus, the confidence that the answers for the question are biased is the confidence with which the null hypothesis may be *rejected*. If a confidence threshold is specified, sometimes called the **critical level** in statistical analyses, the probability of a Type I error can also be computed; this is the probability that a decision that there is a bias, will be wrong. A bias confidence that exceeds the critical level implies that the bias is so strong that it will significantly distort any meaningful responses to the question. A typical critical level might be about 95%, which corresponds to a 5% Type I error probability, but each research project must impose its own critical level based on a thoughtful risk analysis. More on this subject is presented in Chapter 13 (Validation).

Finding a bias is one thing. But once a bias has been identified, what can be done about it? How can the bias be negotiated? If the bias-confidence threshold is *not exceeded,* the distributions of the answer counts from the external control experiment for this question may be used as the **control baseline** for the test results from the same question. When the test experiment is complete, the multinomial proportion variance may be used to compute the standard difference between two proportions (control and test) and estimate the (two-tailed) confidence that the test cohort indicated a statistically significant preference for or against each answer.[9] For example, considering the first of the three hypothetical questions posed previously for the task *Evaluate GUI Effectiveness* (assuming, once again, that the bias confidence threshold was not exceeded for this question), the final test results might read as follows:

Preferences Ranked by Strength
(positive confidences = for; negative confidences = against)
1.   Blue   47% of the responses   (+99% confidence)
4.   Pink   8% of the responses   (−91% confidence)
3.   Red   20% of the responses   (−28% confidence)
2.   Green   25% of the responses   (+7% confidence)

Clearly, results like this could be neatly summarized in a compact table for the entire questionnaire, including, if appropriate, some summary metrics, such as the averages and standard deviations of the confidences of related questions or answers. This entire process of reducing the results and computing the performance metrics can be accomplished with ease in an Excel workbook. Once again, more on this subject is presented in Chapter 13 (Validation).

If the confidence threshold for a bias is *exceeded,* there are three options for negotiating the bias. First, the question can simply be removed from the questionnaire. This assumes, of course, that the information captured by the question is not essential to achieve the task objective. If it is, then several versions of each question should be included when the questionnaire is put through the control task, just to cover your bets. Second, the question may be able to be reworded so that the bias is removed or sufficiently minimized. The disadvantage of this solution is that the question must again be put through a set of control trials to find out if the rewording actually worked, which may be impractical. Third, if the questionnaire does not have to be a forced-answer questionnaire, then a *No Opinion* or *Don't Know* answer may be included for each question. As mentioned previously, this will not necessarily remove the bias. However, it may weaken it

---

[9]There are other statistical methods by which this can be accomplished. Although it is beyond the scope of this book to present explanations of these statistical methods, they can be found in any standard text on descriptive statistics, several of which are listed in the Bibliography in Appendix A.

sufficiently, so that your bias-confidence threshold is not exceeded and the control distribution may be used as a baseline as previously described.

Note that the *test* cohort was used in the control task to establish the control baselines for the biases. Normally, the cohort for a control experiment or a pilot is *not* reused for the test task, because the knowledge gained from participating in the control trials **taints** the subjects and disqualifies them as subjects for the test trials. Instead, an equivalent but completely different **naive** cohort is used for the test trials. However, in cases like the example task *Evaluate GUI Effectiveness*, the project team could reason that reading the questionnaire during the control task could provide no information that would improve or degrade the subjects' performance during the test task. This is because the test task does not measure *good* or *bad* performance, but simply the *opinions* of the subjects, based on their qualitative impressions from working with the proposed GUI. The project team might well assume, therefore, that the control trials for the questionnaire would neither influence those impressions one way or the other nor change the way the subjects would answer the (then meaningful!) questions on the questionnaire in its intended role as a sensor during the exit interview. This presumptive statement must, of course, be included as an explicit assumption in the governing propositions.

> **TIP 25**
> Experiment biases are often very subtle, unexpected, and counterintuitive. Conduct a thorough inventory of every source that could conceivably exert an unwanted influence on the task components. Then, based on perceived risk and cost, decide whether these sources and their effects should be further investigated for potential biases. Use psychologists who have experience with experiments involving humans to help you identify and negotiate biases.

The control tasks presented so far have all been *external* control tasks, which are called that because their experiment trials are conducted separately from the test experiment trials, usually before them. (For this reason an external control task is sometimes called a **pretest**.) The subjects in an external control cohort are completely aware of the fact that they are participating in a control task. In fact, they might as well be told this before the trials start.

By contrast, however, it is often extremely important to measure the performance of the subjects in the absence of a required inducer *without their knowledge* to prevent such knowledge from influencing their performance either consciously or unconsciously. A task that has such an objective includes a set of **internal control trials** during which one or more of the required inducers is disabled or omitted. There are two different strategies for achieving this objective. First, the control treatments can be randomly intermixed with the test treatments, and every subject is exposed to some of the control treatments and some of the test treatments. Second, one group of subjects is given the control treatments exclusively, while the rest of the subjects are given the test treatments.

To illustrate the second strategy, consider the example running task *Measure Drug Effectiveness*. During the experiment trials one group of subjects will be given the new drug, while the rest of them will *not* be given the new drug, but rather a placebo that appears (smells, tastes, looks) exactly like the drug, but has no effect. The purpose of these internal control trials with the placebo is to establish a baseline for computing the confidence that the new drug is effective. This is similar to the method just described for computing the biases for the answers on the questionnaire. But there is one important difference. Although it did not matter that the subjects in the questionnaire *knew* they were participating in a control task, the subjects for the new drug trials *must not be told whether they have been given the drug or the placebo.* This is particularly important if the drug is expected to provide only a small improvement in the subjects' illness, not a complete cure. If the subjects are told whether they have been given the drug or the placebo, then the researchers will never know if a change in a subject's health (either an improvement or a deterioration) was a result of the presence or absence of the drug, or simply because the subject *knew* what the treatment was: the drug or the placebo. The mind can exert powerful forces on the body, both positive and negative, and the researchers need unambiguous evidence that the results of the experiments are caused by the drug, not by a psychological bias. When the subjects in a trial are not told whether they are receiving test or control treatments, it is called a **blind trial**.

But keeping just the subjects in the dark is usually not enough to avoid significant biases. Most experiments with intelligent animals require human supervision at times during the procedure: orienting the subjects before the trials begin; administering the treatments to the subjects; caring for and measuring the subjects during the trials; reducing the experiment data; and so forth. As has been described previously, just like the subjects, these human supervisors are also susceptible to conscious and unconscious biases that can influence the course of the experiment. Thus, it is essential that *neither the cohort nor any other human in contact with the subjects* knows which subjects are receiving the test treatments and which are receiving the control treatments (the placebo). When neither the subjects nor the human supervisors are told whether the treatments in a trial are test or control treatments, it is called a **doubleblind trial**.

Doubleblinding must be maintained until the data have been completely reduced and all the required conclusions have been drawn. In the meantime, each subject is identified only by a randomly assigned ID number, and the link between a subject's cohort membership and the ID number is a well-kept secret from *everyone involved in the experiment*. It is revealed only after the conclusions have been formally validated and acknowledged. Computers are handy agents for enforcing such rules, because they are quite good at keeping secrets without exhibiting occasional unconscious winks and nods.

Running internal control trials often raises thorny ethical issues, especially if the trials deal with life and death situations. Consider, once again, the task **Measure Drug Effectiveness**. Half the cohort receives the new cancer drug, and half the cohort receives a placebo. That is good science, but the fact remains that *all* the subjects have cancer! Wouldn't it be better just to give *all* the subjects the new drug, just on the chance that a few lives might be saved? Or is it more important to carefully control the current experiment at the expense of a few, so that sometime in the future, if it works, we can confidently help many more? After all, we cannot simply give every patient every candidate drug; there are good practical and medical reasons for that. On the other hand, that's easy for *us* to say. We're not sick.

Although there are no universal answers for these questions, perhaps the situation is mitigated just a bit by knowing that all the subjects in such experiments have been fully informed of the risks and have explicitly agreed to participate, no matter which treatment they get. Perhaps.

One final point about the subjects in such experiments. The fact that they are not *told* which treatment they get does not mean they do not soon begin to *suspect*. After all, if you were very ill and you stayed that way even after days, weeks, or even months of treatments, three guesses which treatment you are getting (or you *think* you are getting). Of course, no subject really knows (and will not be told) how the other subjects are doing. Nevertheless, the high intelligence of humans, fired by anger and fear, can easily unravel the best of controls. Bias rears its ugly head again. Even if the subjects' speculations are wrong or their reasoning is specious, stress often makes the sick sicker, and despair undermines the medical and lifestyle discipline necessary for the experiment to be successful. That is one reason why the cohorts for testing new drugs for very serious diseases need to be so large (typically thousands of subjects). Many subjects become convinced they are receiving the placebo (or at least are dissatisfied with their progress), violate the necessary conditions of the experiment, and eventually abandon the experiment to seek help elsewhere. Who can blame them?

Back to the more mundane. The case-study project **Measure Character-Recognition Performance** required two control tasks for measuring the biases of the cohort: one with characters, but in the absence of noise; and the other in the absence of characters, but with noise.

The first control task applied a sequence of 20 external control treatments to each subject at the end of the practice session that was conducted just prior to the test trials. The objective of this external control task was to ensure that each subject had sufficient experience with the interface during the practice sessions, so that the effects of the learning curve would be negligible during the test trials. If the conclusions of the external control task revealed a learning bias (*i.e.,* the selection times were not statistically constant, implying that the subject was still getting used to the interface), the bias was negotiated by repeating the practice session. If the subject still had a significant learning bias after two repeats of the practice session (three sessions in all), the subject was excused and removed from the test cohort.

The second control task applied a sequence of internal control treatments randomly intermixed with the test treatments. During these treatments, no character was displayed, although each frame was filled with random noise at the noise level assigned to the subject. The objective of this internal control task was to measure the selection time in the absence of any character, *i.e.,* when it was known for sure that the subject had to guess without any relevant information. The statistics of these selection times were used as the baseline for computing the confidence that the subject had *not* simply guessed correctly during a test treatment, but rather had actually recognized the character correctly. The higher the confidence that the average selection time was significantly less than the control baseline, the lower the associated **cost** of that treatment.

Table 12.11 is an expansion of Table 12.10 to include the number of external and internal control treatments for each candidate block design, along with the experiment durations estimated by the project team for each of the four candidate block designs and the formulas for calculating these numbers and estimates.

As given in row *j,* the number of treatments for each external control trial (20) was chosen for all candidate block designs to provide an adequate sample for the correlation metric. At the same time, it had to be small enough to allow the trial to be repeated several times without boring the subjects with a task devoid of any real challenge, *i.e.,* just clicking buttons that correspond to clearly recognizable characters as fast as possible.

For the internal control trials, each of the 12 noise levels had to be presented at least 30 times over the entire cohort to ensure a sufficient statistical sample of results. It was also necessary that each subject be presented with the same number of control treatments. Thus, as shown in row *k,* the actual number of internal control treatments per subject depended on the size of the cohort that was available.

Table 12.12 presents abbreviated versions of the descriptions of the two control tasks that were included in the project documentation. Also included is a description of the task used by the project team to gain a thorough understanding of the methods and procedures used by the license-plate inspectors on a day-to-day basis as part of their job. To put the information in Table 12.12 in the proper context, it may be useful to take another look at Figure 9.5 and Figure 11.1, which present the task and goal trees for the project.

The final component of the experiment block design is the **protocol**, which is the exact step-by-step procedure to be followed during the preparation and conduct of experiment trials. The major purpose of a detailed protocol is to ensure that the experiment can be accurately and precisely repeated, satisfying the *reproducibility* criterion for the provisional acceptance of knowledge, as postulated in Chapter 5 (An Epistemological Journey). Many researchers have torn their hair out when repeats of their experiments in different venues or as a part of a formal peer review failed to yield the same results. The burden of proof always lies with the original researchers. Strict adherence to well-planned and detailed protocols helps to ease this burden.

Every experiment has its own unique and highly specialized protocol, but a few general guidelines are offered in Table 12.13. Designing a protocol is much like designing a computer program (see the guidelines in Table 12.1). In fact, a computer very often *does* function as the supervisor during an experiment, and the program is its protocol. And just as a computer program must appropriately handle unexpected inputs and unusual conditions during the course of its execution, an experiment protocol executed by a human supervisor must explicitly provide for graceful recovery from mistakes. Of course, not every snafu can be foreseen, and not all unexpected situations are recoverable. If a subject suddenly smashes some equipment, or a meteor hits the laboratory during a treatment, there is no way (and probably no need) to design a procedure for handling such eventualities. As a matter of fact, in such nonrecoverable situations, it does not matter much whether the protocol includes a procedure for handling them, because the results from such trials are probably rendered meaningless and must be ignored anyway. On the other hand, if the subject during the example task ***Evaluate GUI Effectiveness*** hits the F1 key by mistake (or for whatever reason) and the "application quits unexpectedly," it does little good to say to the subject, "Hey, you weren't supposed to do that!" Instead, the keys that are not necessary for the treatment must be disabled so that they will do nothing at all when

Table 12.11 Complete test block designs for Project ***Measure Character-Recognition Performance***

		Randomized Block Designs				
	Experiment Conditions	*A*	*B*	*C*	*D*	*Formula*
	cohort size (number of subjects)	6	**8**	12	12	a = input (less than 14)
factors	character: performance-factor combinations	12	**18**	12	6	b = input
factors	typeface: settings per subject (trial)	1	**1**	1	1	c = 1 (constant per subject)
factors	noise level: levels per subject (trial)	12	**6**	6	12	d = input (6 or 12)
	total test responses by the cohort	864	**864**	864	864	e = abcd
	test responses per typeface for each noise level	36	**72**	72	36	f = e/d/2
	test responses per noise level for each typeface	72	**72**	72	72	g = e/c/12
	test responses per subject (trial)	144	**108**	72	72	h = bcd
	time per trial for test treatments*	12.0	**9.0**	6.0	6.0	i = 5h/60
	external control responses per subject (trial)**	20	**20**	20	20	j = input
	internal control responses per subject (trial)	60	**45**	30	30	k = (30)(12)/a
	time per trial for control treatments*	6.7	**5.4**	4.2	4.2	l = 5(k + j)/60
	total time per trial (includes 10 minutes for practice)*	28.7	**24.4**	20.2	20.2	m = i + l + 10
	total time for all trials (complete block)*	172	**195**	242	242	n = am

*(in minutes): Assumes a response requires an average of 4 seconds followed by a 1-second programmed pause.
**This trial may be repeated twice, as necessary, until the correlation confidence drops below 90%.

Table 12.12 Experiment tasks for Project *Measure Character-Recognition Performance*

Task 1.4:Acquisition of Recognition Measurements

Task 1.4.1: Manual Inspection Procedures

Problem: What are the conditions and procedures currently used by trained inspectors to examine and identify license-plate numbers?

Objective: To establish the appropriate conditions and procedures for the character-recognition experiments

Solution: Interview and observe trained inspectors in the field

Task 1.4.2.1: External Control Experiment Design

Problem: Because of the unfamiliarity of the user interface, without sufficient practice the selection times of the subject during the test trials may systematically decrease as they continue to become accustomed to the interface (learning curve).

Objective: To ensure the constancy of each subject's selection times after a practice period selecting characters displayed without noise

Solution: Before beginning the test trials, each subject was asked to study the display for 5 minutes to become familiar with the configuration of its elements (character frame, buttons, *etc.*) and then to practice selecting characters displayed without noise for 5 minutes. The typeface of these characters was the same as the one that had been randomly selected for the test trial for this subject. . . .

At the end of this practice session, a sequence of 20 randomly selected characters without noise was presented to the subject, who was instructed to click the corresponding button as quickly as possible. The sequence of the selection times for these control trials was correlated with the order of the sequence. If the confidence of a non-zero correlation exceeded 90%, the subject was asked to practice for another 5 minutes. If the confidence still did not exceed 90% after two repeats of the practice session, the subject was excused. . . . (See the protocol in Table 12.14.)

Task 1.4.2.3: Test and Internal Control Experiment Design

Problem: What are the recognition accuracies and response times for trained inspectors?

Objective: To measure the accuracy and response times of character selections as a function of typeface and noise level, accounting for the guess-bias for each character

Solution: For each treatment of each trial, the subject was shown a frame with a noise level chosen randomly from the noise-level set assigned randomly to the subject (see Table 12.9). For each of the 108 test treatments (variable h in Block B in Table 12.11), the frame included a randomly selected character. For the 45 control treatments (variable k in Block B in Table 12.11), no character was displayed. For each treatment the subject was instructed to click the button corresponding to the character judged most likely to be in the frame (forced choice). For the control frames, this selection was made in the absence of any actual character. There was no time limit for a subject's response to each treatment. The doubleblinded sequence of control and test treatments for each trial was generated randomly by the supervisor computer, which also conducted all trials and recorded the experiment results. . . . (See the protocol in Table 12.14.)

Table 12.13 Guidelines for the design of experiment protocols

1 Protocols specify all activities and scripts for preparing the laboratory, conducting the experiment, and recording the data.

2 Check lists can help to assure uniformity in the preparation of the laboratory as a final step before each member of the task unit is admitted to the laboratory and a trial is started.

3 *Everyone* in the laboratory during the conduct of an experiment must have a detailed list of steps to be carried out by them. (If they have no steps to carry out, they do not belong there.)

4 *Everyone* involved in the conduct of the experiment must follow his or her protocol exactly, so that every trial happens in exactly the same way.

5 Printed instructions for human task units can help to ensure uniformity in the laboratory experiences of the subjects, as well as minimize the biases that can easily result from interaction between subjects and human supervisors.

6 If a supervisor finds it necessary to improvise in the face of an unexpected situation, the trial must be marked as suspect. The results of such trials must usually be ignored.

7 Pilots should be used to plan and debug the protocols. An ounce of prevention. . . .

8 Anyone in contact with the humans or other intelligent animals involved in the experiment (including human supervisors) should be told as little as possible about the objectives of the experiment and the associated tasks and project.

9 Protocols can be published as pseudo-code, flowcharts, or lists of numbered steps using GOTO statements to indicate repeats. Choose the method most easily understood by the person who must *execute* the protocol, such as nurses, interviewers, test-car drivers, *etc.*

10 It is unreasonable to expect humans to follow the protocols perfectly; it is easier and safer to design the experiment so that protocol violations are handled automatically (fail-soft).

11 Whenever possible, the protocols for handling emergencies in dangerous experiments should be automated and fail-safe.

pressed, thereby avoiding the problem entirely. If the experiment designers do not know how to handle unwanted actions like this (*e.g.,* they do not know how to disable specific keys on the keyboard or out-of-bounds mouse clicks), then their skills are simply not up to snuff for such a task.

Next time you are at a computer exhibition, and some software salesperson is extolling the virtues of a new application program you are interested in purchasing, you might try running an informal OP pilot of your own right on the spot. While the software is running and the user interface is turning somersaults, reach down and begin pressing keys (especially combinations with the ALT, OPTION,

SHIFT, and CTRL keys). The results can be very enlightening. By discovering such fallibilities in the user interface, you will be doing yourself and the software company a real favor (although you probably should not expect to receive their heartfelt gratitude right then and there).

This story about the task ***Demonstrate New Software*** emphasizes two important requirements: 1) a protocol should include explicit procedures to recover from unwanted events that are predictable and recoverable; and 2) rather than expecting human participants to obey a protocol to the letter, it is much smarter and safer to design the experiment such that protocol violations have no significant effect or the recovery is handled automatically (fail-soft). Dangerous industrial and professional task domains routinely enforce these requirements. Automatic gates at railroad grade crossings are limited by "fail-safes," so that if the control system fails, the gate comes down automatically under the natural force of gravity. If caught in an unusual attitude (like a spin) most small fixed-wing civilian aircraft today are designed so that the machine will right itself and automatically resume straight and level flight under the natural forces of gravity and aerodynamics (providing it has enough altitude). If the human supervisor of a high-speed assembly line releases the pressure on the "dead-man's switch" in his hand for any reason, the power to the assembly line is immediately switched off and alarms are sounded.

Table 12.14 shows the protocol for the experiment technician for Task 1.4.2 in the project ***Measure Character-Recognition Performance*** (see Table 12.8 and Figure 9.5). Appendix F presents the experiment protocol for an industrial research project conducted at Robert Bosch corporate research department in Stuttgart, Germany, in 2001.

12.2.3 DATA MANAGEMENT

The final component of the experiment design is data management. Anyone with extensive R&D experience knows that managing data, whether acquired in the domain of a task (input data) or generated in the range of a task (output data), is perhaps the single most critical and frustrating task in a research or development project. Data is the currency of the information age, often collected and coveted with the same harsh emotional and political motivations as money itself, *i.e.,* for power and control. To an insurance company, losing money to expensive claims is an annoyance; losing *data,* however, can be a catastrophe. And the same is true from the lofty heights of pure intellection. The single most tragic setback for the epistemological maturation of human civilization was the destruction of the Great Library at Alexandria early in the fifth century of the modern era, just as the Roman Empire was entering the final stage of its disintegration. Over its 700-year lifetime, this astonishing repository of human knowledge had accumulated more than 500,000 manuscripts, amounting to one book for every 600 people on the planet (about the same ratio as today!). And then, in a single night, it was gone, and the Western World descended into the Dark Ages. Not a single manuscript from this library exists today.

Table 12.14 Protocol for Task 1.4.2 for Project *Measure Character-Recognition Performance*

PROTOCOL 142: TECHNICIAN PROTOCOL FOR EXPERIMENT TRIALS (TASK 1.4.2) REV 7.6 04/06/96		
Step	*Action*	*Check*
1	Prepare laboratory for the experiments (Protocol P142a).	❑
2	Make sure the laboratory is in its proper initial state (Protocol P142b).	❑
3	Launch and/or initialize the experiment software (CharRec v3.1).	❑
4	Click NEXT SUBJECT button on the display, and note the Subject ID of the next subject.	❑❑❑
5	If the display message is "NO MORE SUBJECTS," discard this check list, and exit the protocol.	❑❑❑
6	Open the waiting-room door, and call the next subject by the Subject ID number from Step 4.	❑❑❑
7	If the selected subject is not present in the waiting room, close the door, and go to Step 4.	❑❑❑
8	Ask the subject to enter the laboratory.	❑❑❑
9	Return to the display, and click the START button next to the Subject ID field on the display.	❑❑❑
10	Return to the waiting-room door, and close it after the subject has entered.	❑❑❑
11	Ask subject to sit down on the workstation chair.	❑❑❑
12	Ask the subject to enter his/her Subject ID on the display input field.	❑❑❑
13	If the display message is "INVALID SUBJECT," dismiss the subject (script S142a). Go to Step 4.	❑❑❑
14	Introduce yourself, the experiment, and the laboratory to the subject (Script S142b).	❑
15	Help the subject to adjust the chair, position the mouse, and adjust the display parameters.	❑
16	Tell the subject about the process for this experiment (Script S142c).	❑
17	Ask the subject to click the PRACTICE button on the display to begin a practice session.	❑❑❑
18	Observe and assist the subject throughout the practice session.	❑❑❑
19	Leave the laboratory and enter the control booth, closing the door behind you.	❑❑❑
20	Using the intercom, ask the subject to click the CONTROL button to begin the control trial.	❑❑❑
21	When the subject is finished with the control trial, note the control decision on the display.	❑❑❑
22	Return to the laboratory, closing the door behind you.	❑❑❑
23	If the displayed control decision was "DISMISS SUBJECT" go to Step 28.	❑❑❑
24	If the displayed control decision was "REPEAT PRACTICE SECTION," go to Step 17.	❑❑❑
25	Leave the laboratory and enter the control booth, closing the connecting door behind you.	❑
26	Using the intercom, ask the subject to click the TEST button on the display to begin the test trial.	❑
27	When the subject is finished with the trial, return to the laboratory, closing the door behind you.	❑
28	Thank the subject for participating in the experiment (script S142d).	❑
29	Open the door to the waiting room, and escort the subject out of the laboratory.	❑
30	Reenter the laboratory, closing the door behind you.	❑
31	File this check list in the protocol envelope. Get a new blank check list. Go to Step 2.	❑

This book can only scratch the surface of the important data management issues for modern day R&D projects. It is well to remember, however, that data is what the entire project is all about, its *raison d'être*. Data management issues should therefore be handled with commensurate care.

Avoid the evolution of data empires. Organizations usually regard data as intellectual property of considerable worth, and rightly so. Understandably, they go to great lengths to ensure the integrity of this data and to prevent unauthorized access to this data, especially from those outside their organization. Those who are charged with the responsibility of protecting the data are also held responsible for any and all data losses. They quickly learn that the easiest way to protect the data is to provide as little access as possible to *anyone,* including their own fellow employees. Preservation takes precedence over utility. Access to the data soon requires jumping enormous bureaucratic hurdles and providing unreasonable justification, *i.e.,* demonstrating an abiding "need to know." Those in charge of the data are promoted (often by themselves) from data administrator to Director of Information Resources (read: Keeper of the Faith). Slowly but surely, the data management staff evolves into a priesthood to guard the sacred parchments, rather than to maintain, protect, and provide access to a valuable knowledge resource. In the end, of course, this renders the data useless, albeit thoroughly protected.

It is an age-old dilemma between access and control, borne of the inherent fragility of data media. Books and documents are easily ruined and will eventually deteriorate all by themselves from natural causes. In the Middle Ages, because there was such a shortage of virgin parchment, monks often took an ancient Greek or Roman scroll, scraped off the old ink, and wrote their daily prayers on the erased parchment. Recently, using sophisticated technology, some of the original works of the Greek mathematician Euclid were detected under the religious scribblings of a minor thirteenth century monk. Thus, librarians (data managers) have always been very reticent to allow people to come into direct contact with the precious documents under their care. It can be, however, a short distance from legitimate concern to paranoia.

Today, however, the problem of preserving the *integrity* of the data, which historically was quite real and severe, is largely moot. Data can be stored directly on CDs without fear of deterioration over time. The original CDs can be carefully archived to protect them from mechanical damage. Copies of these CDs can be distributed very inexpensively with complete confidence that the data is *exactly* the same as on the original. Eliminating one of the most important reasons for the tight control of data (the protection of its integrity) greatly weakens the excuse usually used by ambitious data administrators for developing powerful data empires.

However, because CDs can be copied and distributed with such ease and so inexpensively, the risk of data *theft* is greatly increased, especially with universal access to the Internet.[10] Upwardly mobile data administrators often use this burgeoning threat as a justification for tightening access to the data and, once again, building data empires. But make no mistake; the threat of data theft is very real! The only question is: should a staff of data administrators (bureaucrats) have the centralized authority to control *access* to the data, or simply protect the *integrity* of the data (which really is no longer in question)? And if they are not allowed to control access, who *will* be charged with this responsibility? One possibility is simply to place this responsibility in the hands of those who generate the data, *i.e.,* the members of the associated project teams. Many high-level corporate executives, however, are not comfortable with the oversight problems that can arise with such decentralization of authority. Measuring the sense of responsibility (and trustworthiness) of every employee in a large organization is very difficult and highly subjective. Top management thus often decides that the inconvenience of data access that results from the establishment of data empires is simply a price that must be paid for reliable protection of the intellectual property of their organizations. The trick is to find the right balance.

Protect data from unauthorized access with public-key encryption. Simple password protection is *not* sufficient to protect data from unauthorized access. Passwords are easily broken; students in computer science routinely cut their teeth on cracking password-protected systems. Moreover, most users are understandably unwilling to changing their passwords very often, and when they do, they resist choosing unique and nonobvious strings; they have trouble enough remembering all the different passwords they must use every day. Brittle rules designed to prevent users from indulging in such frivolous violations of security are easily circumvented.[11]

---

[10]Occasionally I travel across the Atlantic Ocean as a passenger on a commercial freighter. When I do so, I often telephone my wife from midocean using the ship's single-side-band shortwave radio. The first time I telephoned her in this manner, I warned her that the whole world could listen to our conversation; all they needed was an inexpensive shortwave receiver. For the same reasons, it must be understood that when you send an unencrypted message or file to someone via the Internet, you are effectively broadcasting it to the entire world. Consider yourself warned.

[11]I am unconcerned about the security of my data files in my university computer account. I would never dream of archiving any sensitive data (*e.g.,* student grades) on such vulnerable media. Therefore, to maximize convenience, my password has always been my first name (capitalized) followed by a dot and a two-digit number. Currently I am up to Peter.37. My students and colleagues always know how to access what they need from me.

The best way to protect data from unauthorized access (other than locking it up and letting the data administrators decide who gets access to what) is to encrypt it. Data encrypted using the sophisticated technologies available today simply cannot be decrypted without knowing the key. In particular, public-key encryption allows a reasonably comfortable way to share data without compromising confidentiality, authentication, and authorization.

Here's how it works. Every person in an organization is given two encryption keys: a *public* key, which the person is free to give out to anyone; and a *secret* key, which the person reveals to no one. When the project team completes its experiments, the raw results (data) are stored unencrypted on a CD, which is then stored in the company vault. Each project team member is given a copy of the CD encrypted with his or her public key. To view or use this data, a team member must apply his or her *secret* key to the data. If a team member's CD is misplaced or lost (or stolen), it cannot be read without knowing the corresponding team member's secret key. To share data with someone outside the team, a team member decrypts the appropriate files, re-encrypts them with the intended recipient's public key, and sends the encrypted files to the recipient. When the files arrive at the intended destination, the recipient applies his or her secret key to view the shared data.

With very little additional overhead, the files can also be encoded so that the recipient can authenticate whether the received files originated directly from the expected sender (as opposed, for example, from an unauthorized intermediary). Many other clever combinations of public and secret keys allow most of the important data security risks to be avoided or significantly reduced. All of these capabilities can be automated in the appropriate software packages (*e.g.,* FTP applications and CD protection utilities), so that the users need not memorize their public and secret keys (which are necessarily very long and complex).

Maintain logs of every copy of important data files. Because data files are often shared with different departments in a company, as well as interested parties outside the company, management requires an audit trail of the distribution of these files. Should questions arise about the appropriateness of a transaction or the possibility of data theft, the ability to audit the copy-transaction history is essential. Maintaining records of all *electronic* file transfers, which often is done automatically, is not enough. Provision must also be made to automatically log all copies made on removable media (disks, CDs, tapes, *etc.*). This suggests that the computer operating system itself should automatically log all such copies, recording the date, filenames, and media type (all of which the operating system knows at the time the copy is made). This software can also be configured to ask the user to supply the identity of the *intended* recipient of the copy (and other information, as necessary), which is then also recorded in the copy-transaction log.

Keeping such logs does not guarantee that copies of specific data files can be traced with certainty. It is straightforward to change the name of a file before copying it to removable media or transmitting it electronically across a network. Moreover, the transaction log itself is vulnerable to tampering. Additional precautions can be taken, but no system is perfect. A person dedicated to the theft of data can always find a way to circumvent the security barriers. Eventually, as with most things, it comes down to trust and, unfortunately, access limitations.

Copyright and publish all experiment data promptly. As an alternative to keeping new knowledge secret, intellectual property can also be protected by: 1) establishing the fact that the project team generated the knowledge *first* before anyone else; and 2) hindering other people from using this knowledge without explicit permission.

The first step in this process is to make the date when this knowledge was first generated a matter of public record. This is easily accomplished by publishing it in the open literature, such as in a professional journal or in the proceedings of a professional conference. Of course, the knowledge must be accompanied by a technical paper that explains what you did to generate it; without these explanatory details, the paper will probably not be accepted by any credible journal or conference. This, however, is not difficult; thousands of such papers are published every year. When such knowledge is published in this manner, it becomes eligible for recognition as *prior art,* which can be a strong legal basis for deciding the ownership of the intellectual property, should it ever be challenged in a court of law.

The second step in the process is to copyright the data. Almost always, the *paper* just mentioned will be copyrighted by the publisher. However, it is also a good idea to copyright the *data* itself in your own (or your company's) name. This is a simple and inexpensive process that will help to discourage anyone from copying the data for any purpose without your written consent.

When considering how to protect the ownership of intellectual property, it may seem a bit ironic that an alternative to keeping it a tightly held secret is to publish it in the open literature. In fact, probably the *only* sure way that doctoral students can protect their research ideas is by publishing their solutions and results as early as possible. This helps to ensure that their contributions will be recognized and acknowledged as both original and significant by their defense committees when they meet to examine the student several years later.

In addition to these high-level strategic issues for data management, there are several important low-level tactical recommendations for ensuring the longevity and utility of experiment data.

Record ALL experiment data. A project team that spends time trying to decide which of all the possible results from the experiment should be recorded is wasting its time. The answer is simple: record *everything.* In the case of a task based on a computer simulation, this should be taken even one step further. Don't bother computing the performance metrics directly in the simulation program. Just record all of the raw results, even if this requires many gigabytes of storage. Thereafter, another program (or an Excel workbook) can be used to reduce these results to the required performance measures. This provides the invaluable capability to postulate and compute new performance metrics without reprogramming and rerunning the simulation, which can be both expensive and risky. (Does the changed program really do the same thing as the original program?)

In the old days, when experiment results were recorded by hand on paper, it made sense to limit the human effort and amount of paper by trying to limit the recorded results to those deemed essential in advance. But today, when the results of experiments are usually stored automatically on high-density media, this makes no sense. Media is cheap; repeating experiments is very expensive! ("From what I can remember about each of the subjects, there seems to be a strong bias in the results as a function of the duration of the trial. Why didn't we simply record this duration? We knew it at the time!")

There are exceptions to this seemingly profligate strategy. For example, it is not reasonable to ask human subjects to fill out a 20-page questionnaire about their backgrounds simply because such data *might* turn out to be useful. In such cases, somehow (perhaps using pilots) the team must carefully decide in advance which demographic data about the cohort is most important and must be captured. Even so, when all is said and done, you will wish you had asked additional or different questions. As stated previously, running experiments with humans and other intelligent animals is very difficult and frustrating, and normally you get only one shot at them.

It is generally useful to store the raw results of experiments as delimited text files, so that Excel (or other similar applications) can read such files directly to reduce the results and compute the performance metrics. However, if it is *really* important to use as little media as possible to store the raw results, then store them as binary data files. Files that store numeric values as text strings are five times larger than their binary equivalents, which are stored as 32-bit real or integer values (20 bytes per value *vis-à-vis* 4 bytes per value). When access to a binary data file is required to reduce some results, a short program can be used convert it to a delimited text file, so that it can then be read by Excel.

Do not preprocess the raw experiment data in any way before recording it. Often the data files generated by an experiment or used as input to an experiment from another task have been preprocessed in some way before you get it. As long as the recipient knows about this preprocessing and has approved its application, that's fine. But too often this preprocessing takes place without the knowledge of the recipient, leading to mysterious results that cannot be explained and defy expectations. For example, suppose an experiment team makes the unilateral decision to compress the images that result from an image-processing experiment, using a lossy method, such as JPEG, before sending them to the analysts responsible for reduction of the results and computation of the performance metrics. If this compression is not known to the analysts (the effects of such compression are usually undetectable by the human eye; after all, that's the point of such compression), they may be bewildered by the strange artifacts that crop up here and there in their analysis, seemingly without rhyme or reason. When the analysts finally figure out what has happened, the experimenters will, of course, be apologetic: "Sorry. We thought it would help you if the files we sent you were a good deal smaller than the originals. So we compressed them to save you some space and time." In fact, an enormous amount of time was wasted.

Below is a partial list of preprocessing functions that are sometimes applied to data to "enhance" it before delivering it to the intended recipients (with the best of intentions):

- Low-pass filtering to eliminate small "artifacts" or reduce "noise"
- High-pass filtering to "accentuate" certain signals
- Quantization (reduction in precision) to eliminate "insignificant" differences in values
- Normalization into "sensible" dynamic ranges
- Removal of "unreliable," "extreme," or "unrepresentative" instances (decimation)

Once such functions have been applied to the data, the distortions they cause may be irreversible, or worse yet, undetectable. Only proactive roles by both parties involved in the exchange of data can avoid such problems: 1) when you send data to someone, do not preprocess input or output data (domain or range knowledge) in any way without first discussing the processes with them; and 2) when you request data from an external source, be sure to determine exactly what pre-processing functions have been applied to the data prior to sending it to you. Robust communication and cooperation is the only way to avoid serious data pathologies.

Establish clear organizational and documentation conventions for data files. The numbers and sizes of both the input files and the experiment result files for a project can be extremely large. For example, a recent experiment conducted by a research institute used a cohort of 50 subjects to capture more than 80 gigabytes of data composed of 2500 individual files of results. This is equivalent to 30 *million* pages of text. Unless such data files are carefully organized in some sort of hierarchy and accompanied by clear and complete documentation, accessing the data will become a chaotic nightmare. It is well to remember that these input and result data files include a significant fraction of the domain and range knowledge of a task, intended not only for immediate use by the project team itself, but perhaps months or years later by complete strangers, when some abiding need for this knowledge reemerges.

The files of input and result data files for a project should be organized hierarchically in *exactly* the same way as the task trees for the project, using the same numbering scheme. The highest-level folder, corresponding to the primary task, contains subfolders for each of the subtasks. This hierarchy is extended until every lowest-level (terminal) task is represented by its own folder. However, although *every* task in a project has at least one experiment associated with it, many of these experiments do not generate results that are stored as digital data files on magnetic or optical media. Such results are often simply verbal statements in the project documentation, such as "After having been tuned, the piano was checked by the performer and declared to be in satisfactory condition"; or "The user manual for the software application developed for this task was reviewed by the client." Nonetheless, folders for such results are *not* eliminated from the hierarchy, because in the future when other engineers and scientists are reexamining the project, missing folders may be regarded as conspicuous by their absence, triggering a long (and fruitless) search for the "missing" input or result data files. It is better, therefore, to include a folder for every task at every level of the hierarchy, marking each folder that was intentionally left empty with a special symbol (*e.g.,* an *asterisk*), which indicates that there were no digital data files as results or inputs for this task. Including empty folders in the project data hierarchy is a minor waste of computer resources, easily tolerated for the sake of observing one of the three criteria for the provisional acceptance of knowledge, *completeness.*

The folder-naming conventions for the project data hierarchy should be defined so that the folders are sorted automatically in directory listings in an appropriate way. A good method is to begin each folder name with a string of digits to reflect the associated task number, followed by a very short textual abbreviation of the task name. Although the digits of a task number are normally separated by periods, the number in the name of a folder need not contain periods; this shortens the folder names so that they can easily be displayed in a directory listing in a reasonably narrow window. Note that removing the periods in a task number causes no ambiguity in the hierarchical numbering system (as long as no more than 9 sublevels are used.)

To make sure the folder names sort correctly in directory listings, all task numbers in the folder names should be expressed with the same number of digits, padded to the right with nonsignificant zeroes as necessary. Be sure to include plenty of nonsignificant zeroes at the end of the numbers, so that expanding project hierarchy by adding new tasks below the current lowest level does not cause the names to be sorted incorrectly in directory listings. Table 12.15 presents the hierarchical organization of the data files for the project *Measure Character-Recognition Performance*, based on the project task tree presented in Figure 9.5. All the task numbers have been expressed to six digits (padded with zeros), even though the task tree in Figure 9.5 has a maximum of only four levels in its hierarchy, requiring only four digits. As recommended previously, retaining six digits ensures that the directory listing would still sort correctly, even if additional levels for new subtasks were added to the bottom of the project hierarchy.

Each terminal numbered folder that is not empty (*i.e.,* whose name is *not* terminated with an asterisk) contains two folders, one for the input data files and one for the result data files. Within each input folder and results folder (or within the previous folder referenced by an alias) are the associated data files themselves (or aliases that refer to them), organized as needed into sub-folders. Each result data file is given a name that uniquely identifies the specific experiment(s) to which it belongs. As just described, even if an input folder or results folder is

Table 12.15 Data files for Project *Measure Character-Recognition Performance*

```
📁 100000 CharRecExperimentData
    📁 CharacterInputs
        📄 SerifCharacters
        📄 SansSerifCharacters
    📁 110000 RelatedWork
        📁 111000 VideoCond
            📁 Inputs★  asterisk (★) = folder intentionally left empty
            📁 VideoCondResults
                📄 DOTLicensePlateImages
        📁 112000 NoiseModel★
    📁 120000 SolutionMethod
        📁 121000 InspectionCond★
    📁 130000 GoalHypo
        📁 131000 NoisePilot
            📁 CharacterInputs italics = alias for a previous file or folder
            📁 PilotResults
                📄 RealNoiseCond
        📁 132000 SetGoals★
    📁 140000 RecogMeas
        📁 141000 ManInspecProc★
        📁 142000 RecExp
            📁 142100 DesControlExp★
            📁 142200 ConControlExp
                📁 CharacterInputs
                📁 ControlResults
                    📄 SerifBaseline
                    📄 SansSerifBaseline
            📁 142300 DesTestExp★
            📁 142400 ConTestExp
                📁 CharacterInputs
                📁 TestResults
                    📄 SerifRecognition
                    📄 SansSerifRecognition
    📁 150000 ValidateObjective
        📁 151000 PerformComp
            📁 Inputs
                📁 ControlResults
                📁 TestResults
            📁 Results
                📄 AllPerformanceStats
        📁 152000 Conclusions
            📁 Inputs
                📄 AllPerformanceStats
            📁 Results
                📄 Conclusions
        📁 153000 DocProject★
        📁 154000 PeerReview★
```

empty, the folder should nevertheless be included in the hierarchy and marked with the special symbol (★) indicating that it was intentionally left empty.

Often the same input data file is needed for several different experiments in the project, or a result data file from one task is required as an input data file for another task in the project. It makes no sense, however, to put identical copies of this file in every one of the folders that requires it. This would introduce a maintenance nightmare; if the file is routinely modified during the course of the project (which is not unusual), there is a real risk that the changes will not be propagated consistently to all the copies. Instead, a single master copy of the data file should be located in a folder created at a convenient and sensible level in the data file hierarchy. This allows *aliases* of the master data file to be placed in any folder that requires access to it. This way, whenever the master data file is modified, the aliases will all still refer to the same and most recent version of the master data file.

Finally, every result data file must include a **slate** (an informational header) that provides at least the project task number, the date and time of the associated experiment, and the identity of the human supervisors (if any).

Exercise 12.3
For those tasks in the Synthesis Phases of a project you are currently planning or undertaking that require complex experiments, specify the laboratory, block designs, and data management regimes.

12.3 Conduct Experiments

The plans have all been laid. Now is the time to put them to the test. The only guideline that can be offered at this point is this: *follow the plans you have so carefully devised to the letter.* Resist the temptation to improvise. If an experiment does not run smoothly or successfully, stop it, and reenter the Scientific Method to prepare a new and improved set of plans that solve or avoid the problem. Regard a failed experiment as a pilot.

On 21 June 1996, a full set of control and test trials for the experiments of the project *Measure Character-Recognition Performance* was conducted. As specified by the experiment block design in column B of Table 12.11, eight inspectors were used for the primary experiment cohort.

12.4 Reduce Results

Often the performance values from which the conclusions of the task will be drawn cannot be computed directly from the raw results of the experiments. First, the results must be *reduced,* combining and/or transforming them in some way to generate values suitable for the domains of the performance metrics. To illustrate the difference between raw results and reduced results, let us consider two of our example running tasks from Part II.

In the project *Identify Faster Traffic*, the input data file provides estimates of the speeds of the individual vehicles on each highway. These are the *reduced* results from the preceding task that measured these speeds, *not* the *raw* results, which were measurements of the signals generated by a rubber hose laid across the highway. A car that crossed this hose would generate two pulses of compressed air, one from the front wheels and the other from the rear wheels. Each pulse was recorded as an absolute *time*. Then, after a sufficient sample of vehicles had gone by, the list of times in the raw data file had to be *reduced* to a list of vehicle speeds. The transformation that reduced these raw data was based on pairs of successive pressure pulses that were *assumed* to correspond to the front and rear wheels of the *same vehicle*. Selecting these pairs of successive pulses had to be made without knowing the distance between the axles of any particular vehicle or the distance between any two successive vehicles. This is by no means a straightforward logical process; trucks with more than two axles, unusually long or short cars and trucks, and the wide range of vehicle speeds would certainly cause a great deal of confusion when transforming the *raw* results into the *reduced* results, the speed estimates.

The archeologist in the task *Measure Fossil Age* might decide to estimate the distances between the sedimentation layers by photographing the cliff face from a carefully calibrated distance using a camera with carefully calibrated optics. Once the photographs were developed and the likely boundaries of the sedimentation layers carefully identified, precision measurements between these layers on the photograph could be easily transformed into the corresponding distances on the cliff face itself. Thus, the *raw* results are a set of photographs, which must then be transformed into the *reduced* results, the estimated distances between sedimentation layers.

In fact, there are really only two kinds of values that can ever be measured directly: distances and counts. When time is read from a clock, for example, we either measure the *distance* between the position of a hand on an analog clock (test value) and the origin of the time scale on the face of the clock (baseline value); or we count the number of times a counter (escapement) has changed state during the interval between an initial event (baseline value) and a final event (test value). Similarly, to measure the weight of an object, we either measure the distance between the position of the pointer on the scale with the object in the pan (test value) and its position without any weight on the pan (baseline value);

or (given sufficiently sophisticated technology) we count the number of mole-
cules of known weight that the object contains.

It can be (and often is) argued that only *one* of these values, distance or
count, can really be measured directly, *i.e.,* that one metric is simply a transformation of the other, more fundamental metric. This argument is a close relative
of the spirited battle that has been raging in the field of physics since the early
twentieth century: the quantum theory of matter (based on count) *vs.* the wave
theory of matter (based on distance).

Be that as it may, the raw results of an experiment must usually be *reduced* to
values that are suitable for the domain of the required performance metrics: times
to distances; counts to times; distances to speeds; distances to weights, *etc.* The
raw results for the case-study project *Measure Character-Recognition Performance*
were no exception. First, the raw character selections made by the subjects for
each actual character had to be reduced to a list of logical values indicating simply
whether each test response was correct or incorrect, as required by the benefit
metric (percent of selections recognized correctly). Second, the raw selection
times between the successive responses by a subject, which were measured in *days*
(the units of the NOW() function in Excel), had to be converted to seconds, as
required by the performance metrics. Note that the times could have been converted to seconds by the supervisor program, but imagine how unfortunate it
would have been if the conversion function had been programmed incorrectly,
and this error was not discovered until all the experiments had been run and the
cohort dismissed.

The raw and reduced results from the experiments for the project *Measure
Character-Recognition Performance* were then recorded in the result data file.
Every raw and reduced–result file was slated with a header specifying the experiment date and the technician ID. The data entries included relevant trial information, factor values, and the raw or reduced results for each treatment in all
eight trials. Sample entries for both files are shown in Table 12.16.

Why should both the raw and result data files be recorded? Why not just let
the supervisor software simply compute the reduced results (or even the performance metrics) directly, record them, discard the raw results, and be done with it?
The reason is that most transformations that reduce the raw results are not
reversible; once applied, the raw results cannot be recovered without running the
experiments again, which is often expensive and time consuming, and sometimes
impossible. If it is discovered that different methods for data reduction might be
more appropriate, or if unusual behavior in the reduced results suggests examining
the raw results for possible sources of error, the raw results must be available. It is
usually *much* less expensive to record *both* the raw and reduced results, just in case.

> **TIP 26**
> Because the data reduction methods and/or performance
> metrics may have to be modified or corrected, record both
> the raw *and* reduced results of experiments.

Table 12.16 Sample result file entries for Project *Measure Character-Recognition Performance*

Task 1.4.2 06/21/96 Technician 27	Subject ID	Trial Type	Typeface	Noise Level	Correct Character	Selected Character	Selection Time (Days)
Examples of	02	ExtCntrl	*serif*	0	G	G	0.0000248
Raw Results	03	IntCntrl	*serif*	210	n/a	A	0.0000541
	05	Test	*sans serif*	150	R	B	0.0000797

	Subject ID	Trial Type	Typeface	Noise Level	Character Selection	Selection Time (Sec)
Examples of	02	ExtCntrl	*serif*	0	correct	2.14
Reduced Results	03	IntCntrl	*serif*	210	n/a	3.67
	05	Test	*sans serif*	150	incorrect	5.89

CHAPTER 13

Validation

By the time the Terrible Two's draws to a close, the child has gathered considerable experience from its operating-point pilots that tested the limits of the home environment. Ready to be introduced to the challenges of a larger world, the child is enrolled in preschool, and life begins for real. No longer is the child just an inducer or a sensor, but often a subject of other experiments. Some experiments are conducted by the teacher, but many more are mounted by the other children. The benefits and costs of competition and cooperation are evaluated everyday, as the child measures its performance in formal exercises in the classroom and informal games during recess.

As the years slowly pass, and the seemingly never-ending, never-changing routine of school drones on, mind and body mature, and plans are made for graduation and independence. Parents wring their hands while their misguided progeny break bones, wreck cars, and dream of the day when they can stay out as late as they want and change the world their parents have so badly neglected. Lesson after lesson, the results come fast and furious and blur together, misunderstood individually, waiting patiently to be understood collectively.

Finally, a semblance of order is perceived. The young adults come home from college and are astonished how much their parents have learned while they were away. Pacts and deals are made. Tasks are undertaken. Successes and failures are put in perspective. Problems change, goals and hypotheses are defined, experiments are run, results are gathered, and, with the help of friends and colleagues and lovers, conclusions begin to emerge from the murky confusion. Bits of knowledge fall into place like a giant jig-saw puzzle. Still, the big picture is elusive.

On her deathbed, Gertrude Stein is reputed to have been asked by a friend, "Gertrude, you've lived your life in search of truth. What is the answer?" "Answer?" she replied. "My goodness, I still haven't figured out what the *question* is!" In search of **validation**, life comes full circle.

The root of the word "validation" is "valid," which derives from the Latin word *validus,* meaning "strong." The synonyms given in dictionaries (*e.g.,* sound, convincing, telling, conclusive, well-grounded, incontestable) refer to the capacity of knowledge to resist challenge or attack, based on the way in which this knowledge was inductively or deductively acquired. Thus, validation is *earned,* not bestowed.

> **Definition 13.1** The objective of the **Validation Phase** of the Scientific Method is to decide whether the objective of the task has been achieved, based on formal conclusions about its goals and hypotheses and a rigorous peer review of the task methodology.

13.1 Compute Performance

By definition, the reduced results files recorded in the last step of the Synthesis Phase are the basis for computing the performance of the task unit using the performance metrics postulated in the Hypothesis Phase. Because stepwise refinement is not only allowed, but *expected* as part of the application of the Scientific Method to the task, these performance metrics may have to be revised. They may turn out to be unsuitable in the face of the actual results acquired from the experiments, or because the goals and/or hypotheses cannot be validated and must be redefined. Both outcomes may require postulating a different or modified set of performance metrics.

For example, suppose that during the planning of the task *Identify Faster Traffic*, before having access to the actual measurements, the engineer assumes that the distribution of the vehicle speeds will be normal (Gaussian), allowing the use of straightforward parametric statistical tests based on the averages and standard deviations of the speed samples. However, now suppose that when the data is received, a QD pilot reveals that this assumption is not justifiable; the distribution is clearly nonnormal. In this case, the engineer will have to postulate different performance metrics and statistical methods for drawing the conclusions of the task. Although descriptions of the parametric and nonparametric statistical methods available for performance data analysis are outside the scope of this book, Figure 13.1 lists some of them. Complete descriptions of them can be found in any comprehensive text on descriptive statistics.

Before any performance measures are actually computed, it is important to take a good long look at the reduced results to get a feel for the behavior of the data. It may even be necessary to apply some preliminary statistical tests as QD pilots to determine whether the postulated metrics are appropriate, and if not,

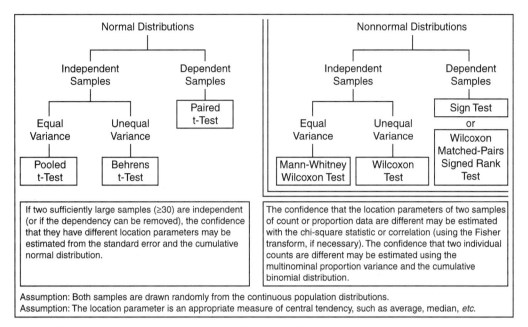

Figure 13.1 Summary of common statistical methods for comparing location parameter values

what metrics might be more appropriate. Histograms and Pareto plots can reveal unusual distributions. Two-dimensional scatter plots of the factors *vs.* the results can reveal subtle functional relationships that will be ignored and obscured by performance metrics that cause heavy reductions of the data, such as the average and standard deviation.

Searching for trends and unusual behavior in the results not only can assist in selecting appropriate performance metrics, but may also detect possible pathologies in the experiments that acquired the results. Consider the results obtained by the Group 1 students who were measuring the length of Room A in the task described in Chapter 8 (Limits of Knowledge) in Section 8.2.2 (See Table 8.4). You will recall that this sample of measurements appeared to be normally distributed, as shown in Figure 8.2, and so we felt comfortable in computing confidence intervals based on the normal areas. However, before any performance metrics are computed, a quick look at the results is in order, as a precautionary measure to make sure there are no surprises lying in the weeds out there, such as an obvious bias in the experiment method or some strange outliers. Figure 13.2 is a plot of these room-length measurements in the same order in which they were acquired (not a Pareto), as listed top-to-bottom and left-to-right in Figure 8.4.

Once again, the students who measured the room have a surprise buried in their results! For some strange reason, the measurement values exhibit a clear trend (slope = +0.434 millimeters per measurement) as a function of measurement order. Unless the room was getting larger during this task, the students somehow managed to introduce a subtle, but systematic bias into their methodology. As

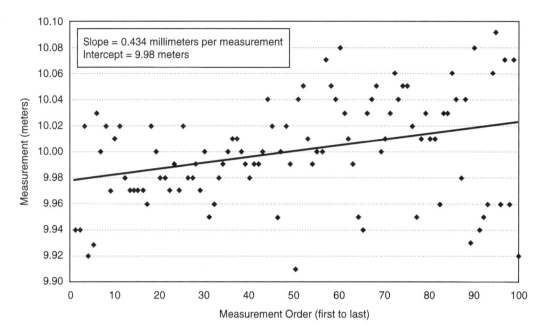

Figure 13.2 Measurements by Group 1 for Room A in the order of their acquisition

discussed in Section 8.2.2, the bimodal distribution exhibited by the histogram of the measurements made by the students in Group 2 was easy to spot (See Figure 8.3), but the clear systematic bias in the results of the students in Group 1 was undetectable in both the histogram and Pareto plot of the measurements (See Figures 8.2 and 8.6).

A statistical analysis based on the variance of such biased measurements would be almost meaningless, because a good deal of the variance is a direct result of the spurious slope, which will significantly distort the accuracy of any conclusions. Moreover, it would be very difficult to negotiate this bias without rerunning the experiment after having identified and corrected the cause of the bias (assuming the cause can be found, which is highly problematic).

> **TIP 27**
> Before computing performance, conduct a QD pilot to ana-
> lyze the raw or reduced results to detect any unusual arti-
> facts or unexpected trends stemming from biases in the
> task methods and to confirm the appropriateness of the
> postulated performance metrics.

Selecting and computing the performance metrics is a process that is heavily task-dependent. Nonetheless, to provide a meaningful illustration of this process, let us compute the performance for the case-study project *Measure Character-Recognition Performance*, using the reduced data files carefully stored in the complete hierarchy of result folders (see Table 12.15). The performance computations follow the top-down organization of the goals and hypotheses listed in Table 11.3 and the goal tree shown in Figure 11.1 (with the exception of the OP Pilot in Goal G1.1).

Goal G1.2: To decide whether a subject is qualified for participation in a test trial. Each subject participated in a set of no more than three practice sessions with the graphical user interface (GUI) of the software that was used to measure character-recognition performance. At the end of each practice session, an external control trial was conducted to measure the subject's character-selection times. If these times revealed no significant residual learning curve, the subject was assumed to be qualified to participate in the experiment trials for Goal G1.3.

Hypothesis H1.2.1: The subject has had sufficient practice with the interface. To make the decision whether a subject was qualified to proceed with the internal control and test trials, the project team asserted research Hypothesis H1.2.1. It will be recalled from Section 11.2 (Set Goals) that a research hypothesis is a declarative sentence asserting a desired, expected, or possible conclusion of the application of the solution (or part of the solution) to achieve the task objective (or part of the objective). In all likelihood, the project team did not have a particular desire or expectation that a subject had had sufficient practice. They had no vested interest in a particular outcome; they merely wanted to *know*, one way or the other, so that they could take the appropriate action. Thus, it is probably fair to assume that Hypothesis H1.2.1 was simply a *possible* conclusion.

In fact, there were only two hypotheses that the project team could have reasonably asserted:

1) The subject requires more practice (absolute correlation > 0); or
2) The subject has had sufficient practice (absolute correlation = 0).

The decision criterion for the first hypothesis is open-ended: it includes an *infinite number* of correlations whose absolute values are greater than zero. Thus, the question arises: when this hypothesis is tested, what absolute correlation value should be deemed large enough to accept the hypothesis that the subject requires more practice? Is 0.1 large enough? Must it be at least 0.5? In fact, there is no meaningful objective basis for establishing this threshold, *i.e.,* a maximum value, beyond which the correlation is just too large.

The decision criterion for the second hypothesis, on the other hand, refers to a single, unique point in the decision space: a correlation of exactly zero. There is no confusion or ambiguity in this value. Zero is zero. Although investigators may assert a *research* hypothesis that expresses a specific desired or expected conclusion referring to an open-ended region in the decision space, it is *always possible and necessary* to restate this hypothesis so that is refers to a single, unambiguous value in the decision space. This is called the **null hypothesis**, usually designated H_0. Table 13.1 lists some sample research hypotheses and their corresponding null hypotheses.

Most of the null hypotheses in Table 13.1 are expressed in two different versions. Both are appropriate, and in fact, there are additional ways in which null hypotheses could be expressed. Whatever wording is used, the statement should be brief, accurate, complete, and clear.

There is no way we can test the *research* hypotheses given in Table 13.1, because we don't know the meaning of such phrases as "better," "more intelligent," or "effective." On the other hand, it is quite straightforward (at least in principle) to test the corresponding *null* hypotheses, which always refer to a single, unique value. Suppose, for example, that we have obtained a sample of measurements of the fuel efficiency for the new engine mentioned in the first example. These measurements will have a distribution with an average and an associated uncertainty (standard deviation). Regardless of the value of the average fuel efficiency, we cannot conclude that it is statistically greater than 40 mpg, because we do not know how much greater it must be to support this claim. However, we *can* compute the confidence of rejecting the null hypothesis that the average measurement is *exactly* 40 mpg. This is called the **rejection confidence**.

To illustrate the application of the rejection confidence, let us suppose that the distribution of the fuel efficiency measurements is approximately normal and we have more than 30 independent measurements in the sample. We can compute the rejection confidence as the normal area corresponding to the standard error (the difference between the measured error and 40 divided by the standard deviation of the measurements) using a one-tailed test. If the average is greater than 40 mpg, the confidence that the null hypothesis may be rejected will be greater than 50%. If the average is less than 40 mpg, the rejection confidence will be less than 50%. If the average turns out to be exactly 40 mpg (at the allowed precision limit), this means that there is a 50/50 chance that we can reject the null hypothesis.

If the research hypothesis is not a null hypothesis, then it is called an **alternative hypothesis**, usually designated H_A. In every example in Table 13.1 the research hypothesis is an alternative hypothesis, not the null hypothesis. However, quite often the research hypothesis is intentionally stated as the null hypothesis, either because the null hypothesis just happens to be the desired or expected conclusion, or simply because there is no desired or expected conclusion, and so the research hypothesis might as well be the null hypothesis. This is precisely the case for Hypothesis H1.2.1 of Task 1.4.2.2 for Project *Measure Character-Recognition Performance*: "The subject has had sufficient practice with the interface." Note that if Hypothesis H1.2.1 had been worded the other way around, "The subject has *not* had sufficient practice with the interface," it would have been necessary to assert and test the null hypothesis to make the decision.

Actually testing the null hypothesis and making the decision whether a subject requires additional practice takes place in the next step of the Validation Phase, Draw Conclusions, which is discussed in Section 13.2.

Goal G1.2.1.1: To measure character-selection times without noise (external control). To acquire the results necessary to test Hypothesis H1.2.1, the external control experiments for Task 1.4.2.2 were designed and conducted (See Figure 9.5). At the end of each practice session, each subject was presented

Table 13.1 Sample research hypotheses and their corresponding null hypotheses

Research Hypothesis, H_A	Null Hypothesis, H_0
The average fuel efficiency of the new engine exceeds 40 mpg.	The average fuel efficiency of the new engine is 40 mpg.
Most people think our widget is better than their widget.	There is no difference between people's opinions about our widget and their widget. Any differences in what people think about our widget and their widget may be attributed to chance.
Women are more intelligent than men.	There is no difference in intelligence between women and men. Any difference between the intelligence of women and men may be attributed to chance.
Drug XYZ is an effective therapy for disease ABC.	Drug XYZ is ineffective as a therapy for disease ABC. Any observed improvement in disease ABC when treated with drug XYZ may be attributed to chance.
The average speed of traffic on Highway A is higher than the average speed of traffic on Highway B.	The average speeds of traffic on Highways A and B are the same. Any difference in the average speeds of traffic on Highways A and B may be attributed to chance.
The measured heat loss exceeds the manufacturer's specification.	The measured heat loss meets the manufacturer's specification. Any difference between the measured heat loss and the manufacturer's specification may be attributed to chance.
Benefit varies inversely with noise level.	There is no relationship between recognition benefit and noise level. Benefit is independent of noise level.

with a sequence of 30 characters displayed without noise and was asked to click the corresponding button as soon as possible after recognizing each character. The sequence of these selections and the corresponding selection times were recorded by the supervisor computer. At the end of the trial, the computer immediately computed the correlation between the sequence of selection times and their order in the sequence, which was reported to the technician. If there was no significant correlation, it was assumed that the subject had had sufficient practice with the interface and could proceed to the test and internal control trials (Goal 1.3). If there was a significant correlation, however, this was assumed to indicate that the subject was still learning to deal with the interface, and more practice was necessary.

The performance metric postulated for Goal G1.2.1.1 was the Pearson product-moment coefficient of correlation, which estimated the strength of a linear relationship between the 30 selection times of the subject and the order of the sequence of these selections. The correlation coefficients for all eight subjects are reported in Table 13.2, along with the corresponding rejection confidences, which are computed directly from the size of the sample (36) and the value of the correlation coefficient.

None of the confidences in Table 13.2 was of interest *per se* to the project team. They were necessary only to validate Hypothesis H1.2.1 and thereby to decide whether a subject had had sufficient practice. Table 13.2 was included in the project documentation merely for completeness and is presented here only for the interest of the reader and to illustrate the underlying methodology.

Goal G1.3: To measure selection times and accuracies (internal control and test). This goal uses the reduced results of the experiment trials for Task 1.4.2.4, defined by randomized block design *B* in Table 12.11, to compute the performance metrics postulated in Table 11.3. These measures of performance were necessary to achieve the primary research objective of the project, *i.e.,* to estimate the peak recognition performance of the cohort of trained license-plate inspectors. As a first step in the process of computing these performance values, Figure 13.3 plots the measured selection times (reduced results) over all characters acquired during the internal control and test trials as a function of the two research factors: noise level and typeface.

Note that the data plotted in Figure 13.3 includes *all* the selection times, whether the subject selected the correct character or not. As expected, the selection times increase quite uniformly as the noise level increases, and there seems to be some evidence that the test selection times for the *serif* characters (dashed line) are a bit lower than the selection times for the *sans serif* characters (thick solid line). Moreover, the selection times for the internal control trials, during which no characters were displayed and the subject had to guess without any relevant information (thin solid line), are much higher than the selection times for the test trials. This seems to suggest that the subjects were puzzled by the apparent absence of a character (which was true, of course), and spent a good deal more time trying to decipher the identity of the (nonexistent) character.

Such first impressions are not valid conclusions, but simply informal observations. The conclusions for this project (like all formal projects) had to be based

Table 13.2 External control performance for Project *Measure Character-Recognition Performance*

Subject ID	Typeface	Number of Practice Sessions	Correlation Coefficient	Confidence (%) Correlation ≠ 0
1	sans serif	3	−0.142	56
2	sans serif	3	−0.300	89
3	sans serif	2	+0.037	16
4	sans serif	1	−0.004	2
5	serif	2	+0.032	14
6	serif	1	+0.052	22
7	serif	3	−0.065	27
8	serif	3	−0.112	45

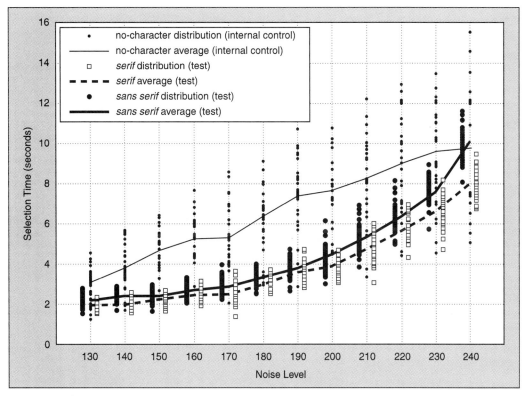

Figure 13.3 Character selection times for Project *Measure Character-Recognition Performance*

on the set of performance metrics postulated by the project team (although these metrics certainly could have been changed if the results showed that they made no sense or if other performance metrics made more sense). The values computed by the postulated performance metrics, which are reported in Table 13.3, became the basis for testing the null hypotheses that were derived from the research hypotheses asserted for this goal (H1.3.1 through H 1.3.6). Table 13.4 lists these research hypotheses and their corresponding null hypotheses.

Table 13.3 Performance values for Project *Measure Character-Recognition Performance*

Performance Metric		NOISE LEVEL											
		130	140	150	160	170	180	190	200	210	220	230	240
Fraction of characters	*sans serif* benefit	97	94	86	78	67	56	44	33	25	17	11	8
identified correctly (%)	*serif* benefit	97	97	92	86	81	72	61	47	28	14	8	8
Type I error	*sans serif* cost	18	11	2	1	7	3	4	5	11	13	21	54
probability (%)	*serif* cost	11	6	1	1	5	1	3	3	7	8	12	29
Product of benefit and	*sans serif* payoff	80	84	85	77	62	54	43	32	22	14	9	4
(100 − cost) (%)	*serif* payoff	86	92	91	85	77	71	59	46	26	13	7	6

Table 13.4 Research and null hypotheses for Project *Measure Character-Recognition Performance*

	Research Hypothesis, H_R	Null Hypothesis, H_0
H1.3.1	Benefit varies inversely with noise level.	Benefit is independent of noise level.
H1.3.2	Cost varies directly with noise level.	Cost is independent of noise level.
H1.3.3	Payoff varies inversely with noise level.	Payoff is independent of noise level.
H1.3.4	Benefit for *serif* characters is higher than for *sans serif* characters.	Any difference in benefit for *serif* and *sans serif* characters may be attributed to chance.
H1.3.5	Cost for *serif* characters is lower than for *sans serif* characters.	Any difference in cost for *serif* and *sans serif* characters may be attributed to chance.
H1.3.6	Payoff for *serif* characters is higher than for *sans serif* characters.	Any difference in payoff for *serif* and *sans serif* characters may be attributed to chance.

The benefit metric (See Table 11.3) was postulated as the percent of the total number of test treatments during which a subject selected the correct character, expressed as a percentage [0, 100%]. This straightforward metric is easily justified (which, as a postulate, it *must* always be).

The reasoning and justification for the cost metric is carefully delineated in the project documentation:

> The results of the internal-control treatments were reduced to the average of the selection times when the subject was guessing. Because the cost metric was intended to reflect the effect of guessing on performance, it was postulated to be the probability of mistakenly claiming that the selections were not guesses, *i.e.,* the Type I error probability. For example, if the average test selection time was much less than the average internal-control selection time (very high rejection confidence), the probability of mistakenly claiming that the subject was *not* guessing is very low, implying a *low cost*. On the other hand, if the average test selection time was almost the same as the average internal-control selection time (very low rejection confidence), the probability of mistakenly claiming that the subject was not guessing is *very high,* implying a *high cost*.

The cost metric, then, was postulated as the probability of committing a Type I error[12] when rejecting the null hypothesis: there is no difference between the average test selection time and the average internal-control selection time. For consistency with the other metrics, the value of the cost was represented as a percentage [0, 100%].

---

[12]A Type I error occurs when you mistakenly reject the null hypothesis, *i.e.,* when you decide the null hypothesis is false, but in fact it is true. This is also known as a false negative.

By Definition 7.14, **payoff** is an arithmetic combination of one or more benefits and costs that represents the impact of their interaction by increasing in value with increasing benefit and decreasing in value with increasing cost. This yielded (at least) two possibilities for combining the benefit and cost performance values in the postulate of the payoff metric for the project *Measure Character-Recognition Performance*. On the one hand, the payoff metric could have been the ratio of the benefit to the cost, so that as the benefit increased for constant cost, the payoff increased, and as the cost increased for constant benefit, the payoff decreased. This choice, however, would have produced a payoff whose values were bounded on the low end by zero, which is fine, but were unbounded on the high end, which is not so fine. In fact, if the cost had ever turned out to be zero (not likely, but possible), the payoff would have been undefined because of an attempt to divide the benefit by zero. The project team wanted the ranges of all the performance metrics to be in the interval [0, 100%] to allow meaningful comparisons with the results and conclusions of future R&D projects. Therefore, a more well-behaved and bounded function was selected by the team for the pay-off metric: the *product* of the benefit and (100 − cost) divided by 100. Thus, if the cost was zero, the payoff was equal to the benefit, and if the cost was 100%, the benefit was equal to zero. Note that the expression (100 − cost) is nothing more than the confidence of rejecting the null hypothesis that there is no difference between the average test selection time and the average internal–control selection time. One may logically interpret the payoff metric as the confidence that the benefit was high AND the cost was low.

Hypotheses H1.3.1 through H1.3.3: Performance as a function of noise level. To decide what statistical measures will be necessary to test Hypotheses 1.3.1 through 1.3.6, the project team first plotted the values in Table 13.3 to observe their general behavior as a function of the relevant factors: noise level and typeface. This plot is presented in Figure 13.4.

On inspecting the behavior of the performance curves in Figure 13.4, it was clear to the project team that no sophisticated statistical tests were necessary to establish the validity of the first three of the six hypotheses under Goal G1.3, Hypothesis H1.3.1, H1.3.2, and H1.3.3, which assert how the three metrics (benefit, cost, and payoff) vary with noise level. The relationships were obvious, and thus the required conclusions for these three hypotheses could be based confidently on mere observations of the performance curves in Figure 13.4. These conclusions are drawn in the next section of the Validation Phase, Draw Conclusions.

Hypotheses H1.3.4 through H1.3.6: Performance as a function of type-face. By contrast, Figure 13.4 reveals that mere observations of the performance curves would *not* serve to validate the last three hypotheses under Goal G1.3, H1.3.4, H1.3.5, and H1.3.6, which assert that performance with the *serif* typeface exceeds performance with the *sans serif* typeface. Although the values along the benefit and payoff curves for the *serif* typeface (solid line with squares and white columns) seem to be consistently higher than those for the *sans serif* typeface (solid line with circles and black columns), the uncertainty of the measurements used to compute these performance values could well render these differences statistically

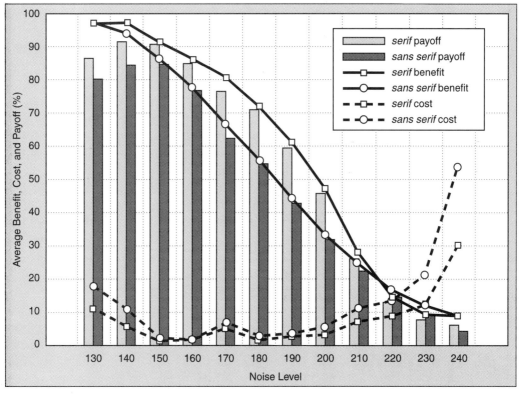

Figure 13.4 Average performance for Project *Measure Character-Recognition Performance*

insignificant. Likewise, the apparent lower cost for the *serif* typeface (dotted line with squares) compared with the *sans serif* typeface (dotted line with circles) could easily be lost in statistical uncertainty of the measurement samples.

Because each of the metrics (benefit, cost, and payoff) is expressed as a confidence or probability derived from a sufficiently large sample of measurements, the project team used the standard difference of two proportions as the statistical basis for testing the three null hypotheses given in Table 13.4:

$H_0$1.3.4: Any difference in benefit for *serif* and *sans serif* characters may be attributed to chance.

$H_0$1.3.5: Any difference in cost for *serif* and *sans serif* characters may be attributed to chance.

$H_0$1.3.6: Any difference in payoff for *serif* and *sans serif* characters may be attributed to chance.

The only assumption needed to apply this statistic to measure the rejection confidence was that the samples are independent and random, and the project team had gone to considerable trouble to ensure these requirements were satisfied by means of careful planning and rigorous control of the experiments. Table 13.5 and Figure 13.5 present the rejection confidences for each of these null hypotheses as a function of noise level.

Table 13.5 Rejection confidences for null hypotheses H1.3.4, H1.3.5, and H1.3.6

Null Hypothesis	NOISE LEVEL											
	130	140	150	160	170	180	190	200	210	220	230	240
H1.3.4: *serif* benefit = *sans serif* benefit	50	72	77	82	**91**	**93**	**92**	89	61	37	35	50
H1.3.5 *serif* cost = *sans serif* cost	78	77	58	57	63	61	55	63	62	60	63	73
H1.3.6 *serif* payoff = *sans serif* payoff	76	82	77	80	88	88	84	78	57	47	47	55

rejection confidences ≥ 90% are shown in **boldface**

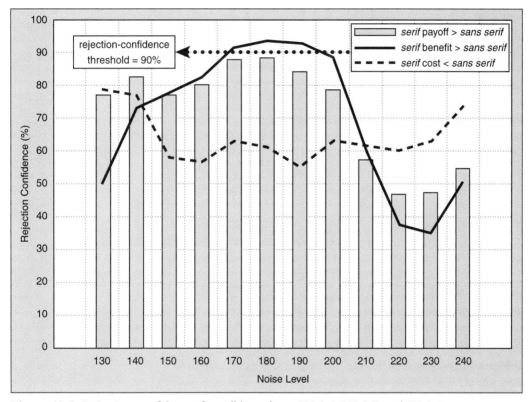

Figure 13.5 Rejection confidences for null hypotheses H1.3.4, H1.3.5, and H1.3.6

The performance values in Tables 13.2, 13.3, and 13.5, supported by Figures 13.3, 13.4, and 13.5, constitute a sufficient basis for drawing the conclusions for the project, as required by its primary research objective (See Table 10.5).

13.2 Draw Conclusions

Once the results for a task have been assembled and the required measures of performance have been computed and summarized in an appropriate set of tables, charts, and graphs, the conclusions for the task may be drawn. As defined in

Section 6.4, it is these conclusions that directly determine the extent to which the objective of the task has been achieved. Two kinds of conclusions may be drawn: **formal conclusions** and **informal observations**.

Formal conclusions are those conclusions that are required to achieve the objective of the task and that are based on demonstrably rigorous methods for acquiring and reducing the results of the task. As required by the task objective, each conclusion may be a decision or an estimate, derived from tests of a set of explicit hypotheses or from the results of a set of explicit goals, respectively. If the task objectives allow, a conclusion drawn from estimates can be inherently soft, suggesting a range of values, rather than a single value, and may be accompanied by qualifiers, such as statistical moments, uncertainty estimates, and/or confidence estimates: "There is a 95% confidence that the temperature of the room lies between 62 and 68°F." Decisions also need not be hard, if not required by the objective. For example, if a research hypothesis for a task asserts that "Customers consider our widget to be better than their widget," the most accurate possible conclusion may be limited to a numeric equivalent of the phrase "it depends," such as "If you ask men, then 89% of the customers agree that our widget is better, but if you ask women, then 22% of the customers agree that our widget is better." For the case-study project *Measure Character-Recognition Performance*, however, the decision about the research hypothesis for Goal 1.2 had to be a hard decision; either the subject was qualified to participate in the test trials, or not. Clearly, it was not acceptable to validate the hypothesis for this goal simply by concluding that there was an 87% confidence that the subject required more practice.

Take a moment to recall (or review) the three criteria for the provisional acceptance of knowledge set forth at the end of Chapter 5 (An Epistemological Journey): reproducibility; completeness; and objectivity. If a great deal rides on the acceptance of the validity of the conclusions of the task, it may be necessary to ask the project team to recuse itself and conduct this critical step of the Scientific Method as a double-blinded task conducted by an independent agent, *i.e.,* someone who is neither a member of the project team nor aware of the task methods. If the measures of performance for a task are made available to this person, along with the explicit goals and *null* hypotheses, it should be possible for this person to draw all the required conclusions of the task without any additional information. If there is a *very high* risk associated with a faulty conclusion, it may even be advisable to ask *several* appropriately qualified persons to undertake this task independently. Although expensive and time-consuming, making this effort will go a long way toward ensuring that the final step of the Scientific Method, Peer Review, will result in the certification of the conclusions as reproducible, complete, and objective. This is exactly the espoused purpose of the laws of the United States that afford all criminal defendants the right to a public trial by a jury of their peers, who are supposed to be carefully selected for their lack of bias, who may be sequestered during the trial, and who are permitted access to only certain relevant information.

Even though it has been said that the methods for drawing conclusions must be demonstrably rigorous, this does not always mean that statistical methods must be used. When the associated risk is not very high, a conclusion may be drawn based on obvious trends in the performance values. For example, suppose an elementary school teacher is trying to decide which flavor of ice cream is preferred by the school children in preparation for a party. To ascertain this preference, the teacher asks a sample of 20 children, and 19 indicate a preference of chocolate over vanilla. It is probably unnecessary for the teacher to apply the standard difference of two proportions to these results to compute the confidence with which she can reject the null hypothesis that the children have no strong preference. Chocolate seems a sure bet.

However, a word of warning must accompany this seductive invitation to abandon rigor. It is often impossible to decide with reasonable certainty whether the estimates of the location parameters of two samples are significantly different without taking into account the uncertainty of these estimates, as discussed in Chapter 8 (Limits of Knowledge). Consider, for example, the statistician who must decide which highway has the higher traffic speed in the running example task *Identify Faster Traffic*. If the average traffic speed on Highway A is 80 mph and the average traffic speed on Highway B is 60 mph, it may seem obvious that Highway A has a significantly higher average traffic speed. After all, a 20 mph speed difference seems considerable when you imagine two cars passing each other with this speed difference. But this perspective does not tell the whole story. Remember that the statistician is trying to characterize the entire *population* of vehicles on both highways from *samples* of vehicles, and it is entirely possible the difference of 20 mph between the average speeds of the two samples would not hold if the entire population were measured. Moreover, the averages may include cars that are traveling at a snail's pace or seem to be auditioning for the Indianapolis 500. Clearly, just knowing the location parameters (average speeds) is not enough. At the very least, the standard deviations of the samples must be taken into account. Better yet, the actual distributions of the speeds should be inspected to make sure the difference in the average speeds is an indicative and appropriate metric and is not being swallowed up by outliers.

> **TIP 28**
> Before abandoning formal rigor in the face of what seem to be obvious conclusions, check the distributions of the underlying performance values. Lacking estimates of the uncertainty of these values, when asked if their averages (or other location parameters) are significantly different, your answer must be "I don't know. I do not have enough information."

Once all the required conclusions have been drawn, the project team is free (in fact, encouraged) to draw additional conclusions. Some may be based on rigorous and formal methods, but are only identified after inspecting the results and are often somewhat tangential to the stated objective. Others, however, may simply be speculations based on insufficient evidence. Even though speculative knowledge was explicitly *proscribed* for the governing propositions of an R&D task, it is entirely appropriate to report informal observations in the conclusions of a task. Such knowledge must never be used to support other formal conclusions, of course, and must always be clearly identified as speculative, qualified by such phrases as *"It may be speculated that. . . . ,"* or *". . . can be tentatively identified as . . . ,"* or *". . . may be suggested, subject to formal experiments."*

To illustrate how conclusions are drawn from the performance values, we will draw the conclusions for the case-study project *Measure Character-Recognition Performance*, starting with the *low-level* hypotheses and goals in the goal tree (See Figure 11.1), and then climbing up to the next level of the goal tree to draw these conclusions at this level, and so forth, until we have drawn the conclusion for the primary project goal G1.

Goal G1.2.1.1: Measure character-selection times without noise (external control). No conclusions were intended to be drawn directly from the information gathered by this goal. As pointed out previously, the results of the external control task (Task 1.4.2.2) that accomplished this goal served no purpose other than to provide the knowledge necessary to test the parent hypothesis H1.2.1 that the subject had had sufficient practice with the interface.

Hypothesis H1.2.1: The subject has had sufficient practice with the interface. The confidence with which the project team could reject Hypothesis H1.2.1 for each subject is reported in the last column of Table 13.2. However, just knowing the rejection confidence is not enough. In addition, a **rejection-confidence threshold** was needed to specify the maximum rejection confidence above which the null hypothesis would be rejected, thus *disqualifying* the subject from participation in an internal control and test trial. Such a rejection-confidence threshold may also be expressed as $1 - \alpha$, where α is the maximum allowable probability of a Type I error, also known as the **critical level**.

Consider, for example, the null hypothesis asserted for the previous example of the new engine in Table 13.1: "The average fuel efficiency of the new engine is 40 mpg." Suppose that the measurements from a series of experiments reveal that the null hypothesis can be rejected using a one-tailed test with a 92% confidence. This means that 92% of the time the average fuel efficiency will be greater than or equal to 40 mpg, and 8% of the time it will be less than 40 mpg. If the critical level $\alpha = 10\%$, then the rejection-confidence threshold $(1 - \alpha)$ is 90%, and we can reject the null hypothesis, because the rejection confidence of 92% is larger than the rejection-confidence threshold of 90%. On the other hand, if the rejection confidence is only 88%, then we cannot *quite* reject the null hypothesis and must acknowledge that the average fuel efficiency is not significantly different from 40 mpg *for the specified rejection-confidence threshold of 90%*. If the rejection-confidence threshold is lowered to 85%, however, then both cases would allow the null hypothesis to be rejected, concluding that the fuel efficiency is significantly greater than 40 mpg. Of course, by lowering the rejection-confidence threshold from 90% to 85%, we have increased the probability of committing a Type I error from 10% to 15%. The risk associated with this trade-off must always be taken into account.

Picking an appropriate rejection-confidence threshold is very problem dependent, depending heavily on the amount of risk associated with making a wrong decision and the amount of benefit to be accrued from making a correct decision. However, a rule of thumb may be offered: all things being equal (which, of course, they seldom are), most statisticians agree that a rejection-confidence threshold of at least 90% is required to reject a null hypothesis.

If we decide to reject the null hypothesis, *i.e.,* the rejection confidence exceeds the rejection-confidence threshold $(1 - \alpha)$, what conclusions can we draw, if any, about the *alternative* hypothesis (which might also be the same assertion as the research hypothesis)? If the null hypothesis is rejected, may the alternative hypothesis be accepted? These are murky waters, sometimes patrolled by sharks. If the cost of committing a Type II error[13] is high, then the most cautious decision is to "reserve judgment," which is a fancy way of refusing to comment, one way or the other. If the cost of committing a Type II error is low, then an appropriate decision might be to "suggest the validity of the alternative hypothesis." Either way, drawing such conclusions in the absence of any knowledge about the probability of committing a Type II error is often very risky and must be done with great care. Unfortunately, it is usually very difficult to estimate the Type II error probability.

Consider, for example, the radiologist who must examine a series of mammograms to decide if the patient has breast cancer. Note that the null hypothesis asserts that the patient is completely free of cancer. A Type I error occurs when the doctor rejects the null hypothesis and decides to have further tests done, but *in fact* the patient does *not* have cancer. A Type II error occurs when the doctor does not reject the null hypothesis and decides there is no cancer, but *in fact,* the patient has breast cancer. The commission of a Type I error is, at worst, annoying and expensive: more tests are performed to no purpose, because the patient is *in fact* cancer free. On the other hand, the commission of a Type II error is profoundly tragic: the greatly relieved patient is sent home and dies of breast cancer several months later. Clearly, the most humane strategy is to set the rejection-confidence threshold very low, triggering additional tests even when there is only very marginal pathological evidence of disease in the mammogram. However, this is also the most expensive strategy, which insurance companies often resist.

To draw the conclusion for Hypothesis H1.2.1 in the project *Measure Character-Recognition Performance,* the project team had to establish a rejection-confidence threshold for testing the null hypothesis: the subject has had sufficient

---

[13]A Type II error occurs when you mistakenly accept the null hypothesis, *i.e.,* when you decide the null hypothesis is true, but in fact, it is false. This is also known as a false positive.

practice, *i.e.,* the correlation equals zero. (Note that this also happened to be the research hypothesis.) A low rejection-confidence threshold, say 90%, would impose a fairly strict threshold, implying that it was very likely that subjects would have to repeat the practice session several times. A rather high rejection-confidence threshold of 99%, on the other hand, would impose a fairly lenient threshold, implying that it was more likely that subjects would be able to proceed with the test and internal control trials without exceeding the three-practice-session limit. Because the project had such a small cohort and could ill afford to have any subject's results biased by the vestige of a learning curve, the team felt compelled to impose a fairly strict rejection-confidence threshold of 90%.

Goal G1.2: Decide whether each subject is qualified to participate in a test trial. To accomplish Goal G1.2, the rejection-confidence threshold of 90% that had been selected by the project team was programmed into the software of the supervisor computer. At the end of each external control trial, the computer calculated the required correlation coefficient for the subject and reported one of three possible decisions on the display of the supervisor computer in the control booth: *BEGIN TEST TRIAL, REPEAT PRACTICE SESSION,* or *DISMISS SUBJECT.* This decision was noted and executed by the technician, as described in the protocol in Table 12.14.

As it turned out, all eight subjects were qualified to proceed to the test and internal control trials after no more than three practice sessions, as shown in Table 13.2. Note, however, that the subject with ID number 2 barely squeaked by with a rejection confidence of 89%, and it took all three practice sessions for the subject to accomplish this confidence and be qualified.

We are now ready to proceed to Goal 1.3, starting with the lowest level hypotheses, as shown in Table 11.3 and Figure 11.1. Note that no rejection-confidence threshold is given as a factor for any of the six hypotheses. This implies that the project team wanted a soft decision that simply specified the confidences with which the corresponding null hypotheses could be rejected, without subjecting them to a rejection threshold.

Hypotheses H1.3.1 through H1.3.3: Performance as a function of noise level. The null hypotheses for these three hypotheses are given in Table 13.4, which assert that benefit, cost, and payoff performance are independent of the noise level, regardless of typeface. This turned out to be one of those rare cases when the validity of null hypotheses can be accurately estimated with sufficient precision simply by inspecting the performance values. One glance at Figure 13.4, leads to the unambiguous and clear conclusion that the null hypotheses for benefit and payoff can be rejected with extremely high rejection confidences (much higher than 99%). No formal statistical tests are necessary. Thus, accepting the research (alternative) hypotheses for these performance measures incurs little risk: noise level has an obvious and unambiguous effect on benefit and payoff, exactly as described by the research hypotheses.

There is, however, one minor qualification. Note that at both extremes of the range of noise level, between 130 and 140 and between 230 and 240, there is little or no change in benefit with noise level for both typefaces. The project regarded this behavior as testimony to their careful selection of the dynamic range of the noise levels selected for the project in Tasks 1.3.1 and 1.3.2 (Goal G1.1). They speculated that benefit was constant and uniformly good (94 to 97%) at the low end of the range, because there was practically no room for further improvement; and that recognition performance was constant and uniformly poor (8 to 11%) at the high end of the range, because there was practically no room for further degradation. Thus, the dynamic range was neither wider nor narrower than it needed to be.

The conclusion for the null hypothesis about the cost is, however, less clear. The cost, it will be remembered, was postulated to be (approximately) the probability that the subject was guessing. However, as shown in Figure 13.4, at the low end of the noise-level range, the cost for both typefaces *decreased* with increasing noise level, which seems counterintuitive. Why is the cost going down, as the noise goes up? This counterintuitive behavior, the project team speculated, was probably an artifact caused by the fact that the total response time when there is little noise (the character is easy to recognize) is *dominated* by the time required simply to *find* the right button and press it. When asked a simple question, it may take you a moment to get your mouth in gear and blurt out the answer, even though the correct answer flashes into your mind almost instantly. This delay in articulation may mistakenly be misinterpreted as hesitation, *i.e.,* that you are not quite sure and are just guessing. The cost metric postulated for this project ignored that mental articulation delay between recognition and action (pressing the intended button).

After the initial decrease, the cost remains fairly constant from noise level 150 to noise level 190. In this range, the time delay caused by a delay in articulation remained a significant component of the overall delay, but was comparable to the time delay measured by the internal control treatments. Above a noise level of 190, the subject apparently had to spend more and more time studying the display to identify the character. As a result, the articulation delay quickly became an insignificant fraction of the overall delay, and the cost increased monotonically with noise level, as asserted by the research hypothesis. At the highest noise level (240) the cost was quite high, 30% for the *serif* typeface and over 50% for the *sans serif* typeface. This, the project team further speculated, was strong evidence that the subject was truly guessing as a result of uncertainty, *i.e.,* at noise levels greater than 190, the cost metric was measuring what it had been intended to measure.

Note that these conclusions about benefit, payoff, and cost as a function of noise level are *soft* conclusions. The project team was content with such conclusions, because they provided an approximate model with which they could compare the performance of the automatic recognition engine they hoped to develop as a product for their company. For example, given values for the benefit of the automatic engine as a function of noise level, they could assert and test the specific research hypothesis that their automatic system performed as well as or better than trained experts, and validate this hypothesis statistically.

Hypotheses H1.3.4 through H1.3.6: Performance as a function of typeface. The null hypotheses for the last three research hypotheses assert that any difference in performance between the two typefaces may be attributed to chance. The rejection confidences for benefit, cost, and payoff as a function of noise level are reported in Table 13.5 and are plotted in Figure 13.5, which are the soft conclusions. In addition, the project team set a rejection-confidence threshold of 90% and concluded that any advantage of the *serif* typeface over the *sans serif* typeface was significant only for benefit and only for noise levels from 170 through 190. Elsewhere, typeface yielded no significant difference in performance.

Goal 1.3: Measure selection times and accuracies (internal control and test). The conclusions drawn for the six hypotheses, accompanied by the performance reported in Tables 13.4, Table 13.5, Figure 13.4, and Figure 13.5, are the major conclusions for Goal 1.3. Quite understandably, however, the project team could not resist analyzing the errors made by the subjects when they were selecting the characters. Which characters were incorrectly identified most frequently? Which pairs of characters were confused most frequently? Although they did not have a sufficient sample of each of the incorrectly selected characters to draw any formal conclusions based on a rigorous statistical analysis, they did offer some informal observations, which are summarized in Table 13.6 and Figure 13.6.

As it turns out, Figure 13.6 and Table 13.6, which were, of course, included in the project documentation, caught the interest of some high-ranking politicians in the government of the country where the project was conducted. They arranged for a follow-on research project whose objective was to recommend characters that should *not* be used for license-plate numbers to reduce the error rates of both the human inspectors and the new automatic license-plate inspection systems that were being developed. One thing leads to another.

Well, that about wrapped it up. The conclusions just described for the case-study project *Measure Character-Recognition Performance* provided all the information required to achieve its primary research objective. Nothing was left but the shouting. And that is exactly what the last two steps of the Validation Phase of the Scientific Method are all about: documentation and review.

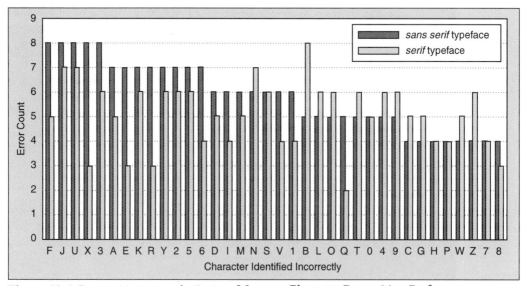

Figure 13.6 Recognition errors for Project *Measure Character-Recognition Performance*

Table 13.6 Character confusions for Project *Measure Character-Recognition Performance*

Selected Character (columns); Actual Character (rows)

Actual \ Selected	A	B	C	D	E	F	G	H	I	J	K	L	M	N	O	P	Q	R	S	T	U	V	W	X	Y	Z	0	1	2	3	4	5	6	7	8	9
A								1										1													8					
B				3																										3			5			
C															5												8									
D															3												9									
E		4				5																	4													
F																	3	3		4																
G															2												1						9			
H										3				5								3														
I										3		3															7									
J											2										5						1									
K																		5				4														
L										3																	7									
M														2								8														
N								5					5																							
O			3														4										7									
P				5														8																		
Q															5												6									
R				3								3					4																			
S																																5		6		
T										6																	4									
U															3							2	4													
V																					7		4													
W														7								5														
X						4					8																									
Y																					6	2														
Z																				2									11							
0			2												9																					
1										7																							3			
2																							7										3			
3		7															2																			
4	3																																			8
5																	11																			
6		3				4																										3				
7																								8	3											
8		6																					3													4
9															5		6																			

13.3 Prepare Documentation

If the methods and knowledge of the task remain locked in the research notebooks and heads of the project team, all the planning and hard work go for naught. Failure to create comprehensive documentation for the project violates the *completeness* criterion for the provisional acceptance of knowledge defined in Chapter 5 (An Epistemological Journey), and threatens the criterion of *reproducibility* as well.

Assuming they have been faithfully maintained and updated, the research notebooks of the team members will contain a complete record of all the processes and knowledge used and created by the project. Such records, however, are necessarily and understandably fragmented and disorganized, and cannot be regarded as sufficient documentation of a project, any more than a diary or appointment calendar can be regarded as a coherent biography of a person's life. To assemble and generate the complete and well-organized **primary project documentation** that explains *every detail of every task* is a critical task in every project. Although the responsibility for accomplishing this task lies with the project team, management must (but all too often does not) acknowledge that this task cannot simply be tacked onto the end of a project as a kind of self-evident overhead to be undertaken with minimal cost. In truth, the preparation of such reports is difficult and time-consuming and, as such, is a significant task in the project milestone chart. Sufficient time and resources must be explicitly allocated to fulfill this critical responsibility.

The irony is that, once written, the primary project report may never see the light of day again. Like an insurance policy, although it is seldom needed, everyone understands the necessity of protecting an important investment. Like the Library at Alexandria, although most of the books simply gathered dust over the centuries, the world was thrown into the abyss of ignorance when it burned. Like a bed of oysters, although most inhabitants live out their short lives at the bottom of the food chain, every now and then one delivers an exquisite gift into the hands of the patient fisherman. Thus, perhaps, the expression: "pearls of wisdom."

There is, however, another much more immediate need for generating the primary project documentation. It becomes the single, official source of material for generating **secondary project documentation**, including oral management briefings and conference presentations; and written technical reports, professional journal articles, and conference proceedings. Each of these venues may need to express a different type and level of detail, tailored to the concerns and backgrounds of the specific audience, while making no compromises in the accuracy of the knowledge, only in the precision. To accomplish this, the author(s) of the report must be thoroughly familiar with the material; understand the limitations, perspective, and background of the audience; know how to abstract the material without turning it to mush; and possess trained skills in oral and written communication, both textual and graphical. This is an extraordinary array of required skills, talents, and sensitivities.

Suppose, for example, the traffic engineer for the running example task *Identify Faster Traffic* must present an oral report to her boss, who is the head of the Department of Public Roads. He will be using her presentation to formulate his recommendations to the County Transportation Committee, which has already committed itself for political reasons to widening the highway that has the higher average traffic speed. The traffic engineer, who has been given little or no information about the political issues of the parent project, must necessarily limit her report to sterile conclusions about the traffic speeds. The boss, who is uninformed about and mystified by the alchemy of statistical analysis, must accept these conclusions at face value. The conversation between the traffic engineer and her boss goes something like this:

BOSS:	So, which highway has the higher average speed?
TRAFFIC ENGINEER:	Highway A.
BOSS:	Are you sure?
TRAFFIC ENGINEER:	Yes, at least 99%.
BOSS:	Well, that seems clear enough. I'll recommend widening Highway A.

At this point, because the traffic engineer is unaware of the issues involved in the parent project, there is no reason for her to go into any more detail; and because the boss thinks a statistical moment is some arcane unit of time, there is no reason for him to ask any more questions. This is a recipe for faulty decision-making in the bliss of ignorance. So, let us expand the horizons of these two individuals a bit by giving the boss a brief introduction to statistical data analysis, and by informing the traffic engineer about the critical issues of the parent project. Now the conversation would probably go a bit differently.

BOSS:	So, which highway has the higher average speed?
TRAFFIC ENGINEER:	Highway A.
BOSS:	What is the certainty of this conclusion?
TRAFFIC ENGINEER:	At least 99%. But that may not be the whole story. The average speed is not be a useful metric for deciding which highway should be widened.
BOSS:	I take it the distributions are weird.
TRAFFIC ENGINEER:	One of them is, yes. Highway A, which is the highway with the higher average speed, has a single mode, so the average is probably a useful metric.
BOSS:	But Highway B?
TRAFFIC ENGINEER:	Take a look at this histogram of the speeds on Highway B. The distribution has two principal modes! The traffic is either going very fast or is stopped dead in its tracks. There are very few

vehicles around the average speed. The average is a misleading metric.

BOSS: Sounds like we should widen the highway with the slower traffic in this case, Highway B. I don't know if I can persuade the committee to change their minds. They're already committed to widening the highway with the faster traffic. The funds have been set aside for this.

TRAFFIC ENGINEER: I know, but I don't think widening either highway will fix the problem. They might spend the money and get no benefit at all. The traffic jam may be caused by something amiss farther down the highway.

BOSS: Maybe it's an inadequate left-turn lane. Whatever it is, I think we need to try to persuade the committee to do some additional investigation. I need you to put together some visuals that show why the results are contradictory and misleading. Nor more than three slides. In English.

TRAFFIC ENGINEER: You got it. Short and sweet. No technobabble.

In this scenario, each player is respectful and mindful of both the strengths and limitations of the other player. To illustrate the critical importance of this mutual understanding, consider a third version of this conversation, in which such sensitivities are completely ignored.

BOSS: So, what's your confidence that the two highways have the same average speed?

TRAFFIC ENGINEER: I cannot conclude anything about that. I can only estimate the statistical confidence with which the null hypothesis can be rejected and the associated Type I error probability.

BOSS: I can't tell the council about Type I error confidences. Politicians don't want to hear about errors.

TRAFFIC ENGINEER: You can't fight statistics. I can only tell you what I know.

BOSS: Okay, so what's the confidence of the null hypothesis?

TRAFFIC ENGINEER: That's an ill-posed question.

BOSS: You're fired.

In this last scenario, a lack of acknowledgment and respect for each other's limitations and areas of expertise leads to frustration and confrontation. In the middle scenario, however, two reasonably well-informed people are cooperating to extract just the information they need to make a recommendation that turns out to be something that was *not* mandated by the objective of the parent project. They understand and complement each other's strengths and weaknesses. Based on the slides that will be generated by the traffic engineer, the expertise of the boss will come into play in the artful design of a concise, clear, and convincing executive presentation to persuade the powers-that-be to change the rules of the higher-level game, keeping in mind that his audience has little knowledge of traffic engineering or statistics, but are skilled politicians whose cooperation is essential to solve the problem. Once again, the boss must be reasonably well-informed about his audience and know how to communicate effectively with them, only this time they are his superiors.

> **TIP 29**
> In addition to the objectives of the current task, make sure the project and management personnel are reasonably well informed about the objectives of both the parent and the descendant tasks. Establish efficient channels of communication between adjacent levels in the project task hierarchy.

After all of the effort that has gone into planning and conducting the project following the four phases of the Scientific Method, it seems eminently logical to organize all the oral and written project reports in like fashion. The primary project documentation requires complete and detailed descriptions each of the four phases and their internal steps, as well as some introductory and concluding material and a bevy of appendices. Secondary project documentation, on the other hand, can be anything from a very long technical report like a dissertation or monograph, to a paper published in an archival journal, to a 20-minute oral presentation at a professional conference, to a 5-minute briefing for a department meeting, to a 30-second response to a question about the project posed by an executive who has just dropped by your office for a short chat. While a general format may be specified for the necessarily comprehensive primary project documentation, the format and content of secondary documentation clearly depend heavily on the interest and background of the intended audience. Many accomplished scientists and engineers are equally comfortable describing their work to professional colleagues or secondary school children. The eminent astrophysicist Carl Sagan was a superb example of this versatility.

13.3.1 PRIMARY PROJECT DOCUMENTATION

The four phases of the Scientific Method and their internal steps (See Figure 9.1) are a sensible basis for defining a general-purpose format for the primary project documentation. The topical outline in Table 13.7 is an expansion and adaptation of these phases and internal steps. Note that the names of the steps have been changed from the imperative forms required for task names to equivalent substantive forms, which are perhaps more appropriate as section titles in a report.

Many of the phrases given in the first column are the precise terms associated with phases and steps of the *methodology* presented in this book and, as such, may not be familiar to the audience for which the report is intended. Thus, the actual section titles used in the report should use terminology that is more readily understood. The section titles may be either terms that are generic to R&D processes, examples of which are given in the last column of Table 13.7; or terms that are specific to the subject matter of the project. In the latter case, a topical version of the generic section title "Solution Description" might be "Design of the Proposed Hypercharged Four-Barrel Carburetor" or "Numerical Integration Using Legendre Polynomials" or "Design of the Dynamically Allocated Data Structure" or "Interview Questionnaire Design." In like fashion, the generic section title "Performance" might be specifically expressed as "Performance of the Hypercharged Carburetor" or "Residual Errors in the Numerical Integration" or "Dynamic Allocation Performance" or "Statistical Measures of the Cohort Responses."

The project documentation need not overemphasize the specialized nomenclature that has been devised for the project organization and methodology presented in this book. Meta-references like *Hypothesis Phase* and *required concurrent inducer* are "inside-the-beltway" terms, as it were, and are important only to those who are designing and conducting R&D projects, not to the readers of the project documentation. After all, someone who sits down to read a book generally has no interest in the vocabulary of the publishing profession, such as leading, gutters, em-spaces, and the like. When doctors talk to their patients, they speak in English, even though much of the vocabulary of their profession is in Latin or the arcane expressions of biochemistry (aspirin = *acetylsalicylic acid* = $CH_3COOC_6H_4COOH$). For similar reasons, section titles and other references in the documentation should have direct and clear meanings in the language of the project itself. On the other hand, the specialized terms for knowledge representations and assertions should always be used to avoid ambiguity, such as *definition, assumption, postulate, accuracy, precision, objective, goal, hypothesis, etc.* Even if the audience is unfamiliar with the strict definitions of these terms, they will quickly surmise their meanings from the context.

Although the outline in Table 13.7 specifies the recommended order of the topics for the project documentation, the arrow between the two columns indicates that there are two alternative orders for documenting the steps that span the Synthesis and Validation Phases. As you may recall from Section 11.2 (Set Goals), each of the goals set in the Hypothesis Phase spawns an explicit task in the Synthesis Phase, which in turn specifies an experiment to acquire the knowledge required by the goal, from which the appropriate conclusions are drawn in the Validation Phase. This raises an interesting question: Should the order of the documentation be *step-by-step,* describing the progress of all the goals at every step; or should the order of the documentation be *goal-by-goal,* following each goal individually through all the steps?

The first alternative implies that the documentation for the steps bracketed by the arrow should be grouped together: Describe all the solution implementations for all the goals; then describe all the experiments for all the goals; then present all the results for all the goals; then report all the performance measures for all the goals; and, finally, state all the conclusions for all the goals. If the goals are very closely related, this may be the clearest way to document these steps,

Table 13.7 General-purpose topical outline for the primary project documentation

Section	Examples of Common General Section Titles
Title	(none)
Abstract	Abstract
Table of Contents	Table of Contents
List of Illustrations	List of Illustrations | List of Tables and Figures
Introduction	Introduction | Summary | Overview
Analysis	Problem Domain
Problem Description	Problem Description | Motivation
Performance Criteria	Performance Criteria | Design Criteria | Requirements Analysis
Related Work	Related Work | Prior Work | Background Information
Objective	Objective | Research Objective | Development Objective
Hypothesis	Technical Approach | Proposed Solution
Solution Description	Solution Description | Solution Method | Design Specification
Goals	Goals | Research Goals | Design Goals | Goals and Hypotheses
Factors	Factors | Research Factors | Design Factors
Performance Metrics	Performance Metrics | Performance Measures
Synthesis	Implementation | Experiments | Experiments and Results
Solution Implementation	Solution Implementation | Software Construction | Equipment
Experiment Design	Experiment Design | Verification Method
Experiments	Experiments | Experiment Trials | Verification Process
Results	Results | Experiment Results
Validation	Conclusions | Performance and Conclusions
Performance	Performance | Performance Measures
Conclusions	Conclusions
Recommendations	Recommendations | Future Work
Appendices	Appendices
Bibliography	Bibliography | References | Bibliographical References
Governing Propositions	Glossary | Governing Propositions
Task Tree	Task Tree | Project Task Tree | Project Task Hierarchy
Task Inventories	Task Inventories | Task Component Inventories
System Diagrams	System Diagrams | Task Architectures | System Architectures
Milestone Chart	Milestone Charts | Project Schedules | Schedules and Costs
Result Files	Result Files | Experiment Data Files
Support Materials	Support Materials | Support Documents
Index	Index

(The Synthesis and Validation sections are bracketed with the note: SEQUENTIAL OR INTERLEAVED)

such as when two different sensors are being used to measure the same task unit, or when two similar task units are being measured under different conditions, *e.g.,* a comparison of oranges grown in California or Florida. This is called the **sequential documentation method**.

If the goals represent a sequence of progressive investigations, however, it will be much clearer if each goal is followed individually all the way through both phases, documenting the progress step-by-step from the solution implementation to the conclusions. This is called the **interleaved** or **multiplexed documentation method**.

It is also possible, and often advisable, to mix the two methods, documenting some of the goals sequentially, with the rest of them interleaved in related groups. Deciding which method to use should be based on a thoughtful analysis of the relationships among the goals, leading to the clearest explication of the Synthesis and Validation Phases.

Regardless of the documentation method, because the lower-level goals are necessary to achieve the higher-level goals (by definition), the order in which the goals themselves are documented must always be bottom-up, starting with the terminal goals in the goal tree and proceeding upwards, level by level, concluding with the primary goal in the goal tree. Note that this bottom-up order for documentation is the reverse of the top-down order in which the goals were *defined*.

Table 13.8 illustrates the bottom-up orders of the topics in the Synthesis and Validation Phases for both documentation methods, applied to the goals of the case-study project *Measure Character-Recognition Performance*. The goals in the first column are listed in the order in which they were defined in the Hypothesis Phase (along with their factors and performance metrics). The numbers in the last two columns indicate the order in which the topics are presented in reports for each documentation method. Remember that these orders apply only to the project *documentation*. The order of events during the *conduct* of the project (as given in the milestone chart) may be entirely different.

Conspicuous by their absence from the topical outline in Table 13.7 are the last two steps of the Validation Phase: Prepare Documentation and Solicit Peer Review. In general, it hardly seems appropriate or useful to include an explanation of the preparation of the documentation in the selfsame documentation. Although there may be special instances where such information should be included, normally the primary project documentation is *itself* the best representative of this step of the Validation Phase. The last step of the Validation Phase, peer review, is excluded from the primary project documentation because it cannot even *begin* until some documentation for the project has been completed, such as a technical report for in-house review or a paper submitted to a refereed conference or journal, both of which will be derived from the primary project documentation. Thus, no comments and suggestions obtained from formal reviews of the project can be included in the primary project documentation, because such reviews have not yet taken place. On the other hand, if an informal in-house review of the project occurs before the primary project documentation has been prepared, the results of this review certainly should be reported in the Recommendations section of the primary project documentation.

Table 13.8 The orders of the topics in the sequential and interleaved documentation methods

Goal	SEQUENTIAL METHOD						INTERLEAVED (MULTIPLEXED) METHOD					
	Solution Implementation	*Experiment Design*	*Experiments*	*Results*	*Performance*	*Conclusions*	*Solution Implementation*	*Experiment Design*	*Experiments*	*Results*	*Performance*	*Conclusions*
G1	5	10	15	20	25	30	25	26	27	28	29	30
G1.1	3	8	13	18	23	28	13	14	15	16	17	18
G1.1.1	1	6	11	16	21	26	1	2	3	4	5	6
G1.1.2	2	7	12	17	22	27	7	8	9	10	11	12
G1.2	4	9	14	19	24	29	19	20	21	22	23	24

A brief explanation of the purpose, content, and format of every section in the recommended outline in Table 13.7 is given in what follows. The four sections that correspond directly to the four phases of the Scientific Method, however, have already been thoroughly defined and explained, and only a few additional guidelines for their documentation are provided.

Title. The title of the primary project documentation (or any report, for that matter) is the least precise description of the project. However, as has been stated previously, a reduction in precision need *not* imply a significant compromise in accuracy. Thus, the title can and should be an accurate and complete synthesis of the four principal steps of the Scientific Method, one from each phase: the objective; the solution; the experiments; *and* the conclusions. Even to be able to include this minimal amount of information at low precision in the title requires a lot of words, often as many as 20 or 30. So be it. The objective of this documentation task (*Invent Meaningful Title*) is to compose a brief, but accurate description of the project, so that when the literature on the subject is perused, the researcher may make a fairly reliable initial decision whether or not the report is of interest, simply by reading the title. By way of illustration, below are some possible titles for the primary project documentation for most of the example running tasks introduced in Part II of this book:

> *The Successful Insertion into Low Earth Orbit of a Manned Mercury Spacecraft Launched with a USAF Atlas Rocket* (18 words)

> *The Inappropriate Use of the Average Traffic Speed as the Sole Criterion for Deciding Whether to Widen Highway A or B* (21 words)

> *A Rigorous and Complete Mathematical Proof of Fermat's Last Conjecture* (10 words)

The Confirmed Discovery of a New Planet in its Predicted Solar Orbit Based on Blink Comparisons of Photographic Images Captured by the Lowell 13-inch Telescope (25 words)

An Estimate of the Geologic Age of a Fern Frond Fossil at 780 Million Years Based on Measurements of the Position and Type of the Sedimentation Layers in the Cliffs of Sagitawe (32 words)

The High Statistical Certainty of the Superiority of TCH-Aminohydroxy- dase-6 over Existing Therapies for Stage III Liver Cancer Based on Large- Scale Doubleblinded Human Experiment Trials (24 words)

The Poor Effectiveness of the Proposed Graphical User Interface of the Find- ItNow! Web Browser Based on the Opinions and Behavior of a Small Cohort of Typical End Users (28 words)

A Database of Summaries of 372 Scholarly Articles, Books, and Reports on Multiple-Personality Disorder Published since 1850 (17 words)

An Estimation of the Date of the Death of Organic Sample S1287–1995 at 210±50 AD based on Standard Carbon-14 Dating Methods (21 words)

It is reasonable to assume that such long and rich titles increase the probability of success and decrease the number of false alarms for the process of searching the literature for relevant information, such as that which takes place during the Related Work step of the Analysis Phase. The advantage of this extra information is especially important as more and more people attempt to use search engines to find significant sources of information among the millions of reference sites accessible by the Internet. Such supposedly smart search engines, as everyone now knows, are still remarkably stupid, and the only hope of using them productively is if the titles of documents are highly descriptive and as detailed as possible.

Abstract. Like the title, the abstract is a summary of the application of the four phases of the Scientific Method to the project, and it should be organized in this fashion. It is, however, generally about ten times longer than the title, *i.e.,* about 200 to 300 words, equivalent to a long paragraph. The abstract serves as the next level of information used by a researcher to find significant sources of knowledge in the literature, *i.e.,* as a more precise tool to help separate the wheat from the chaff.

Given a total length of 200 to 300 words, each phase of the Scientific Method is generally assigned about 50 to 75 words in the abstract, although one phase may certainly deserve more emphasis than another. Given that a sentence is generally about 15 to 20 words long, one simple and obvious approach is to assign one sentence to each of the first 14 steps of the Scientific Method, omitting the last two steps of the Validation Phase for the reasons mentioned previously. In reality, the necessary *information* about each of the 14 steps should be integrated into a long paragraph that presents a logical progression and a clear summary of the project.

One of the most common errors in abstracts is the lack of a summary of the significant *results and conclusions* of the project. To illustrate the impact of this omission, an incomplete abstract for the example task *Measure Drug Effectiveness* might conclude with the following sentence: "A series of doubleblinded experiments was conducted to measure the effectiveness of this promising new cancer drug." This coy statement simply begs for some statement of the "promising" conclusions, and it is both inappropriate and inconsiderate to omit such a statement! Some researchers actually defend the omission of such information as an "incentive" to force people to actually read the entire article. This marketing strategy may be effective for projects on Madison Avenue, but it has no place in science and engineering. Suspense is a mechanism reserved for fictional works like novels, plays, and short stories. To use it as a ploy with technical reports for captivating the attention of the reader is ethically inexcusable and constitutes a clear violation of the *completeness* criterion for the provisional acceptance of knowledge, as postulated in Chapter 5 (An Epistemological Journey).

Take the time to wordsmith the abstract very carefully, so that it says exactly what you mean, nothing more, and nothing less. Clarity and brevity often compete with each other. Like a piece of fine furniture, a good bit of elbow grease is required to make the abstract shine.

Table 13.9 presents the 25-word title and 280-word abstract of the primary project documentation for the case-study project *Measure Character-Recognition Performance*.

Table of Contents and List of Illustrations. The level of detail provided by the table of contents is a function of the length of the report. As a general rule, headings for a set of subsections should be included in the table of contents only if the page numbers for the subsections all differ from each other. For example, if subsections 4.1, 4.2, 4.3, and 4.4 are all on pages 6 and 7, these headings can be omitted from the table of contents; the subsections can be easily found by looking just beyond the page number listed for section 4 (presumably, page 6). For very short reports, the table of contents and list of illustrations may be omitted entirely; in such cases, the reader can simply thumb through the document to find what is needed.

It is generally not a good idea to use numeric section numbers with a precision greater than 4 digits ($n.n.n.n$). In fact, some publishers will not accept more than three levels. If more precision seems to be necessary, consider promoting a set of subsections to a higher level. For example, sections 4.2.5.1, 4.2.5.2, and 4.2.5.3 become sections 4.2.6, 4.2.7, and 4.2.8. Another solution is to use unnumbered paragraph titles in boldface, as is done in this section.

Introduction. The introduction to the primary project documentation has three objectives: 1) to summarize (yet again) the entire project at a higher level of precision than the abstract; 2) to define the typographic conventions used in the report; and 3) to report the levels of effort and costs of the project. It may seem overdone to summarize the entire project report yet again, but be assured that managers and executives will not agree. They appreciate a short overview that tells them everything they need to know about the project. Often they have no time to read any further. In addition, researchers who are searching the literature for information will welcome a summary in the introduction to help make a final decision on the relevancy of the report before undertaking the much larger task of reading and digesting the entire report. Once again, it is important that all four phases of the project are described in the introduction, and in particular, that the important results and conclusions are summarized. As with abstracts, it is very hard to justify omitting a summary of the significant results and conclusions from the introduction.

The introduction, the majority of which is devoted to the overview of the project (the first of the three objectives), must tell the *whole* story in as much detail as possible in no more than 10% of the total length of the report. If the

Table 13.9 Title and abstract for Project *Measure Character-Recognition Performance*

An Estimation of the Peak Character-Recognition Performance of Trained License-Plate Inspectors for Characters Displayed with Noise that Realistically Simulates the Actual Obscuration of License-Plate Numbers
Abstract
The Acme Corporation plans to develop an automated vehicle-license-plate character-recognition system that is at least as reliable as trained human license-plate inspectors. As a precursor to this development project, a series of carefully controlled experiments were designed and conducted to measure the recognition performance of these inspectors. Because characters on actual video images of license plates are often partially obscured by mud and debris, the correct characters cannot be reliably identified, obviating the use of such images for these experiments. Therefore, realistic simulated noise was superimposed on frames containing computer-generated characters, whose format was the same as actual license-plate characters. Both the noise level and typeface (*serif* and *sans serif*) of the characters were factors in the randomized block design of the experiments. The cohort consisted of eight trained license-plate inspectors. After practicing sufficiently to demonstrate proficiency with the graphical user interface (external control trial), each subject was presented with a sequence of 108 test frames, each with a randomly selected character (in one typeface) obscured with a randomly selected noise level, and 45 randomly interleaved control frames with noise, but no character. Performance metrics included recognition accuracy (benefit), response-selection time (cost), and an arithmetic combination of them (payoff). The results of these experiments yielded a reasonable model of the recognition performance of trained license-plate inspectors. The conclusions also validated the six research hypotheses: briefly, recognition benefit and payoff vary inversely with noise level; recognition cost varies directly with noise level; and recognition benefit and payoff for the *serif* characters significantly exceed the benefit and payoff for the *sans serif* characters for moderate noise levels [170 to 190]. Informal observations identified some of the characters that caused the highest recognition-error rates and confusions.

typical length for the primary project documentation of large project is about 50 pages (not including the appendices), then the introduction should be no more than 5 pages long, which is about 2500 words. Note that the level of detail (precision) consistently increases by an order of magnitude from the title to the abstract to the introduction to the report itself:

Title		*Abstract*		*Introduction*		*Report*
25 words	$\rightarrow$	250 words	$\rightarrow$	2500 words	$\rightarrow$	25000 words
3 lines		1 paragraph		5 pages		50 pages

All other things being equal, in the same amount of time that a single report can be digested, 10 introductions, 100 abstracts, or 1000 titles can be reviewed.

For a very small project requiring, say, a 5-page report for the primary project documentation (about 2500 words), the introduction would only be about 250 words long, the same length as a typical abstract. In this case, the summary in the introduction would provide no more information than the abstract and should be omitted. Only an explanation of the typographic conventions of the report and the resources of the project need be included in the introduction.

The second objective of the introduction is to explain the typographic conventions used in the report. In former days, when reports were typed on standard typewriters, there were very few typographic options available for emphasizing propositions and statements. Such embellishments were restricted to documents that were typeset and published professionally. Today, word processors and desktop-publishing software tools provide easy access to a wide variety of typographic attributes, which makes it *possible* for an author to prepare a report that appears as if it was professionally typeset. Possibility and actuality, however, are two different things. In fact, it requires considerable design talent and skill to use this newfound freedom artfully and wisely. To the inexperienced author, it is often tempting to sprinkle many different colors, typefaces, and fonts throughout the text, creating a mish-mash that ends up being confusing and distracting, rather than helpful. Every typographic attribute should be selected with a specific purpose in mind, applied judiciously, consistently, and, most would agree, sparingly. If someone asks you why you chose a particular typestyle, font, or color, be sure you have a reasonable explanation, something considerably more convincing than "it seemed like a good idea at the time."[14]

---

[14]There are some who believe that the use of different and unusual typographic attributes is frivolous: documents have always been typographically simple, and should remain so. In fact, documents have *not* always been typographically simple; the illuminated documents created by monks centuries ago were marvels of decoration and embellishment. Then, the invention of the moveable-type printing press in the fifteenth century put a stop to all that foolishness, aided and abetted soon thereafter by the austerity and asceticism of the Protestant Reformation. Perhaps now in the twenty-first century, with the invention of the word processor, we can go back to illuminating our documents!

It is well to remember that color, although an outstanding attribute for emphasizing parallel structure and differentiating among important concepts, cannot yet be reproduced inexpensively. Therefore, it is a good idea to make sure that the colors used in the report will still be clear when duplicated on a xerographic grayscale copier. Hopefully, in the not too distant future, such constraints will be eliminated, and full-color documents will be inexpensive and ubiquitous, just as the black–and–white television standard of the 1950s has given way completely to color.[15] More and more publishing is being done on the Internet, which will help to bring this about.

How can the wide variety of available typographic attributes be used most effectively? Although this is largely a matter of personal preference and style, Table 13.10 offers some guidelines. No matter which typographic conventions are selected, it is important that they are defined briefly in the introduction and then applied consistently throughout the document.

The final objective of the introduction is to acknowledge the contributions of the project personnel (including their levels of effort) and to list the project costs (both direct and indirect). This is a good place to reference the milestone chart, which is listed as an appendix in Table 13.7. This information will be useful for management to estimate the time and costs for similar projects in the future.

> **TIP 30**
> Typographic and color attributes can provide powerful means of emphasis and differentiation within text and graphics. Improper or arbitrary use of them, however, can cause considerable distraction and confusion. Get some formal training to learn how to use them sensibly and consistently. Make sure you know how the document will appear when duplicated in grayscale.

Analysis, Hypothesis, Synthesis, and Validation. Parts II and III and the preceding chapters of this part of the book have already addressed these topics in detail. There is, however, an important question about these sections that is typically raised when the documentation for a project is being generated: given that the **governing propositions** comprise the knowledge domain of the project, where exactly should they appear in the documentation? At the risk of sounding

---

[15]Some of the same people who oppose typographic variety are also those who venerate the medium of black-and-white (actually grayscale) movies and photography, scoffing at color as some kind of interloper that violates the ageless purity and sanctity of the grayscale medium. In fact, the rendering of images in grayscale is very new, a direct result of the severe limitations of the early photographic technologies. Before the middle of the nineteenth century, no one ever saw anything in grayscale. Everything was seen and painted in color (except for *chiaroscuro*). Color has dominated human visual perception for hundreds of thousands of years, and to avoid using it in technical reports is wasting a powerful and natural mode of discrimination and emphasis.

flippant, the most accurate (and least precise) answer to this question is: *where they belong*. Not wishing to leave it at that, however, a few heuristics can be offered:

- Place the governing propositions as early as possible in the four phases and as high as possible in the task hierarchy to avoid side-tracking the logical development of the ideas in succeeding or lower-level sections;
- Most definitions and assumptions probably belong in the Analysis Section of the primary task, where the domain knowledge for the entire project is established, leading to a clear and concise statement of the primary task objective;
- Additional definitions and assumptions may have to be introduced in the Analysis Sections of lower-level (descendant) tasks;
- Performance metrics are usually postulated in the corresponding step of the Hypothesis Phase; metrics for computing statistical confidences for the conclusions, however, may be deferred until the first step of the Validation Phase (Compute Performance).

Because of the importance of the governing propositions as the epistemological basis of the entire project, the reader must be able to review them quickly and easily while reading the report. For this reason, they should be titled, num-

Table 13.10 Guidelines for the use of typographic attributes

Attribute	*Function*
boldface	• first occurrence of **words** or **phrases** with special technical meanings (include in glossary) • section headers • important numbers or entries in tables
italics	• emphasis, *e.g.,* "there is *no* significant difference. . . ." or "Only *red* paint may be used." • foreign language words and abbreviations, *e.g., Brandenburgertor, et al., i.e., etc.* • Latin symbols for mathematics, *e.g., H, a, x, f(x)* (not for digits and special characters) • low-level section headers and column headers
typestyle	• underline for emphasis, only if underline breaks for descenders (not <u>poppy</u>) • small caps for special sets of recurring names (*e.g.,* COLORS, DAYS) • true sub- and superscripts (not "computer" names) *e.g.,* 10^{17}, not 1E17; K_{min}, not Kmin • shadow, embossing, outline for special purposes (sparingly)
typeface	• *serif* typeface for text • *sans serif* or *serif* typeface for tables and figures • *sans serif* phrases in *serif* text (or *vice versa*) for recurring acronyms (*e.g.,* RAM)
font	• Times and Helvetica for normal text (preferred by most publishers) • monofont (like Courier) to align character strings in successive rows in tables • unusual fonts for special sets of recurring names (*e.g., **Identify Faster Traffic***)
typesize	• smaller typesize for subscripts and superscripts (not necessarily the default size) • decrease with progression from high-level to low-level section headers
shading	• unused cells in tables (*e.g.,* See Table 13.6) • white text on dark shaded background for special emphasis, *e.g.,* emphasis

bered, and isolated in indented paragraphs, as they are in this book. In addition, to facilitate fast and easy access, they should all be listed in an appendix (Glossary), sorted alphabetically by their associated terms or phrases (not their numbers).

One final point must be made about writing up the sections that describe the four sequential phases of the Scientific Method for your project. Linearizing the description of this inherently nonlinear, hierarchical, and recursive process is a very difficult task. But you have no choice; a written document is, by definition, a linear stream of text, and there is no way around that. It will help to remember that the reader is *not* interested in the false starts and the agonies of failure that actually occurred along the way, even though you felt them all very deeply, and even though resolving these difficulties took most of the project time. Eliminating these war stories will help you to distill it all down to the *essential* process, which then can be described as a seemingly effortless sequence of steps from the problem description to the conclusions. Simply tell the story as if you got everything right the first time.

> **TIP 31**
> When documenting the process of the project, omit the detours and false starts, and distill it down to as simple a story as possible: a linear sequence of successful steps from Analysis to Hypothesis to Synthesis to Validation.

Recommendations. For industrial R&D projects, the recommendations for follow-on efforts are likely to be linked specifically to the objectives of some parent project, *e.g.,* "Based on the encouraging conclusions of this project, it is strongly recommended that the parent development project be fully funded in the next fiscal cycle," or "Based on the discouraging conclusions of the project, it is strongly recommended that the parent development project investigate other methods to achieve the objective of this project." Moreover, because they are usually full-time employees of the company, the researchers who conducted the current research project will probably be around to participate in, or at least to provide guidance for the follow-on projects, if any. The combination of a stable and talented R&D department with low turnover and an archive of comprehensive project documentation significantly reinforces the continuity of industrial R&D efforts.

Although it is difficult to imagine an industrial project that does *not* have a parent project, this happens all the time in academia, where research projects are either not part of a long-term research effort, or follow-on funding is heavily dependent on annual reviews conducted by public funding organizations, which can be very fickle. Individual doctoral research projects also often have no parent project; the student completes the research, defends the dissertation, receives the degree, and disappears. In both cases, the inclusion of complete and detailed recommendations takes on special importance, because these recommendations (and, indeed, the entire primary project documentation) may be the only surviving source of guidance for future researchers who wish to continue the research at some unforeseeable time in an unpredictable future. Typically there is little continuity in doctoral research projects in academia, and to protect the fundamental concept of academic freedom, no such requirement should ever be imposed.

Appendices. The purpose of the appendices to the primary project documentation is to include material that is not appropriate to be included, or physically cannot be included in the body of the report itself. Because it is impossible to list all possible appendices, the last one given in Table 13.7, Support Materials, is meant as a catch-all for things that do not fit under the other categories, such as circuit diagrams, blueprints, source listings of computer programs, data sets, and even hardware prototypes. (Strictly speaking, the Mercury Capsule, the Atlas Rocket, and perhaps John Glenn himself should have been included in the primary project documentation for the example running task *Launch Spacecraft into Orbit*. Certainly their complete specifications were.) Items included as appendices are often not textual material. For example, the appendix listed in Table 13.7 as Results Files may actually be a set of CDs (or even punched cards).

Index. As every researcher knows, one of the most useful components of a long report (or book) is a complete and highly detailed index, which offers many different ways for the reader to access the information in the report. As a rule of thumb, the length of the index should be at least 5% of the length of the report. However, unless the project report is longer than 50 pages, an index is probably unnecessary. Although there are several commercial software tools available for indexing manuscripts automatically (*e.g.,* Sonar Bookends from Virginia Systems), they are often not very helpful; they omit too many significant items (Type II errors) and include too many trivial or irrelevant items (Type I errors). Unfortunately, the process of creating an index, which is an extremely time-consuming task, is still best accomplished manually by a technical editor specially trained for such tasks.[16]

13.3.2 SECONDARY PROJECT DOCUMENTATION

Although the major purpose of the primary project documentation is to record the complete and objective details of a research or development project so that it may be fully understood (and, if necessary, faithfully reproduced) at any time after the project is over without any direct help by members of the project team (who will almost certainly be long gone), it has another equally valuable purpose. It is the single authoritative source of information for generating all additional written and oral reports for the project, that is, the secondary project documentation.

Written Reports. Secondary written reports are always much shorter than the primary project documentation, because they are limited by the publishing venue or the needs and expectations of the intended audience. Most secondary written reports can be derived from the primary project report simply by editing it down to an appropriate length.

---

[16]My wife, who is an experienced technical editor, created the index for my previous book, *The Emergence of Artificial Cognition* (World Scientific Press, 1993). It took her many days of very hard work. I am eternally grateful to her for undertaking this mind-numbing, but essential task for me. Do not attempt this highly specialized task unless you know *exactly* what you are doing.

The length of a written project report can vary from a very short paper in the proceedings of a professional conference, to a short report for an R&D manager, to a very long paper in an archival journal, to a 100-page monograph. Journal publishers normally specify no page limit, although the reviewers or editors may ask you to cut the paper down (or, for that matter extend it). Most industrial managers want short progress reports or final reports for projects, knowing that longer reports for other purposes can always be generated on demand from the primary project documentation. (Note that this assumes that the primary documentation has, in fact, been written and properly archived in the company's library. Be sure to make it so.)

Papers for refereed professional conferences provide a reasonably expeditious way to increase the visibility of your research. Many doctoral students publish their first research papers at conferences to protect their ideas, to obtain useful feedback on a timely basis, and to meet their professional colleagues face-to-face for the first time. Such papers, however, are usually limited to 6 or 8 pages to minimize publication costs. If you can get permission to publish an extra page or two (not always possible), the publishers often charge a hefty fee for each one. On the other hand, if you can afford it, paying the fee may be well worth the advantage of getting as much information as possible out there! Because every square centimeter of each page of a conference paper is worth its weight in gold, Table 13.11 offers some guidelines for exploiting the available real estate effectively.

Deciding what to include in a written report depends completely on the intended audience. You must make the hard decisions about what to delete and what to include from the primary project documentation. Even long journal articles must leave out many details. Governing propositions must usually be abbreviated and embedded in the text, or even omitted entirely. The related work must often be omitted. Sometimes only a few goals can be reported. In reports for managers, very often the description of the solution method should be omitted entirely, and focus placed almost exclusively on results and conclusions. Don't be discouraged if managers show little interest in the details of your clever algorithms or gearbox designs. It is not meant personally, and, as a matter of fact, it may well be a sign of their faith in you, simply trusting that you have all that technical stuff well in hand. Think how you would feel if you asked your boss about the future of the company, and out came a pile of *pro formas* and budget projections. Presumably you would be pretty annoyed (and more than a little bit lost), because what you really wanted was the *big picture*. Well, so does your boss.

TIP 32
Before putting together a project report or presentation, understand the intended audience and its primary interests in the project, both intellectual and political. Let this knowledge guide the design and content of the report.

Table 13.11 Guidelines for the effective use of the limited space in conference papers

1	For full control of spacing, use *exact* line spacing (*e.g.,* 12 points), instead of default *single spacing*.
2	To recover a partial line at the end of a page, decrease the line spacing by a point here and there above it.
3	If small line spacing puts subscripts too low or superscripts too high, manually raise or lower the characters.
4	To avoid wasted space between paragraphs, use first-line-indent instead of block-style paragraphs.
5	Construct (or import) clear, but very small figures with small typesizes, each designed especially for the paper.
6	Place figures at the top or bottom of the page to avoid wasting space above or below them.
7	If there are many figures, put them all on the last page in an efficient spatial arrangement.
8	Enclosing figures in boxes may allow the spacing between figures and text to be reduced without crowding.
9	Eliminate the lines required for figure titles by putting the figure title *inside* the figure area (if there is room).
10	Make the typesize of the references smaller than the text of the body of the paper (if not explicitly disallowed).
11	Place lowest-level section titles in boldface as first words of nonindented paragraph: **Section Header**. Text . . .
12	Embed less important governing propositions in existing paragraphs, rather than giving each its own paragraph.

Note: Make sure your typographic conventions do not flagrantly violate the formatting rules set by the publisher.

Oral Reports. Oral presentations present a far greater challenge than written reports for two critical reasons: they require special design and delivery skills, and they are short and ephemeral.

Oral reports must be designed and delivered using skills for which most people have had *no* specific training. Although much of the time from fourth grade through high school is spent learning how to write papers,[17] little time is spent teaching how to design and deliver an effective oral presentation. On the few occasions in college and graduate school when students must give an oral presentation in the classroom, they must figure it out for themselves, usually with

---

[17]Over the 30 years I have spent as a teacher, predominately at the graduate level, my students have demonstrated adequate writing skills — some much better, some much worse. However, I plead with my colleagues in secondary education to focus more pedagogical energy on two specific requirements of English grammar:

    1) the correct conjugation and use of the intransitive verb *to lie* and the transitive verb *to lay*
    2) the correct use of personal pronouns in the nominative and objective cases

I am tired of hearing students talk about something "laying on the table between you and I." Everyone makes grammar errors, and I am not a purist in this respect. These two particular errors, however, make the speaker sound like a country bumpkin, and they have reached epidemic proportions. A little bit of intellectual vigilance and discipline would soon cure the problem. Wax on, wax off.

mixed results. Ironically, during their subsequent careers as engineers and scientists, they will end up spending far more time designing and giving oral presentations than writing reports, their only guide being the inconstant and disorganized teacher known as on-the-job training.

Preparing an effective oral presentation requires special talents and skills in the graphic and verbal arts. Because oral presentations are so important in professional activities, every engineering and science curriculum at our institutions of higher learning should include a *required* course in the preparation of oral presentations (note that this is *not* a technical writing course, which is also essential). In fact, it makes even more sense for high schools to provide this kind of training for all students who are planning to go to college. If you are already in a professional career and have never had any formal training of this kind, seriously consider taking a professional seminar to develop these special skills. It will pay off handsomely in your ability to communicate effectively on your feet.

The Microsoft product *PowerPoint* is a highly effective tool for preparing slide shows. This elegant software application provides all kinds of aids for enhancing lists, charts, tables, and graphs, as well as animating pieces of the presentation and navigating among the slides. Some of these features, of course, are just too fancy for words, but others are very powerful if used judiciously. While this book cannot provide a tutorial on this product, a few guidelines are offered in Table 13.12. In addition, there are several excellent books on this subject, perhaps most notably *The Visual Display of Quantitative Information,* by Edward R. Tufte, which has recently been updated and re-published by Graphic Press.

The second reason oral reports are much more of a challenge than written reports is that oral reports of reasonable length can express only a fraction of the information that can be expressed in written reports of reasonable length. Moreover, unlike written reports which can be read and reread until comprehension finally dawns, oral presentations are ephemeral; the audience must follow the logical progression and understand everything in real time.

If you were to simply read a 6-page paper at the conference (which unfortunately some people do!), it would take you about 30 minutes, a good deal longer than is normally allowed for oral presentations at conferences. And no matter how much you rehearse, the presentation will be lifeless and boring. An oral presentation is a completely different animal from a written report.

Oral presentations can range in length from a few seconds to 60 minutes. But no matter how long it is, the organization of the presentation must still be based on the 16 steps of the four phases of the Scientific Method. Of course, if very little time is available, then only a few of the steps can be included. The question is: which ones?

Suppose that you were limited to a single sentence that described only *one* of the 16 steps in the four phases of the Scientific Method. Which single step would you select? To explore this question, imagine a young engineer, sitting at his computer, finishing up the performance computations for an important project, when, without warning, the CEO of the company pokes her head in the door with a vice-president in tow:

Table 13.12 Guidelines for the design of slides for oral presentations using *PowerPoint*

1	Before you start using *PowerPoint,* go through all the menus in detail and investigate all the available features.
2	The limit on the number of slides cannot be overcome by jamming as much as possible on each slide.
3	Ignore the default typographic and graphic settings supplied by *PowerPoint.* Make your own design decisions.
4	All text should be at least 18 point, although table entries or chart text may be 14 point if absolutely necessary.
5	Put a 14-point section title and slide number in an out-of-the-way, but consistent place on every slide.
6	Use bullet marks sparingly. Usually indentation alone will serve just fine to separate the items in a list.
7	Avoid paragraphs or prose descriptions. Use lists of short phrases and single short sentences to cue your comments.
8	Remove the chart junk, *i.e.,* objects, characters, and punctuation that add nothing and just confuse the eye.
9	Avoid abbreviations unless space is a critical issue. Use *centimeter,* not cm., *pounds,* not # or lb., and so forth.
10	Only sentences end with periods. Phrases do not. Unneeded periods and punctuation marks are "chart junk."
11	Avoid distracting backgrounds with decorative lines, fancy color schemes, and busy patterns. Keep it clean.
12	Colors should have specific and consistent semantic functions for emphasis, membership, or differentiation.
13	Strive for a consistent graphical and textual style on all the slides. Inconsistencies distract the audience.
14	Animation is very powerful for revealing objects (or groups) one at a time, which helps the audience focus.
15	Dim items that are no longer intended to be the primary and current focus for the audience.
16	Avoid cutesy effects for slide animation and transitions. Most of the time the *appear* effect is all you need.
17	Use custom animation to demonstrate dynamic processes with moving objects and changing titles.
18	To assess the readability of a slide, view it from a distance of 4 times the slide width on the computer screen.

CEO: Good morning, Jack.

ENGINEER: Good morning, ma'am.

CEO: I'm sorry to interrupt your work. I'm just poking around the labs this morning, trying to find out what's going on in the trenches. What are you up to these days?

ENGINEER: The speed and accuracy of our Internet search engine are not very good, and we need to find ways to speed it up and make it more accurate.

CEO: Sounds right to me.

VEEP: Excuse me, ma'am, but we have a luncheon appointment in 5 minutes.

CEO: I've got to run, Jack, but keep up the good work. I'll be looking forward to hearing about the results of your work. Have a great day!

Poof. Well, Jack, you may have just lost the opportunity of a lifetime. The CEO descends from the penthouse to find out what's going on in her research labs, and you tell her about a *problem*.

In all fairness, perhaps the young software engineer thought he was just dutifully following the Scientific Method, whose first step most certainly is *Describe Problem*. But by doing so, he missed the point. Let us be perfectly clear about this: the Scientific Method is a prescription for *planning and conducting* a research or development project. *Presenting a report* on the project is something else entirely. If you are allowed only *one sentence* to describe the project, *please* do not waste it on the problem, or the related work, or even the objective. Cut to the chase: report the *conclusions,* even if only informal results are available at the time. Then, if you capture the interest of your audience and they grant you more time, you can begin to fill in more details. If Jack had been aware of this critical difference between the execution and documentation strategies for a project, he would have responded quite differently to the CEO's question:

CEO: I don't mean to disturb you, Jack. I'm just poking around the labs this morning, trying to find out what's going on in the trenches. What are you up to these days?

ENGINEER: I think I may have found a way to double the speed and accuracy of our Internet search engine.

CEO: My word! That's marvelous!

VEEP: Excuse me, ma'am, but we have a luncheon appointment in 5 minutes.

CEO: You go on and extend my apologies, please. I'm going to spend a few minutes here and see what Jack has figured out.

Well, that did the trick! But what does Jack tell the CEO next? Because the CEO's expertise is (hopefully) marketing and management, Jack is probably well-advised to supply additional details about the performance of the new search engine, including comparisons with competing products, which he certainly would have right at his fingertips if he has followed the Scientific Method in the *planning and conduct* of this project (in the Related Work step). He certainly does *not* want to describe the clever data structures and C++ code he has written to implement the solution, or list the formal goals, factors and performance metrics, or describe the test data bases he is using for his experiments, blah, blah, blah. The CEO will trust Jack and his department head to make sure all that stuff is on track and under control, and it will all be thoroughly documented in the written project reports. Right now, the CEO needs to figure out if there's gold in them thar hills.

This is not a question of office politics or of sucking up to the boss. It is simply a question of providing people with the information they need in the most efficient possible manner. Surely, the CEO did not pose her original question out of idle curiosity. Several months later when he goes to a computer conference to give a paper on his new search engine (after the CEO has secured legal protection for this valuable intellectual property), Jack can focus on the technical details. Although these were *not* of interest to the CEO (or within her field of expertise), they are of paramount interest to Jack and his professional colleagues. In this venue, he can wallow in technobabble and be the geek he loves to be.

Here is the bottom line. If you are asked to report the status of an R&D project in a single statement, *summarize your conclusions*, even if they are only tentative informal observations. If no reasonable conclusions or observations are available yet, then describe the *solution method* if it is being created especially for the project (*i.e.*, not simply acquired); otherwise, simply state the *objective*. If more information is requested, the report should quickly converge to the logical progression of the entire Scientific Method. The order in which these steps are added and the emphasis given to each depends on the knowledge level and interests of the audience.

Most oral presentations are not short conversations that happen on the spur-of-the-moment like the scenario just described. Normally, they are scheduled well in advance and are allocated from 5 to 60 minutes. Vice presidents or sponsors often ask managers to set up a one-hour meeting during which reports on several projects are presented, each lasting no more than 5 or 10 minutes. A presentation at a professional conference is normally allocated 20 minutes, followed by a short discussion period. An expert or well-established researcher may be invited to give a talk that lasts an hour. All of these oral presentations are typically organized around a sequence of slides that illustrate and summarize the information being presented. To help design such oral presentations, Table 13.13 gathers together a set of statements, lists, tables, charts, and diagrams that have been used in this book for summarizing the information about a project. The fact that the order of the topics fits the Scientific Method like a glove is no coincidence.

Important governing propositions may need to be included at various points in the presentation to lay the epistemological foundation for the project. For project proposals or project status reports, only preliminary results from QD pilots may be available. The proposal can include these results as initial performance estimates in the first step of the Validation Phase, followed by some informal observations as preliminary conclusions. If no preliminary results are available yet, then the report can go no further than the *Design Experiments* step in the Synthesis Phase. For proposals, it will be understood that no additional steps can be reported, and the presentation should conclude with a high-level project task tree and/or a milestone chart.

How many slides can be included in an oral presentation? The answer to that question depends, of course, on the length of the talk. As a rule of thumb, each slide requires an average of 3 minutes of narration. This limits a 5-minute presentation to 2 slides, a 20-minute presentation to 7 slides, and a one-hour presentation to 20 slides. If the information about each of the 11 summary items in Table 13.13 was included on one slide (on the average), the entire oral presentation of 11 slides would take more than a half hour.

To those who have only a little experience with designing and giving oral presentations, these numbers may seem too low: "Surely, I can present more than 20 slides in an hour." But be assured: these numbers are realistic, based on hard evidence. A common mistake in oral presentations is the inclusion of too much information on too many slides, causing the speaker to run out of time. Dealing with this planning error on the stage in front of your peers can be very embarrassing. Session chairs will gently try to stop those who have exceeded the allotted time, but some speakers simply refuse to take the hint and continue to plow right on. This intransigence not only angers the audience, it is rude and unfair to the other speakers who follow, because the session generally has a fixed length, and the last speaker gets cheated.[18]

There really is only one way to make sure you don't become one of these bad guys: rehearse your presentation aloud, preferably with an audience. (A friend or colleague will suffice, but muttering to yourself in front of the computer screen does *not* work.) After you have a few papers under your belt, you will have a good feel for approximately how long a presentation will take, based on its design and content. But in the beginning, it takes practice. Wax on, wax off.

Above all, remember to organize the material in the order of the Scientific Method, which has been carefully explained in this book. The only reason to abridge this order, as mentioned previously, is if the presentation is limited to very few slides, which will require omitting most of the steps and focusing primarily on the results and conclusions.

Now you are ready to display your wares, written reports and oral presentations alike. Hopefully, you are prepared for feedback, because that is exactly what you are going to get during peer review, which is the last step of the Validation Phase of the Scientific Method. Listen and learn.

---

[18]In my 30 years of experience as a researcher, I have seen this happen countless times. Worst of all, the most egregious offenders are often experienced professionals, who dismiss the rules with an offhanded arrogance and continue to do so at meeting after meeting throughout their careers. This sets a very poor example for students and gives our profession a bad name in this respect.

13.4 Solicit Peer Review

As described in the brief history of the Scientific Method in Chapter 9 (Overview), before the advent of the Scientific Revolution in the seventeenth century, scientists were content to finish up a piece of research, sit back, and admire the fruits of their labors with a sense of peace and self-congratulation. Few heard about their work, because the laborious process of duplicating manuscripts and the limited means and range of travel simply did not allow for the timely interchange of information. But when the printing press revolutionized the dissemination of the written word and ships began to ply the oceans regularly, scientists began to communicate and congregate. This interaction was not comfortable for all of them; Isaac Newton dug in his heels and grouchily refused to cooperate. But the die was cast: Science had decided to police itself.

Now, four centuries later, knowledge offered without formal peer review is unacceptable, at least to scientists and engineers. Acceptance requires an explicit and unbiased confirmation that the three criteria defined in Chapter 5 (An Epistemological Journey) have been met: *reproducibility, completeness,* and *objectivity.* Even then, the acceptance is only *provisional*; the warranty is only good until something better comes along.

Formal peer review takes on many forms. The most prestigious, and probably the most rigorous and objective form of review is the refereeing process for professional archival journals, which abound in all fields. Unfortunately, the time delay from submission to publication is usually at least one year, and sometimes much longer. This is no longer adequate for the high-paced professions like computer science, genetic engineering, biomedical engineering, and medicine, which are all experiencing an avalanche of scientific and engineering breakthroughs.

Table 13.13 Important information items for an oral project report or proposal

Phase	Summary Item (Format)	Reference and Examples
Analysis	Problem Statement (sentence)	Definition 10.2; Tables 10.1, 10.2
	Performance Criteria (short phrases)	Table 10.3
	Primary Objective (infinitive phrase)	Definition 4.3; Table 10.5
Hypothesis	Solution Description (list or diagram)	Section 11.1
	Goals (tree or outline)	Definitions 11.2, 11.3; Tables 11.2, 11.3; Figure 11.1
	Factors (table with ranges)	Definition 7.8; Table 11.3
	Performance Metrics (formulas)	Definition 11.4; Table 11.3
Synthesis	System and Environment (diagram)	Table 12.4
	Experiment Block Design (table)	Definitions 12.5, 12.6, 12.7; Table 12.11
Validation	Performance (tables or charts)	Tables 13.1, 13.3, 13.5; Figures 13.4, 13.5
	Conclusions (sentences)	Section 13.2
	Informal Observations (sentences)	Section 13.2, Figure 13.6, Table 13.6
	Recommendations (list of phrases)	Section 13.2

Some journals are moving to interim publication on the Internet to reduce the delay, but the reviewing process itself is still as slow as molasses. The reasons are not entirely clear. Most researchers acknowledge that they could review a manuscript in a few days, if asked. Yet editors rarely ask for such a fast response, reconciled, it would seem, to months of delay and procrastination.

Although industry has a high level of respect for and interest in the professional journals, its scientists and engineers seldom publish there, with the exception of those at large industrial research institutes. It may be that the level of detail required to pass muster on the *completeness* criterion during formal peer review presents industrial firms with too great a risk of losing control of valuable intellectual property. That is a shame, because it is quite clear that industry makes enormously valuable and significant contributions to the state-of-the-art on a regular basis. Perhaps a law should be passed that provides legal protection for all intellectual ideas that are published in refereed journals and conferences, at least temporarily until formal patent applications can be filed. Without such protection, there is an unfortunate Catch-22 that stifles industrial participation: no one wants to spend the money for an expensive patent search until they are quite sure the idea is both original and valid, but the most rigorous method for ascertaining this, publication in refereed venues, puts the very same intellectual property at risk.

The researchers who submit papers to refereed *conferences* are a much more eclectic group, including academicians, industrial and professional researchers, and graduate students. Although the formal refereeing processes are usually not as rigorous as those for journals, many regular conferences are very selective and highly respected. This situation continues to improve year-by-year, as the demand grows for expediting the peer-review process without sacrificing quality.

However, the formal review that leads to the acceptance or rejection of papers for refereed conferences is by no means the only peer review to which the research is subjected. Unlike professional journals, conference papers may not be submitted solely for publication in the proceedings; the work must also be presented orally at the conference itself, where hundreds of researchers from all over the world gather to listen to scores of presentations over a period of several days. And this audience is far from passive. Each presentation is followed by a discussion period, during which questions are raised and suggestions are offered. Usually the exchange is tranquil, but sometimes it heats up, especially when serious problems in methodology or underlying formalisms are revealed and disputed. Shouting matches, although unusual, are not unprecedented. Many long-term professional relationships, both friendly and otherwise, are born at these conferences. In almost every field, some of the enmities are legendary.

The importance of this kind of intellectual exchange cannot be overstated. Because it occurs in the presence of a group of researchers at all levels from many different organizations, a basic intellectual honesty is preserved. Of course, much more happens behind the scenes, but there is always at least one moment at the conference, during your presentation, when you must wash your underwear in public. In fact, the presentation of the paper itself should be regarded simply as a necessary introduction to the discussion period (peer review) that follows. For this reason, it is essential that the conference organizers and session chairs ensure that these discussion periods do not fall victim to time crunches. And if this means cutting off a speaker who goes beyond the time limit (regardless of rank or prestige!), so be it. Several professional societies, such as the IEEE, enforce this policy strictly at many of their conferences. Most would agree that the session chair who is courageous enough to do this earns the respect of the entire community.

Peer reviews take place in industry as well. Managers periodically organize meetings in which a small group of scientists and engineers present their work to each other, followed by critical discussions. The objectivity of such reviews may not be as high as at academic conferences, because the eventual objectives of industrial R&D projects are necessarily and understandably driven by the profit motive, which introduces a clear bias. Nonetheless, they are useful and productive. Of course, some might complain that such meetings are too often influenced by interdepartmental and intradepartmental politics. While this is undoubtedly true, it would be *very naive* to imagine that such politics are absent during the review processes for professional conferences or journals. Academia, like industry, is rife with politics.

Note that when project reports are reviewed by managers and executives, these are *not* peer reviews, because managers and executives are not the *peers* of their scientists and engineers, and *vice versa*. Peer review for managers and executives implies a formal critical review of their management projects, conducted by other managers and executives, who are their peers.

Likewise, when teachers grade the work of their students, this is *not* peer review. It is simply the Validation Phase of a project whose objective is to educate the student. When a doctoral student defends his or her dissertation, however, this *is* a peer review. By seeking the highest academic degree that can be earned, the student is formally applying for entrance into the same group to which the professors already belong and is judged by the same standards the academicians use among themselves.

There are several interesting examples of peer review in everyday life. The election process of a democracy is a good example of peer review. After the candidates present their credentials, their accomplishments, and their plans, then their fellow citizens vote. The result of the election is a reflection of the will of the people, the peers of the candidates. Of course, this is only true if there is no systematic disenfranchisement of sectors of the voting population, and if the method of counting the votes is fair and unbiased. Interestingly enough, this seems to imply that the election process itself should always be a viable subject for peer review.

Another example of peer review outside the realm of science and engineering is the fundamental right of every citizen of the United States to be tried by a jury of his or her peers, meaning a group of ordinary citizens whose biases have been negotiated by both the defense and the prosecution. Even though Americans treasure this constitutional guarantee, it is interesting to note that very few other democracies in the world afford their citizens this right.

A free press is an excellent example of peer review in action. Everyday, journalists are busy investigating and reporting the decisions and actions of the public and private sectors, holding their feet to the fire of public disclosure and accountability. The founding fathers of American democracy considered the fourth estate to be a mainstay of democracy, guaranteeing its protection in the very first amendment in the Bill of Rights. Perhaps Charles Evans Hughes, Chief Justice of the US Supreme Court from 1930 to 1941, best summarized the critical importance of a free press as a powerful feedback mechanism for both stability and change:

> The greater the importance of safeguarding the community from incitement to the overthrow of our institutions by force and violence, the more imperative is the need to preserve inviolate the constitutional rights of free speech, free press, and free assembly in order to maintain the opportunity for free political discussion, to the end that government may be responsive to the will of the people and that changes, if desired, may be obtained by peaceful means. Therein lies the security of the Republic, the very foundation of constitutional government.
>
> DeJonge v Oregon [1937]

Finally, there are some linguistic hints that the architects of Christianity intentionally encouraged a peer-to-peer relationship with their prophet and deity. Unlike its polytheistic predecessors, such as the ancient Greek and Roman religions, Christianity invites its followers to address their god in the familiar form. In English, *thou* is used, which was the familiar form of the second-person singular pronoun until early in the seventeeth century. In German, *Du* is used, which is still the familiar form of the second-person singular pronoun. Most of the European languages have both familiar and formal forms of the second-person pronouns, and the familiar form is used to address close friends, family, and the Christian deity. Perhaps this is intended to reassure Christians that Jesus will regard them as his peers in the kingdom of heaven. It is a comforting thought.

Peer review is the final step in the process of the Scientific Method, the Supreme Court, if you will, that issues the *imprimatur* for a research and development project, either recommending the provisional acceptance of its conclusions as reproducible, complete, and objective; or remanding the project back to the R&D team for reconsideration in some specified respects. Either way, the R&D project has been well served by the formal encapsulation of the associated knowledge and rigorous adherence to the Scientific Method.

Quod Erat Demonstratum.

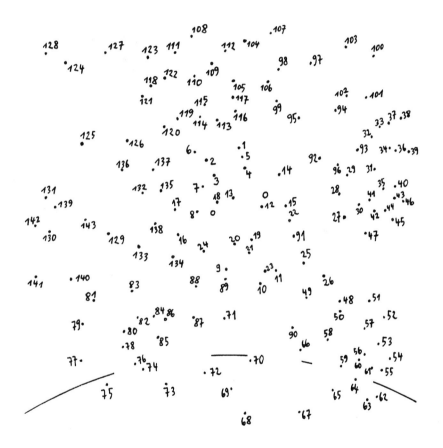

Order and simplification are the first steps
toward the mastery of a subject.

Thomas Mann

Wax on. Wax off.

APPENDICES

APPENDIX A

Bibliography

History and Philosophy of Science

Ayer, A. J. (1985). *Wittgenstein,* New York: Random House.

Bacon, F. (1998). *Essays,* Chicago, IL: NTC Publishing Group.

Bock, P. (1993). *The Emergence of Artificial Cognition,* Singapore: World Scientific Publishing Company.

Bodhi, B. (1988). A look at the Kalama Sutta, *Buddhist Publication Society Newsletter* (cover essay), **9**: Spring.

Birtles, P. J. (1998). *Jetliner for a New Century: The Boeing 777,* Osceola, WI: Motorbooks International.

Bunge, M. (1979). *Causality and Modern Science,* 3rd ed., New York: Dover Publications, Inc.

Bronowski, J. (1973). *The Ascent of Man,* Boston, MA: Little, Brown, and Company.

Cohen, L. J. (1989). *An Introduction to the Philosophy of Induction and Probability,* Oxford: Clarendon Press.

Descartes, R. (1994). *A Discourse on Method, Meditations on the First Philosophy,* translated by J. Veitch, A. D. Lindsay, ed. (reprint edition), Boston, MA: Everyman Paperback Classics.

Galileo, G. (1967). *Dialogue Concerning the Two Chief World Systems: Ptolemaic and Copernican,* translated by S. Drake, 2nd revised ed., Berkeley, CA: University of California Press.

Godwin, R. (compiler) (1999). *Friendship 7: The Nasa Mission Reports,* CG Publishing, Burlington, Ontario, Canada: Apogee Books.

Gregory, R. L. (ed.) (1987). *The Oxford Companion to the Mind,* Oxford: Oxford University Press.

Hofstadter, D. R. (1980). *Gödel, Escher, Bach: An Eternal Golden Braid,* New York: Vintage Books.

Kuhn, T. S. (1970). *The Structure of Scientific Revolutions,* 2nd ed., Chicago: University of Chicago Press.

Levy, D. H. (1991). *Clyde Tombaugh: Discoverer of Planet Pluto,* Tucson, AZ: University of Arizona Press.

Meystel, A. (1995). *Semiotic Modeling and Situation Analysis,* Bala Cynwyd, PA: Adrem.

Nickerson, R. (1999). Basic versus Applied Research, in *The Nature of Cognition,* R. Sternberg ed., Cambridge: MIT Press.

Regis, E. (1987). *Who Got Einstein's Office? Eccentricity and Genius at the Institute for Advanced Study,* Reading: Addison-Wesley Publishing Company.

Scheffler, I. (1963). *The Anatomy of Inquiry: Philosophical Studies in the Theory of Science,* Indianapolis, IN: Bobbs-Merrill Company.

Stuhlinger, E., Hunley, J. D., and Noordung, H. (1995). *The Problem of Space Travel: The Rocket Motor* (The Nasa History Series), Collingdale, PA: DIANE Publishing Company.

Wiles, A. (1993). Modular elliptic curves and Fermat's last theorem. *Annals of Mathematics* **141**:(3), 443–551.

Research Methodology

Babbie, E. R. (1975). *The Practice of Social Research,* Belmont, CA: Wadsworth Publishing Company.

Blalock, H. H., Jr. and Blalock, A. B. (eds.) (1968). *Methodology in Social Research,* New York: McGraw-Hill Book Company.

Cohen, P. (1995). *Empirical Methods for Artificial Intelligence,* Cambridge: MIT Press.

Covey, S. R. (1990). *The 7 Habits of Highly Effective People,* New York: Simon and Schuster (Fireside Book).

Friedman, L., Furberg, C., and DeMets, D. (1996). *Fundamentals of Clinical Trials,* 3rd ed., St. Louis, MO: Mosby-Year Book.

Kaplan, A. (1964). *The Conduct of Inquiry: Methodology for Behavioral Science,* Scranton, PA: Chandler Publishing Company.

Keppel, G., Saufley, W., Jr., and Tokunaga, H. (1992). *Introduction to Design and Analysis: A Student's Handbook,* 2nd ed., New York: W.H. Freeman and Company.

Kirk, R. (1995). *Experimental Design: Procedures for the Behavioral Sciences,* 3rd ed., Pacific Grove, CA: Brooks/Cole Publishing Company.

Mantel, N. (1968). Simultaneous confidence intervals and experimental design with normal correlation. *Biometrics* **24**:(2), 434–437, June.

Martin, D. (1996). *Doing Psychology Experiments,* 4th ed., Pacific Grove, CA: Brooks/Cole Publishing Company.

Ruxton, G. D. (1998). Experimental design: Minimizing suffering may not always mean minimizing number of subjects. *Animal Behavior* **56**:(2), 511–512, August.

Shetty, D. and Kolk, R. A. (1997). *Mechatronics System Design,* Brooks/Cole Publishing Company.

Spector, P. E. (1981). *Research Designs,* Beverly Hills, CA: Sage Publications.

Stump D. A., James, R. L., and Murkin, J. M. (2000). Is that outcome different or not? The effect of experimental design and statistics on neurobehavioral outcome studies. *Annals of Thoracic Surgery* **70**:(5), 1782–1785, November.

Wickens, T. (1999). Drawing Conclusions from Data: Statistical Methods for Coping with Uncertainty, in *Methods, Models, and Conceptual Issues: An Invitation to Cognitive Science,* D. Scarborough and S. Sternberg, eds., Vol. 4, 2nd ed., Cambridge: MIT Press, pp. 585–634.

Young, H. (1962). *Statistical Treatment of Experimental Data,* New York: McGraw-Hill Book Company.

Data Presentation and Visualization

Burn, D. (1993). Designing Effective Statistical Graphs, in *Computational Statistics,* C. R. Rao, ed., Amsterdam: North-Holland.

Hartwig, F. and Dearing, B. E. (1979). *Exploratory Data Analysis,* Beverly Hills, CA: Sage Publications.

Kosslyn, S. (1994). *Elements of Graph Design,* New York: W. H. Freeman and Company.

Schmid, C. F. (1983). *Statistical Graphics: Design Principles and Practices,* New York: John Wiley and Sons.

Tufte, E. R. (1997). *Visual Explanations: Images and Quantities, Evidence and Narrative,* Cheshire, CT: Graphics Press.

Tufte, E. R. (1990). *Envisioning Information,* Cheshire, CT: Graphics Press.

Tufte, E. R. (1983). *The Visual Display of Qualitative Information,* Cheshire, CT: Graphics Press.

Statistics

Bendat, J. and Piersol, A. (1986). *Random Data: Analysis and Measurement Procedures,* 2nd ed., New York: John Wiley and Sons.

Bradley, R. (1984). Paired Comparisons: Some Basic Procedures and Examples, in *Nonparametric Methods,* P. R. Krishnaiah and P. K. Sen, eds., Amsterdam: North-Holland.

Bowker, A. and Lieberman, G. (1972). *Engineering Statistics,* 2nd ed., Englewood Cliffs: Prentice-Hall.

Cohen, J. (1969). *Statistical Power Analysis for the Behavioral Sciences,* New York: Academic Press.

Crow, E., Davis, F., and Maxfield, M. (1960). *Statistics Manual,* New York: Dover Publications.

Efron, B. (1982). *The Jackknife, the Bootstrap and Other Resampling Plans,* Philadelphia, PA: Society for Industrial and Applied Mathematics.

Freund, R. J. and Wilson, W. J. (1993). *Statistical Methods* (revised edition), San Diego, CA: Academic Press.

Gibbons, J. (1993). *Nonparametric Measures of Association,* Newbury Park, CA: Sage Publications.

Harmon, H. J. (1976). *Modern Factor Analysis,* 3rd ed., revised, Chicago: University of Chicago Press.

Hoel, P. (1971). *Introduction to Mathematical Statistics,* 4th ed., New York: John Wiley and Sons.

Hogg, R. and Craig, A. (1978). *Introduction to Mathematical Statistics,* 4th ed., New York: Macmillan Publishing Company.

Ingram, J. (1974). *Introductory Statistics,* Menlo Park, CA: Cummings Publishing Company.

Jackson, J. (1991). *A User's Guide to Principal Components,* New York: John Wiley and Sons.

Johnson, R. and Wichern, D. (1992). *Applied Multivariate Statistical Analysis,* 3rd ed., Upper Saddle River, NJ: Prentice-Hall.

Langley, R. (1970). *Practical Statistics: Simply Explained,* revised ed., New York: Dover Publications.

Lawlis, G. F. and Chatfield, D. (1974). *Multivariate Approaches for the Behavioral Sciences: A Brief Text,* Lubbock, TX: Texas Tech Press.

Mendenhall, W. and Sincich, T. (1992). *Statistics for Engineering and the Sciences,* San Francisco, CA: Dellen Publishing Company.

Papoulis, A. (1991). *Probability, Random Variables, and Stochastic Processes,* 3rd ed., New York: McGraw-Hill.

Salsburg, D. (2001). *The Lady Tasting Tea: How Statistics Revolutionized Science in the Twentieth Century,* New York: W. H. Freeman and Company.

Sincich, T., Levine, D., and Stephan, D. (1999). *Practical Statistics by Example Using Microsoft Excel,* Upper Saddle River, NJ: Prentice-Hall.

Taylor, J. R. (1982). *An Introduction to Error Analysis: The Study of Uncertainties in Physical Measurements,* Oxford: Oxford University Press.

Winer, B. J. and Michels, K. M. (1991). *Statistical Principles in Experimental Design,* 3rd ed., New York: McGraw-Hill Book Company.

APPENDIX B

Glossary

Term	Definition	Location
accuracy	Accuracy is a measure of how close a result is to some baseline.	Definition 8.1
alternative hypothesis	An alternative hypothesis is an hypothesis that refers to a region in the decision space that does not include the value referenced by the null hypothesis.	Section 13.2
Analysis Phase objective	The objective of the Analysis Phase is to gain a thorough understanding of the components of the problem domain, leading to the formulation of a single specific and reasonable task objective.	Definition 10.1
assumption	An assumption expresses an attribute or value of a property or a process that is presumed to be valid for the application domain of the task.	Definition 6.3
axiom	An axiom expresses a fundamental concept or property that is presumed to be universally accepted as valid everywhere in the universe.	Definition 6.1
baseline	A baseline may be defined as any one of several different reference values, such as the correct value, the desired value, or the expected value.	Section 8.1

benefit	A benefit is a performance metric that specifies an advantageous result of applying the solution to the task unit.	Definition 7.11
bias negotiation	Bias negotiation is the process by which the effects of biases are eliminated, minimized, or somehow taken into account.	Section 12.2.2
biased proposition	A biased proposition is a condition or parameter that reflects some limitation, restriction, or exclusion that has been imposed on the task.	Section 7.1
blind trial	A blind trial is an experiment trial in which the subjects are not told whether the treatments are test or control treatments.	Section 12.2.2
channels	The channels in a task domain are the means by which the resources of the task are interconnected to provide the means for exchanging energy and information.	Definition 4.9
cohort	A cohort is a task unit consisting of a set of living plants or animals (called subjects), all of which are assumed to be equivalent for the requirements of the task objective.	Definition 12.3
completeness	The completeness criterion for the provisional acceptance of knowledge requires that the knowledge and methods used to accomplish a task be completely and clearly disclosed using standard terminology.	Chapter 5
conclusion	A conclusion is the final result of a task that states the extent to which the task objective has been achieved.	Definition 6.9
condition	A condition is a specification of the task environment, whose value is either fixed or varied as a factor of the task.	Definition 7.1
constant	A situation is a numerical or categorical quantity that is a parameter; its value is called a level.	Definition 7.4
control trial	A control trial measures the performance of one set of task components in the absence of another set of task components to isolate the effects of the included components on performance.	Definition 12.8
convention	A convention is a rule or statement governed by an accepted standard.	Definition 6.6
cost	A cost is a performance metric that specifies an immediate or long-term disadvantage of an associated benefit.	Definition 7.12

decision task	A decision task requires a decision based on some acquired knowledge to achieve its objective.	Section 11.2
definition	A definition is a statement of meaning or membership.	Definition 6.7
development	Development is a process that applies knowledge to create new devices or effects.	Definition 1.2
doubleblind trial	A doubleblind trial is a trial in which neither the subjects nor the human supervisors are told whether the treatments are test or control treatments.	Section 12.2.2
enumerated block design	An enumerated block design assigns every possible treatment to every member of the task unit.	Section 12.2.2
error	An error is a measure of the distance of a result from some baseline.	Definition 8.2
estimation task	An estimation task requires knowledge, but no decision, to achieve its objective.	Section 11.2
exclusion	An exclusion is a fixed condition or fixed parameter that avoids, minimizes, or ignores an unwanted influence on the task.	Definition 7.7
experiment	An experiment acquires data to measure the performance of the solution under controlled conditions in a laboratory.	Definition 12.2
experiment block	An experiment block is a set of trials that provides a cover of the factor space that is appropriate and adequate for achieving the task objective.	Definition 12.7
fact	A fact is a statement of objective reality.	Definition 6.5
factor	A factor (or performance factor) is a condition or parameter of a task whose value is intentionally varied to measure its impact on the results of the task.	Definition 7.8
feasibility (QD) pilot	Feasibility pilots (also called QD pilots, for Quick-and-Dirty) are simplified and rather informal investigations designed to quickly and inexpensively explore the feasibility of a proposed solution or an experiment design.	Section 9.2.1
goal	Every task has at least one goal to determine the response of the task unit to the application of the solution (or part of the solution), expressed as an infinitive phrase.	Definition 11.2

governing propositions	The governing propositions are those conditions and parameters whose values are fixed during the execution of the task.	Section 7.1
granularity	Granularity is the smallest measurement increment specified for a device or quantity.	Section 8.1
hypothesis	An hypothesis is a declarative sentence asserting a desired, expected, or possible conclusion of the application of the solution (or part of the solution) to achieve the task objective (or part of the objective).	Definition 11.3
Hypothesis Phase objective	The objective of the Hypothesis Phase is to propose a solution to achieve the task objective, a set of goals and hypotheses for this solution, and the factors and performance metrics for testing the validity of the solution.	Definition 11.1
indirect proposition	An indirect proposition is a governing proposition or factor whose value cannot be set directly, but must be set indirectly by observing its value as another condition or parameter is varied.	Section 7.2
inducer	An inducer is a device or mechanism that alters the task unit during or before the task.	Definition 4.5
laboratory	The laboratory for a task is the set of physical regions containing the task unit and all the task resources required during the experiments.	Definition 4.10
law	A law (or theory) is a fundamental relationship or process that is presumed to be universally accepted as valid everywhere in the universe.	Definition 6.2
lemma	A lemma is a conclusion that has been formally proved, often described as a "helping theorem."	Definition 6.10
limitation	A limitation is a fixed condition or fixed parameter that constrains the precision or amount of a task resource.	Definition 7.5
limiting proposition	The limiting proposition is a governing proposition that states the value of the precision limit.	Section 8.1
neutral proposition	A neutral proposition is a condition or parameter that does not impose a limitation, restriction, or exclusion.	Section 7.1
null hypothesis	A null hypothesis is a hypothesis that refers to a single, unambiguous value in the decision space for a task.	Section 13.1

objectivity	The objectivity criterion for the provisional acceptance of knowledge requires that the knowledge propositions for a task be as free of bias and extraneous influence as possible.	Chapter 5
operating-point (OP) pilot	An operating-point (OP) pilot is an informal task to search a factor space for suitable values for the governing propositions and for suitable dynamic ranges and increments for the factors of the formal experiments of the project.	Section 11.3
parameter	A parameter is a specification of the task system, whose value is either fixed or varied as a factor of the task.	Definition 7.2
payoff	A payoff is an arithmetic combination of one or more benefits and costs that summarizes the impact of their interaction by increasing in value with increasing benefit and decreasing in value with increasing cost.	Definition 7.14
performance criterion	A performance criterion is a requirement that any proposed solution to the problem must fulfill.	Section 10.2
performance factor	A performance factor (or factor) is a condition or parameter of a task whose value is intentionally varied to measure its impact on the results of the task.	Definition 7.8
performance metric	A performance metric is a postulate that transforms the results of the task into measures of performance for drawing conclusions about the task objective.	Definition 11.4
placebo	A placebo is a substance or an effect that is used as a neutral or ineffective inducer for control purposes.	Section 12.2.2
policy	A policy is a process, function, rule, or mechanism that is a factor of a task; its value is called a setting.	Definition 7.9
postulate	A postulate is a process, function, rule, or mechanism that is presumed to be valid for the application domain of the task.	Definition 6.4
precision	Precision is the number of states that spans the dynamic range of a quantity, expressing the maximum amount of information that can be expressed by the quantity.	Definition 8.4
precision limit	The precision limit is the minimum of the precisions of all the devices and quantities established by the governing propositions, which determines the maximum allowable precision for the results and conclusions of the task.	Section 8.1

primary goals	The primary goals are the goals of the primary task of a project.	Section 11.2
primary hypotheses	The primary hypotheses are the hypotheses of the primary task of a project.	Section 11.2
primary task	The primary task is the sole task of the highest level, or root, of a project task tree.	Section 3.1
problem statement	The problem statement is an interrogative sentence, a declarative sentence, or an imperative sentence that summarizes a question, complaint, or requirement, respectively.	Definition 10.2
protocol	A protocol is an exact step-by-step procedure for the preparation and conduct of experiment trials.	Section 12.2.2
quality	Quality is a metric that is an arithmetic combination of several benefits, which summarizes the extent to which the task objective has been achieved.	Definition 7.13
randomized block design	A randomized experiment block design assigns a random sequence of treatments to each member of the task unit (or vice versa).	Section 12.2.2
raw error	The raw error is a quantity that expresses the difference between the result and the baseline in the same units as the result itself.	Section 8.1
regime	A regime is a process, function, rule, or mechanism that is a condition; its value is called a setting.	Definition 7.3
representation factor	A representation factor is a parameter or a condition whose values are selected uniformly from its domain and applied to all treatments to assure fair and equal representation of all values.	Section 12.2.2
reproducibility	The reproducibility criterion for the provisional acceptance of knowledge requires that all independent attempts to accomplish a task under the same conditions yield the same results.	Chapter 5
research	Research is a process that acquires new knowledge.	Definition 1.1
resolution	Resolution is the reciprocal of granularity.	Section 8.1
restriction	A restriction is a fixed condition or fixed parameter that constrains the scope of the utility or validity of the products of a task.	Definition 7.6
result	A result is an intermediate measure of performance of a task.	Definition 6.8

sample	A sample is a task unit consisting of a set of nonliving objects or concepts (usually called instances, data points, specimens, items, or cases), all of which are assumed to be equivalent for the requirements of the task objective.	Definition 12.4
scheme	A scheme is a process, function, rule, or mechanism that is a parameter; its value is called a setting.	Definition 7.3
Scientific Method	The Scientific Method comprises four sequential phases—Analysis, Hypothesis, Synthesis, and Validation—which are applied to a task iteratively and recursively to achieve the objective of the task.	Definition 9.1
sensor	A sensor is a device that acquires the required data from the task unit.	Definition 4.6
situation	A situation is a numerical or categorical quantity that is a condition; its value is called a level.	Definition 7.4
supervisor	The supervisor is the set of human and automated agents that operates and monitors the task unit, the concurrent inducers, the sensors, and itself.	Definition 4.8
Synthesis Phase objective	The objective of the Synthesis Phase is to implement the task method (solution and experiments) to accomplish the goals and validate the hypotheses of the task.	Definition 12.1
system	The system for a task is the intersection of the regions altered by the required inducers and the regions sensed by the sensors, as long as this intersection contains the task unit.	Definition 4.7
systematic block design	A systematic experiment block design uses a deterministic algorithm to assign different treatments to different subsets of the task unit in a systematic way, covering the entire factor space.	Section 12.2.2
task	An R&D task applies a specified method to the domain of the task with the objective of obtaining a satisfactory result in the range of the task.	Definition 4.1
task domain	The task domain comprises the task unit and the resources necessary to achieve the task objective.	Definition 4.2
task method	The task method comprises the solution specified to achieve the task objective and the experiments designed to determine the effectiveness of the solution, that is, everything required to transform the task domain into the task range.	Definition 4.11

task objective	The task objective is a statement of what the task is intended to achieve, expressed as an infinitive phrase.	Definition 4.3
task range	The task range comprises all the products of the task, including knowledge, devices, and effects.	Definition 4.12
task unit	The task unit comprises the set of objects or concepts that undergoes some required alteration before or during the task and measurement during the task.	Definition 4.4
test trial	A test trial measures the performance of the task unit under the influence of the required inducers.	Section 12.2.2
theorem	A theorem (or lemma) is a conclusion that has been formally proved.	Definition 6.10
theory	A theory (or law) is a fundamental relationship or process that is presumed to be universally accepted as valid everywhere in the universe.	Definition 6.2
treatment	A treatment is a combination of one level or setting for every factor.	Definition 12.5
trial	A trial is a complete set of treatments assigned to a member of the task unit during the experiment.	Definition 12.6
uncertainty	Uncertainty is a measure of the inconsistency in a sample.	Definition 8.3
variable	A variable is a numerical or categorical quantity that is a factor of a task; its value is called a level.	Definition 7.10

APPENDIX C

Tips

TIP 1 Provide every professional with a comprehensive set of high-quality state-of-the-art tools for planning, designing, and implementation, as well as the time and resources to update and maintain them.

TIP 2 Use pilot tasks throughout a project to identify unexpected consequences, forestall failures, avoid brittle procedures, determine operating ranges, and estimate the feasibility of proposed solutions.

TIP 3 Use task trees for the top-down design of projects; use milestone charts for planning the logistics, schedules, and costs of these projects as they evolve.

TIP 4 Inventory the external and internal alterations that the task unit undergoes before or during the task, carefully deciding whether each is required, unwanted, or irrelevant, and whether it is artificial or natural.

TIP 5 Be cautious with tasks using humans or other animals. They are usually very difficult and expensive to manage and control, and often introduce complex ethical and legal issues that must not be overlooked.

TIP 6 For a given task, carefully define the system, the environment, and the laboratory, and then examine them thoroughly and methodically to identify:

  a) unwanted influences generated by and acting on the task resources;
  b) detrimental ways in which the system boundary can be violated; and
  c) extraneous elements within the system that influence the task unit.

TIP 7   Combining the individual benefits into a summary metric (quality) or combining the benefits and cost into a single summary metric (payoff) is always possible, but has the potential of hiding meaningful information.

TIP 8   As each task is planned, make a brief note of each knowledge proposition in your research book as it occurs to you. Periodically collect the notes into a master list, wordsmith them into *bona fide* knowledge propositions, organize them by category, type, and role, and make sure there are no conflicts.

TIP 9   To minimize the net uncertainty of the results of a measurement task: 1) Measure things as directly as possible, minimizing the number of arithmetic combinations of intermediate results; and 2) ensure that samples of measurements are as independent and random as possible.

TIP 10   For each knowledge proposition that includes a number, state or estimate the precision of the number whenever possible, either directly in the proposition itself or in an accompanying proposition. For example:

Fact: The video camera supports 256 gray levels.

Fact: Speeds in the range 0 to 120 mph were measured to the nearest 0.5 mph.

TIP 11   Below is an Excel formula that can be used to express a result with the correct precision, given the precision limit and the raw result:

```
=FIXED(rawResult, IF (LOG10(precisionLimit/ABS(rawResult)) < 0,
TRUNC(LOG10(precisionLimit/ABS(rawResult))) − 1,
TRUNC(LOG10(precisionLimit/ABS(rawResult)))), TRUE)
```

Note that the result of the FIXED Excel function is text, not a numeric value.

TIP 12   Feasibility or QD pilots are useful for exploring the feasibility of proposed task components. Every QD pilot, although perhaps only minimally planned and not usually reported in the final project documentation, *must* be fully recorded in the cognizant team members' research notebooks. Once a QD pilot has yielded a potentially useful result, it *must* be confirmed and validated by planning and executing a formal task.

TIP 13   In general, too little planning time incurs much heavier penalties in overall task time than too much planning time.

TIP 14   Because the project task tree and milestone chart present an overview of every step within every phase of the Scientific Method for a project, it is particularly useful as one of the primary planning documents, a current copy of which should always be in the hands of every team member.

TIP 15   For R&D scientists and engineers: Resist the urge to "make it up as you go along." Plan the project thoroughly in advance, but allow the plan to evolve as experience and pilots reveal the flaws and potential improvements in the plan. Be prepared to give management time and cost estimates whose accuracy they can trust. If you discover you will not be able to make a deadline, notify your manager immediately; don't wait until the last moment. Understand that managers see and deal with a larger perspective than you do. Accept that the workplace is not a democracy.

TIP 16   For R&D managers: Support your scientists and engineers with the time and resources necessary to plan and stepwise refine these plans throughout the lifetimes of projects. Limit the work of each person to one or two concurrent projects. Don't steal time from project scientists and engineers for "fire drills"; hire special staff members for that purpose. Don't micromanage. Don't nickel-and-dime. Try not to burden your scientists and engineers with bureaucratic tasks; that's *your* job. If you have no choice, then explicitly acknowledge that the associated time and costs of the bureaucratic task are not a part of the ongoing project budgets, and extend all deadlines appropriately.

TIP 17   The task objective is a *contract* between the task team and management. It often has a political component. Word it carefully.

TIP 18   From the very beginning of a project, start building a list of knowledge propositions. Update it whenever a new proposition is established or a previous proposition is modified.

TIP 19   Use OP (operating-point) pilots to search for suitable values for fixed parameters and conditions, and dynamic ranges and increments for factors.

TIP 20   If it meets the needs of the task, using an existing solution is almost always more cost-effective than manufacturing it yourself.

TIP 21   Arrange to be given complete, raw, and unpreprocessed data. Expect to deal with political and ownership issues, as well as unsolicited "help."

TIP 22   Before beginning the design of the experiments, carefully inventory all the components of the task.

TIP 23   Make sure every item in the laboratory has a specific purpose in the experiments. If an item is not on the inventory of required task components, remove it. If it cannot be removed, deal with it proactively. (See Table 12.3)

TIP 24   Deny access to the laboratory for everyone but essential personnel, post and publish this restriction, and strictly enforce it. Disable the adjustment of all fixed parameters and conditions. If necessary, monitor the laboratory continuously during untended periods. (See Table 12.7)

TIP 25   Experiment biases are often very subtle, unexpected, and counterintuitive. Conduct a thorough inventory of every source that could conceivably exert an unwanted influence on the task components. Then, based on perceived risk and cost, decide whether these sources and their effects should be further investigated for potential biases. Use psychologists who have experience with experiments involving humans to help you identify and negotiate biases.

TIP 26   Because the data reduction methods and/or performance metrics may have to be modified or corrected, record both the raw *and* reduced results of experiments.

TIP 27   Before computing performance, conduct a QD pilot to analyze the raw or reduced results to detect any unusual artifacts or unexpected trends stemming from biases in the task methods and to confirm the appropriateness of the postulated performance metrics.

TIP 28   Before abandoning formal rigor in the face of what seem to be obvious conclusions, check the distributions of the underlying performance values. Lacking estimates of the uncertainty of these values, when asked if their averages (or other location parameters) are significantly different, your answer must be "I don't know. I do not have enough information."

TIP 29   In addition to the objectives of the current task, make sure the project and management personnel are reasonably well informed about the objectives of both the parent and the descendant tasks. Establish efficient channels of communication between adjacent levels in the project task hierarchy.

TIP 30   Typographic and color attributes can provide powerful means of emphasis and differentiation within text and graphics. Improper or arbitrary use of them, however, can cause considerable distraction and confusion. Get some formal training to learn how to use them sensibly and consistently. Make sure you know how the document will appear when duplicated in grayscale.

TIP 31   When documenting the process of the project, omit the detours and false starts, and distill it down to as simple a story as possible: a linear sequence of successful steps from Analysis to Hypothesis to Synthesis to Validation.

TIP 32   Before putting together a project report or presentation, understand the intended audience and its primary interests in the project, both intellectual and political. Let this knowledge guide the design and content of the report.

Summaries and Guidelines

Table 1.1 Sources of chaos and confusion in typical R&D groups

- Research results cannot be reproduced because of poor methodology and documentation.
- Speculations are not identified as such and are intermixed with supported conclusions.
- Knowledge is precarious, locked up in the heads of individuals.
- Data collection is haphazard and confounded with political issues.
- Experiment methods are chaotic, dominated by a "try this, try that" mentality.
- Experiment processes cannot be audited or reviewed due to a lack of logs and records.
- Reports are too long (or too short), poorly organized, incomplete, and confusing.
- Project documentation is sparse or nonexistent.
- Statistical analysis of results is missing or naive.
- Oral presentations are disorganized, confusing, and emphasize the wrong things.
- Data visualization techniques are poor and the presenters are hard to follow.
- Project activities are isolated and inbred.

Table 2.1 Tools and resources necessary for all professional scientists and engineers

- a research notebook that contains a complete record of *all* your activities, updated faithfully and regularly
- a thorough understanding of descriptive statistics to avoid blind reliance on statistical packages
- thorough knowledge of all functions and capabilities of a powerful spreadsheet program (*e.g., Excel* or *MatLab*)
- thorough knowledge of all functions and capabilities of a powerful slideshow design program (*e.g., PowerPoint*)
- access to a wide variety of specialized software packages for data analysis and visualization
- very fast and unconstrained Internet access
- a very fast and large-RAM-capacity desktop workstation at home and at work, replaced every three years
- convenient access to a well-stocked technical library, including books and professional journals
- a small collection of professional books, including handbooks and trade references
- regular attendance at and participation in professional conferences and workshops
- subscriptions to a few professional journals and magazines in your technical area
- the services of a professional technical editor for in-house technical reports and published papers

Table 2.2 Guidelines for the use of the research notebook

- Put EVERYTHING in your notebook — use it like a diary and a scrapbook, both professional and personal.
- Keep your notebook with you at all times so that you are always ready to record your thoughts and ideas.
- Your notebook is your private record. No one else need ever see it, although you may choose to share items.
- Ask other people to sketch or write suggestions or information for you in your notebook, not on loose paper.
- Resist using loose sheets of paper for "casual" notes and ideas. Use your research notebook.
- Use paginated notebooks with high-quality quadrilled paper so that erasures will not wear out the paper.
- Put your telephone number and a request for return on the cover; loss of this record can be very serious.
- Number and date each notebook. You will fill one every few months. Archive the filled books in a safe place.
- Slate every new subject, and date every new entry. Specify the sources of nonoriginal information.
- Use right-hand pages for regular entries, left-hand pages for revisiting topics (or *vice versa,* if more convenient).
- Use preprinted (or manually entered) page numbers to cross-reference related entries.
- Paste or tape copies of important and useful documents into your notebook, including computer printouts.
- Do not erase major "mistakes." Cross them out and explain them. You may need to reevaluate them later.
- When a notebook is full, put copies of important pages in a looseleaf binder using sheet protectors.

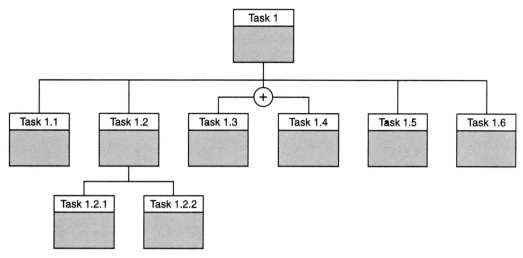

Figure 3.1 Sample project task tree

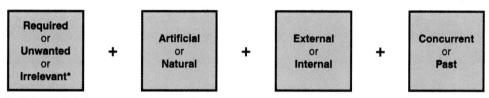

*not included in the task resources

Figure 4.1 Types of alterations and inducers that can affect the task unit

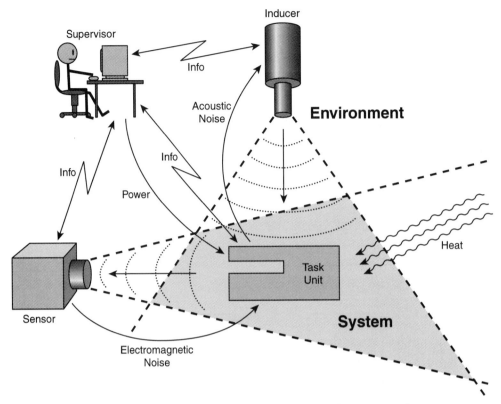

Figure 4.3 A simple system connected to its environment by information and energy channels

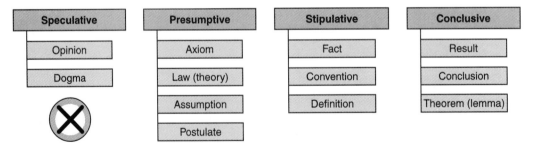

Figure 6.1 Taxonomy of categories and types of knowledge propositions

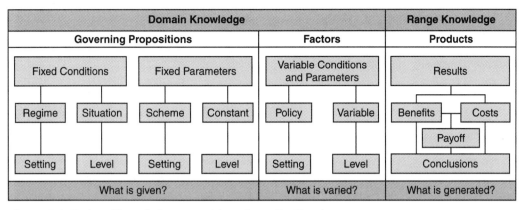

Figure 7.1 Roles of knowledge propositions for an R&D task

Table 8.1 Common error metrics

Formula	Type	Formula	Type
$\varepsilon = x - \beta$	signed raw	$\varepsilon = \dfrac{\Sigma(x-\beta)}{n\lvert\beta\rvert}$	average signed fractional
$\varepsilon = \dfrac{x-\beta}{\lvert\beta\rvert}$	signed fractional	$\varepsilon = \dfrac{\Sigma\lvert x-\beta\rvert}{n}$	average absolute raw
$\varepsilon = \lvert x-\beta\rvert$	absolute raw	$\varepsilon = \dfrac{\Sigma\lvert x-\beta\rvert}{n\lvert\beta\rvert}$	average absolute fractional L_1 norm
$\varepsilon = \dfrac{x-\beta}{\lvert\beta\rvert}$	absolute fractional	$\varepsilon = \sqrt{\dfrac{\Sigma(x-\beta)^2}{n}}$	root-mean-square (rms) L_2 norm
$\varepsilon = \dfrac{\Sigma(x-\beta)}{n}$	average signed raw	$\varepsilon = \dfrac{1}{\lvert\beta\rvert}\sqrt{\dfrac{\Sigma(x-\beta)^2}{n}}$	fractional root-mean-square (rms) L_∞ norm

x = measured value β = baseline value n = number of measurements ε = error value

Table 8.6 Summary of uncertainty metrics

1) If there is no need to normalize the measure of uncertainty of a sample to allow comparison with other samples with different averages, then the **uncertainty** and **fractional uncertainty** are useful metrics.

   uncertainty = u = standard deviation   (Postulate 8.1)

   fractional uncertainty = δu = absolute value of coefficient of variation   (Postulate 8.2)

   Note: If the average is exactly zero, the coefficient of variation is undefined.

2) If a normalized measure of uncertainty is required to allow comparison of two or more samples with different averages, then the **normalized fractional uncertainty**, **scaled normalized fractional uncertainty**, and **certainty** are useful metrics.

   normalized fractional uncertainty   = $\delta \acute{u}$ (Postulate 8.3)

   = ratio of the fractional uncertainty to the square root of the sample size

   scaled normalized fractional uncertainty = $\delta \tilde{u}$ (Postulate 8.4)

   = normalized fractional uncertainty raised to the s power where s = scaling factor $(0,1)$

   certainty = c = unit complement of the scaled normalized fractional uncertainty

   Note: The value of the scaling factor k is arbitrary, but constant across all comparisons.

3) The **confidence** that a measurement will lie in a specified confidence interval and the **confidence interval** that will include measurements with a specified confidence are useful uncertainty metrics.

   Parametric methods (*e.g.,* the standard distance) may be used to compute the values of these metrics *if (and only if) the distribution is known, e.g.,* Gaussian (normal), Poisson, Binomial, *etc.*

   Nonparametric geometric estimation methods can be used to compute the values of these metrics by constructing and analyzing histograms and/or Pareto plots of the samples. Histogram binning will introduce quantization error. For Pareto plots, accuracy of the estimates is only a function of sample size (data sufficiency).

4) To compare the uncertainties of two samples, the $K\%$ uncertainty ratio is a useful metric.

   $K\%$ uncertainty ratio = r_K = ratio of the fractional deviations for two samples

   fractional deviation = ratio of width of the $K\%$ confidence interval to the sample average

   Note: 50% is a common value for K.

   The minimum, average, or maximum r_K over all $K[0, 100\%]$ may also be useful.

Table 8.7 Propagation of uncertainty in arithmetic combinations of sample results

Random and Independent Samples	*Dependent Samples*
Given a set of measurement sample results $u, v, \ldots, z$ with uncertainties $\delta u, \delta v, \ldots, \delta z$:	

IF the results are combined as:	**IF** the results are combined as:
$q = u \pm v \pm \ldots \pm z$	$q = u \pm v \pm \ldots \pm z$
THEN the uncertainties add in quadrature:	**THEN** the uncertainties add:
$\delta q = \sqrt{(\delta u)^2 + (\delta v)^2 + \ldots + (\delta z)^2}$	$\delta q \leq \delta u + \delta v + \ldots + \delta z$

IF the results are combined as:	**IF** the results are combined as:
$q = \dfrac{x\, y \, \ldots \, z}{u\, v \, \ldots \, w}$	$q = \dfrac{x\, y \, \ldots \, z}{u\, v \, \ldots \, w}$
THEN the fractional uncertainties add in quadrature:	**THEN** the fractional uncertainties add:
$\dfrac{\delta q}{q} = \sqrt{\left(\dfrac{\delta u}{u}\right)^2 + \left(\dfrac{\delta v}{v}\right)^2 + \ldots + \left(\dfrac{\delta z}{z}\right)^2}$	$\dfrac{\delta q}{q} \leq \dfrac{\delta u}{u} + \dfrac{\delta v}{v} + \ldots + \dfrac{\delta z}{z}$

In general, given the function $q(u, v, \ldots, z)$ for combining the measurement sample results:

$\delta q = \sqrt{\left(\dfrac{\partial q}{\partial u}\delta u\right)^2 + \ldots + \left(\dfrac{\partial q}{\partial z}\delta z\right)^2}$	$\delta q \leq \left\|\dfrac{\partial q}{\partial u}\right\|\delta u + \ldots + \left\|\dfrac{\partial q}{\partial z}\right\|\delta z$
	EXAMPLE: for $q = x^n$, $\dfrac{\delta q}{\|q\|} \leq \|n\|\dfrac{\delta x}{\|x\|}$

The Scientific Method

The iteration of four recursive phases for
the planning, conduct, and stepwise refinement
of a research or development task

Analysis	Describe Problem
	Set Performance Criteria
	Investigate Related Work
	State Objective
Hypothesis	Specify Solution
	Set Goals
	Define Factors
	Postulate Performance Metrics
Synthesis	Implement Solution
	Design Experiments
	Conduct Experiments
	Reduce Results
Validation	Compute Performance
	Draw Conclusions
	Prepare Documentation
	Solicit Peer Review

Figure 9.1 The four phases of the Scientific Method and their internal steps

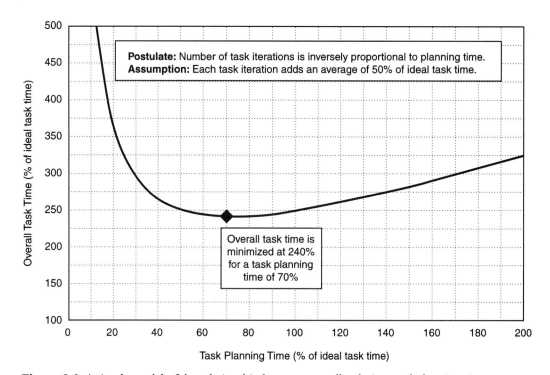

Figure 9.2 A simple model of the relationship between overall task time and planning time

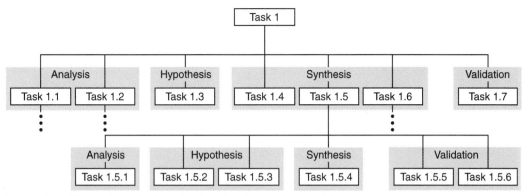

Figure 9.4 Illustration of a three-level project hierarchy (partial)

Table 12.1 Critical software documentation guidelines

Internal Documentation

1   Limit each code module to one or two standard pages for convenient inclusion in reports on facing pages.
2   Introduce each code module with a block of comments that specify:
  • Slate (title, module hierarchy position, programmer, contact info, date, revision number, remarks)
  • Purpose of the module
  • Values of relevant governing propositions
  • Purpose of each major data structure (variable, container, constant, pointer, *etc.*)
  • Intermodule communication parameters
3   Insert a blank line and a row of asterisks before and after each code section to highlight it.
4   Begin each code section with a highlighted comment stating its task function.
5   Clearly indent embedded and nested structures to emphasize the hierarchy of operations.
6   Enter control words in uppercase and boldface (**IF, THEN, ELSE, REPEAT, END**, *etc.*).
7   Invent names that are long enough to clearly communicate their task function.
8   Enter shorthand internal documentation statements DURING programming. Periodically clean them up.

External Documentation

1   Generate and maintain module hierarchy charts in cooperation with all project designers and programmers.
2   Generate and maintain flow charts or pseudo-code for every code module.
3   Keep accurate and up-to-date maintenance and revision logs.
4   Hire experienced documentation specialists to design user manuals and help files.
5   Maintain internal network sites and Internet web pages (if appropriate) for software documentation and help.

Table 12.3 Inventory and planning checklist for the significant components of a task

Component	Reference
DOMAIN	
❏ task objective Section 4.1.1	
❏ task unit	Section 4.1.2
❏ task resources	Section 4.1.3
❏ inducers required or unwanted artificial or natural external or internal concurrent or past	Section 4.1.3.1
❏ sensors quantitative qualitative	Section 4.1.3.2
❏ system and environment	Definition 4.7
❏ supervisor	Section 4.1.3.3
❏ channels (input, output)	Section 4.1.3.4
❏ domain knowledge	Chapters 6, 7, 11
knowledge propositions (presumptive, stipulative, conclusive) conditions (regimes, situations) parameters (schemes, constants)	Sections 6.2–6.4
governing propositions neutral propositions biased propositions (limitations, restrictions, exclusions)	Section 7.1
goals and hypotheses	Section 11.2
factors (policies, variables)	Section 7.2
performance metrics	Sections 6.2, 13.1
benefits cost payoff	Section 7.3
RANGE	
❏ devices and effects (development tasks)	Section 4.3
❏ range knowledge (research tasks)	Section 7.3
❏ results	Sections 6.4, 12.4
❏ conclusions	Section 6.4
METHOD	
❏ solution	Sections 4.2, 11.1
❏ experiments	Section 12.2
❏ laboratory	Section 12.2.1
❏ block design	Section 12.2.2
❏ data management	Section 12.2.3

Table 12.6 Safety tips for the design and use of experiment laboratories

! Do not allow anyone with any of the following items near machines with exposed moving parts: long sleeves or hair; neckties or scarves*; watches and jewelry on the hands, neck, or limbs.

! Do not allow anyone to wear metal hand jewelry when working on high-voltage equipment.

! Limit contact with electrical equipment to one hand; touch only your own body with the other hand.

! Require hard hats when work is performed near apparatus that is suspended overhead.

! Require safety glasses in labs where chemical or biological agents could be released into the air.

! Allow no exceptions for safety rules, regardless of the rank or status of the person in the lab.

! With dangerous equipment use panic switches controlled by technicians who have no other lab duty.

! During dangerous experiments, limit the number of people in the lab to an absolute minimum.

! Make video tapes (with a sound track) of all dangerous or ethically controversial experiments.

! Consult your legal staff about all experiments involving humans or animals. The laws are very strict.

*Many years ago, I had the distinct displeasure of watching (helplessly) as an R&D manager caught his necktie in a rotating tool on a lathe. He spent months in the hospital and was disfigured for life. He is lucky to be alive.

Table 12.7 Options for continuous monitoring of the laboratory

• Video cameras (visible and IR)	• Power line activity	• Temperature
• Sound and vibration	• Electromagnetic radiation	• Humidity
• Computer keyboard and mouse	• Nuclear radiation	• Smoke and gas

Table 12.13 Guidelines for the design of experiment protocols

1 Protocols specify all activities and scripts for preparing the laboratory, conducting the experiment, and recording the data.

2 Check lists can help to assure uniformity in the preparation of the laboratory as a final step before each member of the task unit is admitted to the laboratory and a trial is started.

3 *Everyone* in the laboratory during the conduct of an experiment must have a detailed list of steps to be carried out by them. (If they have no steps to carry out, they do not belong there.)

4 *Everyone* involved in the conduct of the experiment must follow his or her protocol exactly, so that every trial happens in exactly the same way.

5 Printed instructions for human task units can help to ensure uniformity in the laboratory experiences of the subjects, as well as minimize the biases that can easily result from interaction between subjects and human supervisors.

6 If a supervisor finds it necessary to improvise in the face of an unexpected situation, the trial must be marked as suspect. The results of such trials must usually be ignored.

7 Pilots should be used to plan and debug the protocols. An ounce of prevention. . . .

8 Anyone in contact with the humans or other intelligent animals involved in the experiment (including human supervisors) should be told as little as possible about the objectives of the experiment and the associated tasks and project.

9 Protocols can be published as pseudo-code, flowcharts, or lists of numbered steps using GOTO statements to indicate repeats. Choose the method most easily understood by the person who must *execute* the protocol, such as nurses, interviewers, test-car drivers, *etc.*

10 It is unreasonable to expect humans to follow the protocols perfectly; it is easier and safer to design the experiment so that protocol violations are handled automatically (fail-soft).

11 Whenever possible, the protocols for handling emergencies in dangerous experiments should be automated and fail-safe.

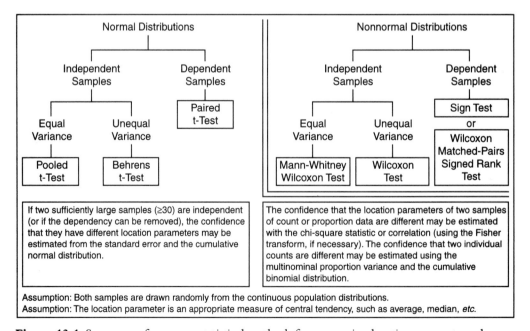

Figure 13.1 Summary of common statistical methods for comparing location parameter values

Table 13.7 General-purpose topical outline for the primary project documentation

Section	Examples of Common General Section Titles
Title	(none)
Abstract	Abstract
Table of Contents	Table of Contents
List of Illustrations	List of Illustrations | List of Tables and Figures
Introduction	Introduction | Summary | Overview
Analysis	Problem Domain
Problem Description	Problem Description | Motivation
Performance Criteria	Performance Criteria | Design Criteria | Requirements Analysis
Related Work	Related Work | Prior Work | Background Information
Objective	Objective | Research Objective | Development Objective
Hypothesis	Technical Approach | Proposed Solution
Solution Description	Solution Description | Solution Method | Design Specification
Goals	Goals | Research Goals | Design Goals | Goals and Hypotheses
Factors	Factors | Research Factors | Design Factors
Performance Metrics	Performance Metrics | Performance Measures
Synthesis	Implementation | Experiments | Experiments and Results
Solution Implementation	Solution Implementation | Software Construction | Equipment
Experiment Design	Experiment Design | Verification Method
Experiments	Experiments | Experiment Trials | Verification Process
Results	Results | Experiment Results
Validation	Conclusions | Performance and Conclusions
Performance	Performance | Performance Measures
Conclusions	Conclusions
Recommendations	Recommendations | Future Work
Appendices	Appendices
Bibliography	Bibliography | References | Bibliographical References
Governing Propositions	Glossary | Governing Propositions
Task Tree	Task Tree | Project Task Tree | Project Task Hierarchy
Task Inventories	Task Inventories | Task Component Inventories
System Diagrams	System Diagrams | Task Architectures | System Architectures
Milestone Chart	Milestone Charts | Project Schedules | Schedules and Costs
Result Files	Result Files | Experiment Data Files
Support Materials	Support Materials | Support Documents
Index	Index

(Between the Synthesis and Validation sections, a bracket spanning the two is labeled: SEQUENTIAL OR INTERLEAVED)

Table 13.8 The orders of the topics in the sequential and interleaved documentation methods

	SEQUENTIAL METHOD						INTERLEAVED (MULTIPLEXED) METHOD					
Goal	*Solution Implementation*	*Experiment Design*	*Experiments*	*Results*	*Performance*	*Conclusions*	*Solution Implementation*	*Experiment Design*	*Experiments*	*Results*	*Performance*	*Conclusions*
G1	5	10	15	20	25	30	25	26	27	28	29	30
G1.1	3	8	13	18	23	28	13	14	15	16	17	18
G1.1.1	1	6	11	16	21	26	1	2	3	4	5	6
G1.1.2	2	7	12	17	22	27	7	8	9	10	11	12
G1.2	4	9	14	19	24	29	19	20	21	22	23	24

Table 13.10 Guidelines for the use of typographic attributes

Attribute	*Function*
boldface	• first occurrence of **words** or **phrases** with special technical meanings (include in glossary) • section headers • important numbers or entries in tables
italics	• emphasis, *e.g.*, "there is *no* significant difference. . . ." or "Only *red* paint may be used." • foreign language words and abbreviations, *e.g.*, *Brandenburgertor, et al., i.e., etc.* • Latin symbols for mathematics, *e.g.*, $H, a, x, f(x)$ (not for digits and special characters) • low-level section headers and column headers
typestyle	• underline for emphasis, only if underline breaks for descenders (not <u>poppy</u>) • small caps for special sets of recurring names (*e.g.*, COLORS, DAYS) • true sub- and superscripts (not "computer" names) *e.g.*, 10^{17}, not 1E17; K_{min}, not Kmin • shadow, embossing, outline for special purposes (sparingly)
typeface	• *serif* typeface for text • *sans serif* or *serif* typeface for tables and figures • *sans serif* phrases in *serif* text (or *vice versa*) for recurring acronyms (*e.g.*, RAM)
font	• Times and Helvetica for normal text (preferred by most publishers) • monofont (like Courier) to align character strings in successive rows in tables • unusual fonts for special sets of recurring names (*e.g.*, **Identify Faster Traffic**)
typesize	• smaller typesize for subscripts and superscripts (not necessarily the default size) • decrease with progression from high-level to low-level section headers
shading	• unused cells in tables (*e.g.*, See Table 13.6) • white text on dark shaded background for special emphasis, *e.g.*, emphasis

Table 13.11 Guidelines for the effective use of the limited space in conference papers

1	For full control of spacing, use *exact* line spacing (*e.g.,* 12 points), instead of default *single spacing*.
2	To recover a partial line at the end of a page, decrease the line spacing by a point here and there above it.
3	If small line spacing puts subscripts too low or superscripts too high, manually raise or lower the characters.
4	To avoid wasted space between paragraphs, use first-line-indent instead of block-style paragraphs.
5	Construct (or import) clear, but very small figures with small typesizes, each designed especially for the paper.
6	Place figures at the top or bottom of the page to avoid wasting space above or below them.
7	If there are many figures, put them all on the last page in an efficient spatial arrangement.
8	Enclosing figures in boxes may allow the spacing between figures and text to be reduced without crowding.
9	Eliminate the lines required for figure titles by putting the figure title *inside* the figure area (if there is room).
10	Make the typesize of the references smaller than the text of the body of the paper (if not explicitly disallowed).
11	Place lowest-level section titles in boldface as first words of nonindented paragraph: **Section Header**. Text . . .
12	Embed less important governing propositions in existing paragraphs, rather than giving each its own paragraph.

Note: Make sure your typographic conventions do not flagrantly violate the formatting rules set by the publisher.

Table 13.12 Guidelines for the design of slides for oral presentations using *PowerPoint*

1   Before you start using *PowerPoint,* go through all the menus in detail and investigate all the available features.

2   The limit on the number of slides cannot be overcome by jamming as much as possible on each slide.

3   Ignore the default typographic and graphic settings supplied by *PowerPoint.* Make your own design decisions.

4   All text should be at least 18 point, although table entries or chart text may be 14 point if absolutely necessary.

5   Put a 14-point section title and slide number in an out-of-the-way, but consistent place on every slide.

6   Use bullet marks sparingly. Usually indentation alone will serve just fine to separate the items in a list.

7   Avoid paragraphs or prose descriptions. Use lists of short phrases and single short sentences to cue your comments.

8   Remove the chart junk, *i.e.,* objects, characters, and punctuation that add nothing and just confuse the eye.

9   Avoid abbreviations unless space is a critical issue. Use *centimeter,* not cm., *pounds,* not # or lb., and so forth.

10   Only sentences end with periods. Phrases do not. Unneeded periods and punctuation marks are "chart junk."

11   Avoid distracting backgrounds with decorative lines, fancy color schemes, and busy patterns. Keep it clean.

12   Colors should have specific and consistent semantic functions for emphasis, membership, or differentiation.

13   Strive for a consistent graphical and textual style on all the slides. Inconsistencies distract the audience.

14   Animation is very powerful for revealing objects (or groups) one at a time, which helps the audience focus.

15   Dim items that are no longer intended to be the primary and current focus for the audience.

16   Avoid cutesy effects for slide animation and transitions. Most of the time the *appear* effect is all you need.

17   Use custom animation to demonstrate dynamic processes with moving objects and changing titles.

18   To assess the readability of a slide, view it from a distance of 4 times the slide width on the computer screen.

Table 13.13 Important information items for an oral project report or proposal

Phase	Summary Item (Format)	Reference and Examples
Analysis	Problem Statement (sentence)	Definition 10.2; Tables 10.1, 10.2
	Performance Criteria (short phrases)	Table 10.3
	Primary Objective (infinitive phrase)	Definition 4.3; Table 10.5
Hypothesis	Solution Description (list or diagram)	Section 11.1
	Goals (tree or outline)	Definitions 11.2, 11.3; Tables 11.2, 11.3; Figure 11.1
	Factors (table with ranges)	Definition 7.8; Table 11.3
	Performance Metrics (formulas)	Definition 11.4; Table 11.3
Synthesis	System and Environment (diagram)	Table 12.4
	Experiment Block Design (table)	Definitions 12.5, 12.6, 12.7; Table 12.11
Validation	Performance (tables or charts)	Tables 13.1, 13.3, 13.5; Figures 13.4, 13.5
	Conclusions (sentences)	Section 13.2
	Informal Observations (sentences)	Section 13.2, Figure 13.6, Table 13.6
	Recommendations (list of phrases)	Section 13.2

APPENDIX E

Case-Study Figures and Tables

To illustrate the evolution of a project plan based on the Scientific Method supported by its knowledge propositions, this appendix is a collection of the excerpts from the case study presented in Part IV: *Measure Character-Recognition Performance*. It is an abbreviated version of the final report for an actual industrial research project conducted several years ago to measure the ability of inspectors to accurately recognize the numbers and letters on video images of automobile license plates. Only short and often incomplete textual descriptions are included in the excerpts of the case study: ellipses (. . .) at the end of sentences or paragraphs are used to indicate incomplete text that was much longer in the original project documentation. This compromise, however, does demonstrate that the information presented using the proposed methodology, albeit in abbreviated form, can be largely self-explanatory.

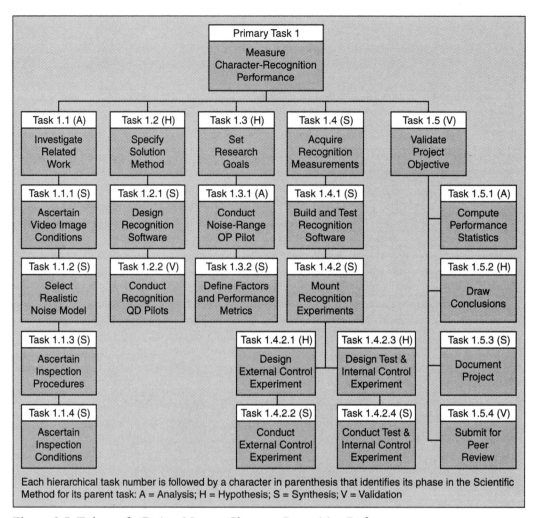

Each hierarchical task number is followed by a character in parenthesis that identifies its phase in the Scientific Method for its parent task: A = Analysis; H = Hypothesis; S = Synthesis; V = Validation

Figure 9.5 Task tree for Project *Measure Character-Recognition Performance*

Table 10.2 Problem description for Project *Measure Character-Recognition Performance*

Primary Task 1: *Problem Description*

As a part of its R&D program to develop innovative highway and traffic surveillance and control systems, Acme Corporation plans to develop an automated vehicle license-plate character-recognition system that is more reliable and at least as accurate as humans are. Currently, license-plate numbers are read manually from images of license plates captured by fixed video cameras overlooking highways and roads. Often the license plates are worn, dented, and spattered with mud and debris that partially obscure the letters and digits. Although it appears that manual inspection can yield high recognition accuracy under these conditions for short times, the task is very labor intensive and incurs high error rates when the inspector gets tired or bored. Clearly, the performance of an automatic character-recognition system, which would avoid these problems, must be at least as good as the *peak performance* of trained human inspectors. Unfortunately, there is very little data available from formal experiments on human character-recognition performance under comparable conditions. . . .

Problem Statement:　How well can trained license-plate inspectors recognize partially obscured characters on license plates?

Table 10.3 Performance criteria for Project *Measure Character-Recognition Performance*

Primary Task 1: *Performance Criteria*

Performance Criterion 1:　The experiment must use computer-generated images of *serif* and *sans serif* characters which have been obscured by a realistic simulated noise model. (exclusion)

To use video images of real license plates was impractical, because the true characters on the plate cannot be ascertained without examining the actual plates, which were not available. . . .

Performance Criterion 2:　The subjects for the recognition experiments must be inspectors who have been trained to read the numbers on video images of license plates.

To avoid learning curves during the recognition experiments, the subjects had to be trained inspectors who were accustomed to the procedures and domain of the job, bringing with them any subtle skills they may have acquired that could assist them in the recognition task. . . .

Performance Criterion 3:　The level-of-effort for the research task team must not exceed one person-year over an elapsed time of six months. (limitation)

Table 10.4a Related work for Project *Measure Character-Recognition Performance*

Task 1.1: Related Work

Most of the research in optical character recognition has attempted to recognize letters that are embedded in words, using a dictionary to assist the final classification of each letter. Contextual information of this kind greatly enhances the probability of accurate recognition (Jones and Tatali 1994). However, the objective of the higher-level parent project in the R&D department requires the recognition of numbers on license plates. In such cases, there is no contextual information available to help with the recognition task, because the character strings on license plates are randomly generated (subject to a few limitations) (USDOT Publication #124–789A3). Thus, the results and conclusions of past character-recognition studies using text (Jones 1982, Green and Johnson 1990, Brown *et al*. 1979) were not useful for estimating the typical license-plate character-recognition performance of human inspectors. . . .

Task 1.1.1: Video Image Conditions

As prescribed by Performance Criterion 1 (Table 10.3), the images of the license plates for this project had to be realistically simulated. Examination of video images of license plates acquired by existing cameras that observe highway traffic (USDOT Publication #106–111T1) revealed that a typical character on a license plate is captured at a vertical resolution of about 9 pixels. The horizontal resolution varies with the width of the character; the letter I is only 1 pixel wide, while the letter W is 8 pixels wide.

There are no uniform international typeface standards or conventions for license plates. Because the Acme Corporation wants to market the proposed automatic license-plate recognition system to many countries that use the Latin alphabet and decimal digits on their license plates (Internal ACME Memorandum AZ119.97 1998), the current project had to measure the recognition performance of human inspectors for both *sans serif* and *serif* fonts of the 26 uppercase Latin characters and 10 decimal digits. Figure CS.1 shows a *sans serif* and a *serif* letter D at the required resolution centered in a 21 × 21 pixel frame without noise. . . .

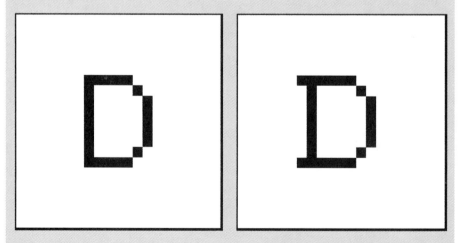

Figure CS.1 Sample simulated video image of a *sans serif* and a *serif* letter D

Table 10.4b Related work (abbreviated) for Project *Measure Character-Recognition Performance*

Task 1.1.2: *Realistic Noise Model*

The open literature reported several research projects that had used noise models to simulate the obscuration of text by dirt, debris, and normal wear-and-tear. The noise model selected for this project was postulated by Jacobi (1993), as illustrated in Figure CS.2. This choice was justified based on. . . .

> **Postulate 1:**   A realistic noise model consists of 32 discrete gray levels uniformly distributed over the dynamic range from 0 (white) to ±255 (black), whose probabilities are normally distributed about an average gray level of 0 (white) with a specified standard deviation (**noise level**).

> **Definition 1:**   The **noise level** is the standard deviation of the normal distribution of the noise generated by the model specified in Postulate 1 and as shown in Figure CS.2.

Task 1.1.3: *Manual Inspection Procedures*

Based on observations of the license-plate inspectors on the job at the license-plate inspection facility and extensive oral interviews with them, a profile of the data-management and analytical procedures undertaken by them during typical inspection tasks was compiled. . . .

Task 1.1.4: *Manual Inspection Conditions*

Based on an extensive inventory of the layout and environment of the license-plate inspection facility, a complete list of recommended specifications for an accurate model of this facility was compiled, including ambient and workspace lighting conditions, temperature regimes, background acoustic noise levels, inspection duty cycles, *etc.* . . .

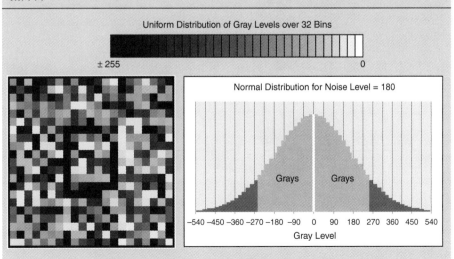

Figure CS.2 Sample image of *sans serif* letter D (noise level = 180) and its noise distribution

Table 10.5 Primary objective for Project *Measure Character-Recognition Performance*

Primary Objective 1:	To estimate the peak character recognition accuracy and response times of trained license-plate inspectors for both *serif* and *sans serif* computer-generated 9-point characters embedded in a 21×21 pixel frame with superimposed noise that realistically simulates the typical range of the actual obscuration of license-plate numbers due to dirt, debris, and normal wear-and-tear. (1 person-year over 6 months)

Table 11.1 Solution specification for Project *Measure Character-Recognition Performance*

Task 1.2: *Solution Specification*

To achieve the primary objective (see Table 10.5), a software application was designed to simulate the video images of license plates inspected by the license-plate inspectors on the job. The simulated images of license plates were presented to a small cohort of trained license-plate inspectors to measure their ability to identify the displayed characters in the presence of simulated, but realistic noise. . . .

Task 1.2.1: *Recognition Software Design*

Each subject in a small cohort of trained license-plate inspectors (*task unit*) was asked to identify each instance in a sequence of randomly selected characters (letters or digits) displayed on a CRT monitor screen capable of supporting at least 256 gray levels (*inducer*). Each 9-point character was centered in a 21×21 pixel frame, partially obscured by a noise level specified by Definition 1 (Table 10.4b). The character was surrounded on all sides by a margin of at least 6 pixels with the same noise level as that which obscured the character. Both the character and the noise were generated as specified by Postulate 1 (Table 10.4b). To the right of the character frame was an array of 36 buttons, one for each of 26 uppercase Latin letters {A through Z} and the 10 decimal digits {0 through 9}, which were used by each subject to select the recognized characters using the mouse (*sensor*). (See Figure CS.3 in Table 12.4.) Each generated and selected character was recorded by the computer, along with the time of the selection. . . .

Task 1.2.2: *Recognition QD Pilots*

Periodically during the software design phase, preliminary versions were evaluated with a series of informal feasibility pilots, using members of the project team as subjects. . . .

Table 11.3 Research goals for Project *Measure Character-Recognition Performance*

Task 1.3: *Research Goals*

Goal G1: To estimate the ability of trained license-plate inspectors to correctly identify characters on simulated video images of partially obscured license plates

    Goal G1.1: To ascertain dynamic range and increment for the noise-level factor. (OP Pilot Task 1.3.1) Based on their experience, a small number of trained license-plate inspectors were asked to select a reasonable and realistic dynamic range and increment size for the noise-level factor for the formal experiments in Task <None>1.4. . . .

    Goal G1.2: To decide whether each subject is qualified to participate in a test trial

        Hypothesis H1.2.1: The subject has had sufficient practice with the interface.

            Factor: Rejection-confidence threshold

            Goal G1.2.1.1: To measure character-selection times without noise (external control Task 1.4.2.2)

                Factor: Displayed characters = {A...Z} and {0...9} (representation factor)

                Performance Metric: Rejection confidence that absolute correlation of selection times with acquisition order is zero.

    Goal G1.3: To measure character selection times and accuracies with noise (internal control and test Task 1.4.2.4).

        Factors: Displayed characters = {A...Z} and {0...9} for test trials; or no character for internal treatments (representation factor)

            Typeface = *serif* or *sans serif* (n/a for internal control treatments)

            Noise level = 130 to 240 in increments of 10 (from Goal G1.1)

    Performance Metrics: Average selection times $\overline{T}_{control}$ and $\overline{T}_{test}$

            Number of characters selected correctly

            Benefit B = percent of characters identified correctly

            Cost C = Type I error probability (%) that $\overline{T}_{test} < \overline{T}_{control}$

            Payoff $P = B\,(100 - C)/100$

    Hypothesis H1.3.1: Benefit varies inversely with noise level.

    Hypothesis H1.3.2: Cost varies directly with noise level.

    Hypothesis H1.3.3: Payoff varies inversely with noise level.

    Hypothesis H1.3.4: Benefit for *serif* typeface is higher than for *sans serif* typeface.

    Hypothesis H1.3.5: Cost for *serif* typeface is lower than for *sans serif* typeface.

    Hypothesis H1.3.6: Payoff for *serif* typeface is higher than for *sans serif* typeface.

        Performance Metric: Confidence that the null hypothesis may be rejected, (H1.3.4 thru H1.3.6) based on the standard difference of two proportions

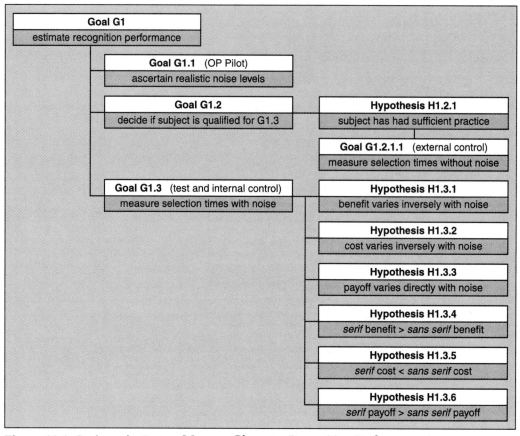

Figure 11.1 Goal tree for Project *Measure Character-Recognition Performance*

Table 11.4 Knowledge propositions for Project *Measure Character-Recognition Performance*

Knowledge Propositions (Role Code)	Number of Factor Values
1 The ambient laboratory conditions have no significant effect on recognition performance. (gcsE)	
2 The CRT screen conditions are constant and have no significant effect on performance. (gcsE)	
3 Test characters are uppercase instances of all 26 uppercase Latin letters and 10 decimal digits. (fcp) ⟶	36
4 The typefaces of the characters on license plates are Roman and either *serif* or *sans serif*. (fcp) ⟶	2
5 The design of the user interface has no significant effect on recognition performance. (gcrE)	
6 The colors of the displayed objects have no significant effect on recognition performance. (gcsE)	
7 A realistic noise model consists of 32 discrete gray levels, uniformly distributed. . . . (gcrL)	
8 The noise level is the standard deviation of the normal distribution of the noise generated by the model . . . (gcs)	
9 Simulated video images of license plate characters yield reliable performance estimates. (gcrL)	
10 Noise levels from 130 to 240 gray levels in increments of 10 represent a typical range . . . (fcvL) ⟶	12
11 Recognition performance metric B (benefit) = percent of selections recognized correctly. (gcr)	
12 Recognition performance metric C (cost) = confidence subject has had sufficient practice. (gcr)	
13 Recognition performance metric P (payoff) = BC = product of benefit and cost. (gcr)	
14 The test cohort is an accurate sample of the recognition skills of trained license-plate inspectors. (gps)	
15 There are no statistically significant differences between the pilot cohorts and the test cohorts. (gpcE)	
16 There are no statistically significant differences in the recognition skills of the subjects. (gpcE)	
17 Personal attributes (age, sex, *etc.*) of the subjects have no significant effect on performance. (gpsE)	
18 There are no statistically significant differences in the typographic biases of subjects. (gpcE)	
19 Testing each subject with 6 to 30 characters per trial assures the independence of responses. (gpcE)	

Role Codes

g = governing proposition	+	p = parameter + s = scheme c = condition + r = regime	or c = constant or s = situation	L = limitation +R = restriction
f = factor +	p = parameter or	c = condition + p = policy or	v = variable	E = exclusion
r = range proposition +	r = result or	c = conclusion + (b = benefit and/or c = cost) or		p = payoff

Table 12.2 Solution implementation for Project *Measure Character-Recognition Performance*

Primary Task 1: Solution Implementation
The software application for acquiring the character-recognition data during the experiment trials was constructed in Microsoft Excel. The appearance of the graphical user interface is shown in Figure CS.4 (see Table 12.8) in the description of the laboratory design. . . .

A field was provided for the experiment supervisor to enter the subject ID; this field disappeared after the ID was successfully entered. No keyboard was provided; all input from the subject was accomplished with mouse clicks. Care was taken to disable and/or ignore all illegal mouse-clicks by a subject during data acquisition. Only clicks on the START button (which disappears after being clicked) and the character buttons {A . . . Z} and {0 . . . 9} were accepted and processed. . . .

Results for each subject were recorded in a tab-delimited text file that could be read directly by Excel for convenient reduction of the results. The recorded data items included timeStamp, technicianID, trialType, subjectID, typeface, noiseLevel, and recognition data {correctCharacter, responseCharacter, responseTime}. . . .

Table 12.4 Task components for Project *Measure Character-Recognition Performance*

Summary of Task Components for the Primary Task

Task Objective: To measure the peak performance of trained license-plate inspectors in recognizing computer-generated 9-point *serif* and *sans serif* characters embedded in a 21×21 pixel frame with superimposed noise that realistically models the actual range of obscuration of license-plate numbers due to dirt, debris, and normal wear-and-tear.

Task Unit: Subject in a small cohort of trained license-plate inspectors

Inducers: CRT display capable of supporting at least 256 gray levels
        Test image (See Figure CS.1 in Table 10.4a)
                Noise [RAEC]
                Characters [RAEC]
        Other visible objects [UNEC]
        Background [UAEC]
      Subject chair [UAEC]
      Cognitive distractions [UNIC]
      Sound, light, and air-conditioning sources [UAEC]
      Smells [UNEC]

Sensors: Mouse used by the subject to select the recognized characters from an array of 36 buttons for the 26 uppercase Latin letters {A through Z} and the 10 decimal digits {0 through 9}

System and Environment: See Figure CS.3

Supervisor: Computer (PC) and technician

Channels: Air for sound, smell, and light (input)
        Air for heat (input and output)
        Subject PNS* (input and output)
        Mouse cable (input and output)
        CRT display cable (input and output)
        Power cables (input)
        Chair (input)

Domain Knowledge: See Table 11.4

Range Knowledge:
        Results: timeStamp
                technicianID
                trialType
                subjectID
                typeface
                noiseLevel
                recognition data
                    correctCharacter
                    responseCharacter
                    responseTime

Figure CS.3 System and environment

        Conclusions:
                Recognition performance as a function of noise level and typeface
                Confidence performance with *serif* typeface > performance with *sans serif*

Solution: See Table 11.1

Experiments: See Tables 12.8 through 12.12 and 12.14

*PNS = peripheral nervous system

Table 12.8 Laboratory design for Project *Measure Character-Recognition Performance*

Task 1.4.2 (see Figure 9.5): Recognition Experiments (Laboratory Design)

The data collection facility for the experiments was a small room (3 by 4 meters) with no windows and blank walls painted a light beige. The room was illuminated by a diffuse 200 watt overhead light whose intensity was set at 80% as a result of the pilot tasks. The temperature of the room was not directly controlled, but varied little from an average temperature of 20°C. The room was furnished with a single table (0.75 meters wide, 1.0 meter long, and 0.72 meters high) and a 5-leg swivel chair on casters with an adjustable back angle and seat height. Data was collected using a Simsong Model 65J467 15-inch CRT display that was bolted to the table, and an Ace Model 56D mouse. A sample of the CRT display objects is shown in Figure CS.4 below. Before each trial began, the character frame was empty (white) except for the START button and the Subject ID field. Both objects disappeared after the Subject ID was entered and the START button was clicked. One second later the question box appeared at the top of the display, and the first frame was displayed. Thereafter, whenever the subject clicked a character button, the question box disappeared, and the subject's response was recorded. One second later, the question box reappeared, and a new frame was displayed. When the trial was over, the words THANK YOU appeared in the middle of the frame. . . .

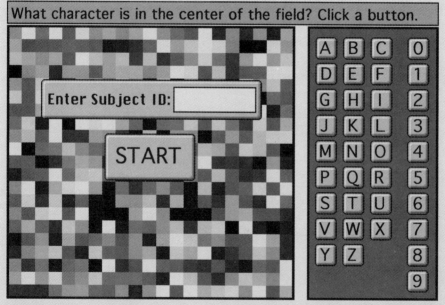

Figure CS.4 Sample CRT display (actual size was 16 cm wide by 11 cm high)

The technician monitored the activity of the subject during the experiment trials via the supervisor computer in an adjoining room. The inducer CRT display was driven by a graphics card in the supervisor computer. The cables connecting the display and the mouse to the supervisor computer passed through the wall separating the two rooms. . . .

Table 12.9 Block design criteria for Project *Measure Character-Recognition Performance*

Factor	Feasible Factor–Value Subsets
character	all 36 characters, applied a whole number of times to each subject, treatment, and block
typeface	*serif* or *sans serif*
noise level	{130, 150, 170, 190, 210, 230} and/or {140, 160, 180, 200, 220, 240}
cohort size	smaller than 14 and yields at least 30 responses for each treatment

Table 12.10 Sample test block designs for Project *Measure Character-Recognition Performance*

Experiment Conditions		RANDOMIZED BLOCK DESIGNS A	B	C	D	Formula
cohort size (number of subjects)		6	8	12	12	a = input (less than 14)
character: performance-factor combinations		12	18	12	6	b = input
typeface: settings per subject (trial)		1	1	1	1	c = 1 (constant per subject)
noise level: levels per subject (trial)		12	6	6	12	d = input (6 or 12)
total test responses by the cohort		864	864	864	864	e = abcd
test responses per typeface for each noise level		36	72	72	36	f = e/d/2
test responses per noise level for each typeface		72	72	72	72	g = e/c/12
test responses per subject (trial)		144	108	72	72	h = bcd
time per trial for test treatments*		12.0	9.0	6.0	6.0	i = 5h/60

*(in minutes): Assumes a response requires an average of 4 seconds followed by a 1-second programmed pause.

Table 12.11 Complete test block designs for Project *Measure Character-Recognition Performance*

Experiment Conditions		RANDOMIZED BLOCK DESIGNS A	B	C	D	Formula
cohort size (number of subjects)		6	8	12	12	a = input (less than 14)
character: performance-factor combinations		12	18	12	6	b = input
typeface: settings per subject (trial)		1	1	1	1	c = 1 (constant per subject)
noise level: levels per subject (trial)		12	6	6	12	d = input (6 or 12)
total test responses by the cohort		864	864	864	864	e = abcd
test responses per typeface for each noise level		36	72	72	36	f = e/d/2
test responses per noise level for each typeface		72	72	72	72	g = e/c/12
test responses per subject (trial)		144	108	72	72	h = bcd
time per trial for test treatments*		12.0	9.0	6.0	6.0	i = 5h/60
external control responses per subject (trial)**		20	20	20	20	j = input
internal control responses per subject (trial)		60	45	30	30	k = (30)(12)/a
time per trial for control treatments*		6.7	5.4	4.2	4.2	l = 5(k + j)/60
total time per trial (includes 10 minutes for practice)*		28.7	24.4	20.2	20.2	m = i + l + 10
total time for all trials (complete block)*		172	195	242	242	n = am

*(in minutes): Assumes a response requires an average of 4 seconds followed by a 1-second programmed pause.
**This trial may be repeated twice, as necessary, until the correlation confidence drops below 90%.

Table 12.12 Experiment tasks for Project *Measure Character-Recognition Performance*

Task 1.4: Acquisition of Recognition Measurements

Task 1.4.1: *Manual Inspection Procedures*

Problem: What are the conditions and procedures currently used by trained inspectors to examine and identify license-plate numbers?

Objective: To establish the appropriate conditions and procedures for the character-recognition experiments

Solution: Interview and observe trained inspectors in the field

Task 1.4.2.1: *External Control Experiment Design*

Problem: Because of the unfamiliarity of the user interface, without sufficient practice the selection times of the subject during the test trials may systematically decrease as they continue to become accustomed to the interface (learning curve).

Objective: To ensure the constancy of each subject's selection times after a practice period selecting characters displayed without noise

Solution: Before beginning the test trials, each subject was asked to study the display for 5 minutes to become familiar with the configuration of its elements (character frame, buttons, *etc.*) and then to practice selecting characters displayed without noise for 5 minutes. The typeface of these characters was the same as the one that had been randomly selected for the test trial for this subject. . . .

 At the end of this practice session, a sequence of 20 randomly selected characters without noise was presented to the subject, who was instructed to click the corresponding button as quickly as possible. The sequence of the selection times for these control trials was correlated with the order of the sequence. If the confidence of a non-zero correlation exceeded 90%, the subject was asked to practice for another 5 minutes. If the confidence still did not exceed 90% after two repeats of the practice session, the subject was excused. . . . (See the protocol in Table 12.14.)

Task 1.4.2.3: *Test and Internal Control Experiment Design*

Problem: What are the recognition accuracies and response times for trained inspectors?

Objective: To measure the accuracy and response times of character selections as a function of typeface and noise level, accounting for the guess-bias for each character

Solution: For each treatment of each trial, the subject was shown a frame with a noise level chosen randomly from the noise-level set assigned randomly to the subject (see Table 12.9). For each of the 108 test treatments (variable h in Block B in Table 12.11), the frame included a randomly selected character. For the 45 control treatments (variable k in Block B in Table 12.11), no character was displayed. For each treatment the subject was instructed to click the button corresponding to the character judged most likely to be in the frame (forced choice). For the control frames, this selection was made in the absence of any actual character. There was no time limit for a subject's response to each treatment. The doubleblinded sequence of control and test treatments for each trial was generated randomly by the supervisor computer, which also conducted all trials and recorded the experiment results. . . . (See the protocol in Table 12.14.)

Table 12.14 Protocol for Task 1.4.2 for Project *Measure Character-Recognition Performance*

PROTOCOL 142: TECHNICIAN PROTOCOL FOR EXPERIMENT TRIALS (TASK 1.4.2) REV 7.6 04/06/96		
Step	**Action**	**Check**
1	Prepare laboratory for the experiments (Protocol P142a).	☐
2	Make sure the laboratory is in its proper initial state (Protocol P142b).	☐
3	Launch and/or initialize the experiment software (CharRec v3.1).	☐
4	Click NEXT SUBJECT button on the display, and note the Subject ID of the next subject.	☐☐☐
5	If the display message is "NO MORE SUBJECTS," discard this check list, and exit the protocol.	☐☐☐
6	Open the waiting-room door, and call the next subject by the Subject ID number from Step 4.	☐☐☐
7	If the selected subject is not present in the waiting room, close the door, and go to Step 4.	☐☐☐
8	Ask the subject to enter the laboratory.	☐☐☐
9	Return to the display, and click the START button next to the Subject ID field on the display.	☐☐☐
10	Return to the waiting-room door, and close it after the subject has entered.	☐☐☐
11	Ask subject to sit down on the workstation chair.	☐☐☐
12	Ask the subject to enter his/her Subject ID on the display input field.	☐☐☐
13	If the display message is "INVALID SUBJECT," dismiss the subject (script S142a). Go to Step 4.	☐☐☐
14	Introduce yourself, the experiment, and the laboratory to the subject (Script S142b).	☐
15	Help the subject to adjust the chair, position the mouse, and adjust the display parameters.	☐
16	Tell the subject about the process for this experiment (Script S142c).	☐
17	Ask the subject to click the PRACTICE button on the display to begin a practice session.	☐☐☐
18	Observe and assist the subject throughout the practice session.	☐☐☐
19	Leave the laboratory and enter the control booth, closing the door behind you.	☐☐☐
20	Using the intercom, ask the subject to click the CONTROL button to begin the control trial.	☐☐☐
21	When the subject is finished with the control trial, note the control decision on the display.	☐☐☐
22	Return to the laboratory, closing the door behind you.	☐☐☐
23	If the displayed control decision was "DISMISS SUBJECT" go to Step 28.	☐☐☐
24	If the displayed control decision was "REPEAT PRACTICE SECTION," go to Step 17.	☐☐☐
25	Leave the laboratory and enter the control booth, closing the connecting door behind you.	☐
26	Using the intercom, ask the subject to click the TEST button on the display to begin the test trial.	☐
27	When the subject is finished with the trial, return to the laboratory, closing the door behind you.	☐
28	Thank the subject for participating in the experiment (script S142d).	☐
29	Open the door to the waiting room, and escort the subject out of the laboratory.	☐
30	Reenter the laboratory, closing the door behind you.	☐
31	File this check list in the protocol envelope. Get a new blank check list. Go to Step 2.	☐

Table 12.15 Data files for Project *Measure Character-Recognition Performance*

```
📁 100000 CharRecExperimentData
   📁 CharacterInputs
      📄 SerifCharacters
      📄 SansSerifCharacters
   📁 110000 RelatedWork
      📁 111000 VideoCond
         📁 Inputs★  asterisk (★) = folder intentionally left empty
         📁 VideoCondResults
            📄 DOTLicensePlateImages
      📁 112000 NoiseModel★
   📁 120000 SolutionMethod
      📁 121000 InspectionCond★
   📁 130000 GoalHypo
      📁 131000 NoisePilot
         📁 CharacterInputs italics = alias for a previous file or folder
         📁 PilotResults
            📄 RealNoiseCond
      📁 132000 SetGoals★
   📁 140000 RecogMeas
      📁 141000 ManInspecProc★
      📁 142000 RecExp
         📁 142100 DesControlExp★
         📁 142200 ConControlExp
            📁 CharacterInputs
            📁 ControlResults
               📄 SerifBaseline
               📄 SansSerifBaseline
         📁 142300 DesTestExp★
         📁 142400 ConTestExp
            📁 CharacterInputs
            📁 TestResults
               📄 SerifRecognition
               📄 SansSerifRecognition
   📁 150000 ValidateObjective
      📁 151000 PerformComp
         📁 Inputs
            📁 ControlResults
            📁 TestResults
         📁 Results
            📄 AllPerformanceStats
      📁 152000 Conclusions
         📁 Inputs
            📄 AllPerformanceStats
         📁 Results
            📄 Conclusions
      📁 153000 DocProject★
      📁 154000 PeerReview★
```

Table 12.16 Sample result file entries for Project *Measure Character-Recognition Performance*

Task 1.4.2 06/21/96 Technician 27	Subject ID	Trial Type	Typeface	Noise Level	Correct Character	Selected Character	Selection Time (Days)
Examples of Raw Results	02	ExtCntrl	serif	0	G	G	0.0000248
	03	IntCntrl	serif	210	n/a	A	0.0000541
	05	Test	sans serif	150	R	B	0.0000797

	Subject ID	Trial Type	Typeface	Noise Level	Character Selection	Selection Time (Sec)
Examples of Reduced Results	02	ExtCntrl	serif	0	correct	2.14
	03	IntCntrl	serif	210	n/a	3.67
	05	Test	sans serif	150	incorrect	5.89

Table 13.2 External control performance for Project *Measure Character-Recognition Performance*

Subject ID	Typeface	Number of Practice Sessions	Correlation Coefficient	Confidence (%) Correlation ≠ 0
1	sans serif	3	−0.142	56
2	sans serif	3	−0.300	89
3	sans serif	2	+0.037	16
4	sans serif	1	−0.004	2
5	serif	2	+0.032	14
6	serif	1	+0.052	22
7	serif	3	−0.065	27
8	serif	3	−0.112	45

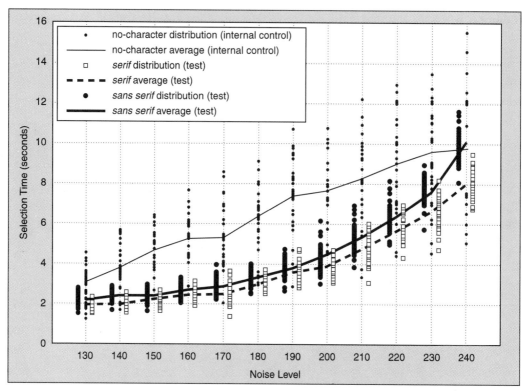

Figure 13.3 Measured internal control and test selection times for all characters

Table 13.3 Performance values for Project *Measure Character-Recognition Performance*

		NOISE LEVEL										
Performance Metric		*130*	*140*	*150*	*160*	*170*	*180*	*190*	*200*	*210*	*220*	*230*	*240*
Fraction of characters	*sans serif* benefit	97	94	86	78	67	56	44	33	25	17	11	8
identified correctly (%)	*serif* benefit	97	97	92	86	81	72	61	47	28	14	8	8
Type I error	*sans serif* cost	18	11	2	1	7⁻	3	4	5	11	13	21	54
probability (%)	*serif* cost	11	6	1	1	5	1	3	3	7	8	12	29
Product of benefit and	*sans serif* payoff	80	84	85	77	62	54	43	32	22	14	9	4
(100 − cost) (%)	*serif* payoff	86	92	91	85	77	71	59	46	26	13	7	6

Table 13.4 Research and null hypotheses for Project *Measure Character-Recognition Performance*

	Research Hypothesis, H_R	*Null Hypothesis, H_0*
H1.3.1	Benefit varies inversely with noise level.	Benefit is independent of noise level.
H1.3.2	Cost varies directly with noise level.	Cost is independent of noise level.
H1.3.3	Payoff varies inversely with noise level.	Payoff is independent of noise level.
H1.3.4	Benefit for *serif* characters is higher than for *sans serif* characters.	Any difference in benefit for *serif* and *sans serif* characters may be attributed to chance.
H1.3.5	Cost for *serif* characters is lower than for *sans serif* characters.	Any difference in cost for *serif* and *sans serif* characters may be attributed to chance.
H1.3.6	Payoff for *serif* characters is higher than for *sans serif* characters.	Any difference in payoff for *serif* and *sans serif* characters may be attributed to chance.

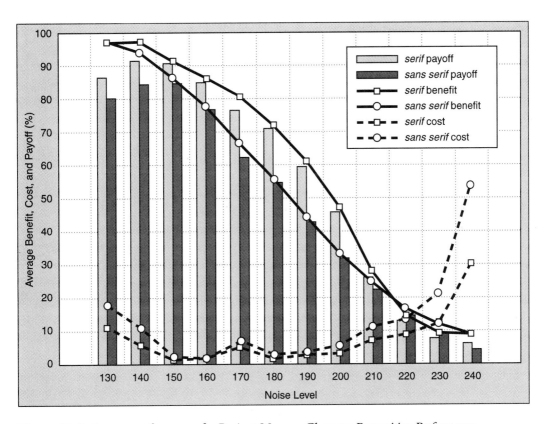

Figure 13.4 Average performance for Project *Measure Character-Recognition Performance*

Table 13.5 Rejection confidences for null hypotheses H1.3.4, H1.3.5, and H1.3.6

	NOISE LEVEL											
Null Hypothesis	*130*	*140*	*150*	*160*	*170*	*180*	*190*	*200*	*210*	*220*	*230*	*240*
H1.3.4: *serif* benefit = *sans serif* benefit	50	72	77	82	**91**	**93**	**92**	89	61	37	35	50
H1.3.5 *serif* cost = *sans serif* cost	78	77	58	57	63	61	55	63	62	60	63	73
H1.3.6 *serif* payoff = *sans serif* payoff	76	82	77	80	88	88	84	78	57	47	47	55

rejection confidences ≥ 90% are shown in **boldface**

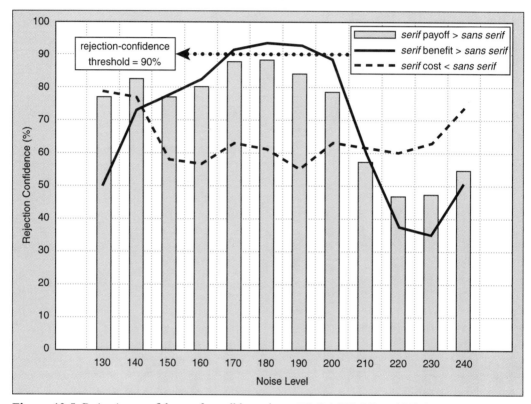

Figure 13.5 Rejection confidences for null hypotheses H1.3.4, H1.3.5, and H1.3.6

Table 13.6 Character confusions for Project *Measure Character-Recognition Performance*

Actual \ Selected	A	B	C	D	E	F	G	H	I	J	K	L	M	N	O	P	Q	R	S	T	U	V	W	X	Y	Z	0	1	2	3	4	5	6	7	8	9
A	▪							1										1													8					
B		▪		3																									3						5	
C			▪												5										8											
D				▪											3										9											
E		4			▪	5																4														
F						▪									3		3		4																	
G							▪								2											1								9		
H								▪			3			5								3														
I								3	▪		3																	7								
J								2		▪											5							1								
K											▪				5							4														
L								3				▪																7								
M													▪	2									8													
N							5							▪	5																					
O			3												▪			4							7											
P			5													▪		8																		
Q															5		▪								6											
R			3								3					4		▪																		
S																			▪													5		6		
T						6														▪					4											
U															3						▪	2		4												
V																					7	▪		4												
W												7											▪	5												
X					4			8																▪												
Y																				6	2				▪											
Z																				2						▪			11							
0		2													9												▪									
1						7																						▪							3	
2																							7						▪						3	
3		7																2												▪						
4	3																														▪					8
5																		11														▪				
6		3			4																												▪			
7																												8	3					▪		
8		6																				3													▪	4
9															5		6																			▪

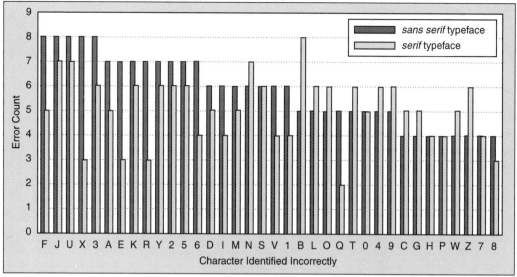

Figure 13.6 Recognition errors for Project *Measure Character-Recognition Performance*

Table 13.9 Title and abstract for Project *Measure Character-Recognition Performance*

AN ESTIMATION OF THE PEAK CHARACTER-RECOGNITION PERFORMANCE OF TRAINED LICENSE-PLATE INSPECTORS FOR CHARACTERS DISPLAYED WITH NOISE THAT REALISTICALLY SIMULATES THE ACTUAL OBSCURATION OF LICENSE-PLATE NUMBERS
Abstract

The Acme Corporation plans to develop an automated vehicle-license-plate character-recognition system that is at least as reliable as trained human license-plate inspectors. As a precursor to this development project, a series of carefully controlled experiments were designed and conducted to measure the recognition performance of these inspectors. Because characters on actual video images of license plates are often partially obscured by mud and debris, the correct characters cannot be reliably identified, obviating the use of such images for these experiments. Therefore, realistic simulated noise was superimposed on frames containing computer-generated characters, whose format was the same as actual license-plate characters. Both the noise level and typeface (*serif* and *sans serif*) of the characters were factors in the randomized block design of the experiments. The cohort consisted of eight trained license-plate inspectors. After practicing sufficiently to demonstrate proficiency with the graphical user interface (external control trial), each subject was presented with a sequence of 108 test frames, each with a randomly selected character (in one typeface) obscured with a randomly selected noise level, and 45 randomly interleaved control frames with noise, but no character. Performance metrics included recognition accuracy (benefit), response-selection time (cost), and an arithmetic combination of them (payoff). The results of these experiments yielded a reasonable model of the recognition performance of trained license-plate inspectors. The conclusions also validated the six research hypotheses: briefly, recognition benefit and payoff vary inversely with noise level; recognition cost varies directly with noise level; and recognition benefit and payoff for the *serif* characters significantly exceed the benefit and payoff for the *sans serif* characters for moderate noise levels [170 to 190]. Informal observations identified some of the characters that caused the highest recognition-error rates and confusions.

APPENDIX F

Sample Experiment Protocol

This appendix presents an actual protocol for a complex experiment that was designed and conducted by a team of scientists and engineers at the Corporate Research Headquarters of the large industrial firm Robert Bosch GmbH in Stuttgart, Germany in 2001. The protocol governed the preparation and execution of the experiment trials for the Driver Monitoring Project, whose objective was to identify reliable physiological indicators of stress and fatigue in car drivers after long periods of uninterrupted driving under a variety of carefully controlled simulated driving conditions.

The experiment trials took place in a specially equipped Driving Simulator laboratory, in which the front half of a small car faced a wraparound screen on which animated highway scenes were projected. The subject (*i.e.,* the driver) controlled the speed and direction of the car from the left front seat of the car using the steering wheel, accelerator, and brake. Every subject drove through the same set of preprogrammed scenarios for 120 kilometers (about 2.5 hours) without interruption.

Each subject (the task unit) was monitored by a wide variety of physiological sensors as he drove the car in the driving simulator. The actual data acquisition period for each experiment trial occurred during Steps 116 and 117 in the protocol given here. All the other protocol steps were concerned with preparation of the simulator and the subject before the trial began (Steps 1–115), and the removal of the physiological sensors, data management, and laboratory shut-down procedures after the trial had ended (Steps 118–186).

The protocol was executed by a trained nurse. For control purposes, no one else was allowed to have contact with the subjects at any time during the entire procedure, although the subject and the laboratory could be observed during the driving trial via closed-circuit television. The first version of the experiment protocol, designed before conducting the first feasibility pilot, was about 30 steps long. Step-wise refinement over a series of subsequent pilots increased the length of the protocol to its final version of 186 steps, which is shown here. The high level of detail in this final protocol ensured that the actual experiment trials were conducted under the same carefully controlled conditions for all subjects in the cohort.

The author is grateful to Robert Bosch GmbH and the members of the Driver Monitoring Project team for translating the research experiment protocol from German into English and allowing it to be presented in this book.

Step	Place	Operation
1	Observer Room	Locate and prepare subject questionnaires and subject folder.
2	Observer Room	Determine subject number and record on the Assistant Form.
3	Observer Room	Label the blank CD for data storage in both versions.
4	Simulator Room	Check the inventory of electrodes and blank CDs.
5	Observer Room	Clean EEG cap.
6	Observer Room	Take blood pressure (BP) equipment from closet, change batteries, and turn it on.
7	Observer Room	Fill syringe with electrode gel.
8	Observer Room	Recharge the previously used storage batteries.
9	Observer Room	Put fabric cover on BP cuff.
10	Observer Room	Turn on monitors and video recorder (VCR).
11	Observer Room	Insert videotape into VCR.
12	Observer Room	Remove empty bottles from the room.
13	Division Office	Get subject's fee from secretary, and turn in the fee receipt from the previous trial.
14	Division Office	Fill out the lunch credit for the subject.
15	Division Office	Store the videotape from the previous trial.
16	Division Office	Take labeled blank CDs from the previous trial to the CD-writer.
17	Hallway	Pick up subject at the main gate.
18	Cafeteria	Take the subject to lunch.
19	Simulator Room	Turn lights on.
20	Simulator Room	Help the subject take off his personal coat, if any.
21	Simulator Room	Turn on the electrical power for the Measurement Computer.
22	Simulator Room	Pull out Measurement Computer drawer and open the access door.
23	Simulator Room	Turn on the Measurement Computer.
24	Simulator Room	Uncover the carbon dioxide sensor.
25	Simulator Room	Turn on the breakers.
26	Simulator Room	Log on to all Simulation Computers.
27	Simulator Room	Load Configuration File into Simulation Computer 1.
28	Simulator Room	Open the external window, if necessary.
29	Simulator Room	Load the driving simulation scenario.
30	Simulator Room	Load simulator software on Simulation Computer 1.

Continued on next page

31	Simulator Room	Turn on the video projectors.
32	Simulator Room	Turn on the electrical power for the force-feedback systems.
33	Simulator Room	Log on to the Measurement Computer.
34	Simulator Room	Help subject into simulator driver's seat and adjust seat position for his comfort.
35	Simulator Room	Hand out Instruction 1 to subject and ask him to read it.
36	Simulator Room	Turn on the simulation-sound amplifiers.
37	Simulator Room	Turn off the lights.
38	Simulator Room	Start the simulator software on Simulation Computer 1.
39	Observer Room	Fetch the BP equipment and the EEG cap.
40	Simulator Room	Hang "No Entrance" sign on the Simulator Room door.
41	Simulator Room	Connect the BP equipment to the Measurement Computer.
42	Simulator Room	Activate the BP recording software on the Measurement Computer.
43	Simulator Room	Press *New* button on the BP recording software user interface.
44	Simulator Room	Fill in patient number and name and press OK.
45	Simulator Room	Press the *TM2430* button.
46	Simulator Room	Press the *Assign PC Time* button and click on OK.
47	Simulator Room	Shut down the BP software on the Measurement Computer.
48	Simulator Room	Disconnect the BP equipment from the Measurement Computer.
49	Simulator Room	Check the BP equipment and correct the displayed time, if necessary.
50	Simulator Room	Move the equipment cabinet, waste basket, and chair over to the driver's seat.
51	Hallway	Take subject to restroom, wait for him, and return with him to Simulator Room.
52	Simulator Room	Collect subject's jewelry, watch, and cell phone; turn off cell phone.
53	Simulator Room	Have subject remove his shirt and undershirt.
54	Simulator Room	Fix BP belt with holding pocket on the subject's right side.
55	Simulator Room	Put BP cuff on subject's left upper arm.
56	Simulator Room	Fasten BP tube on the subject's shoulder with tape.
57	Simulator Room	Put BP instrument in the BP belt pocket and connect the tube to it.
58	Simulator Room	Make a test measurement by pressing the *Start* button on the BP equipment.
59	Simulator Room	Clean the skin at the ECG and ground-electrode positions.
60	Simulator Room	Attach the ECG and ground electrodes.
61	Simulator Room	Measure the impedance between the two electrodes (must be less than 20 Kohms).
62	Simulator Room	Connect the ECG cable and ground wire to the electrodes.
63	Simulator Room	Attach skin-resistance electrodes (no gel).
64	Simulator Room	Help subject put on the lab coat without closing it.
65	Simulator Room	Put EEG-cap on shoulder.
66	Simulator Room	Clean the earflaps.
67	Simulator Room	Put stickers on the reference electrodes.
68	Simulator Room	Put stickers on ECG electrodes. (Make sure sticker tongue does not touch the eye.)
69	Simulator Room	Attach the ECG electrodes after cleaning the skin.
70	Simulator Room	Route the cable around the subject's ear.
71	Simulator Room	Put the EEG cap on the subject, which subject may adjust to feel comfortable.
72	Simulator Room	Attach motion sensor on cap so that the cable is directed downward.
73	Simulator Room	Put respiration sensor between skin and belt.
74	Simulator Room	Tape skin-temperature sensor to the left side of the subject's abdomen.

Continued on next page

75	Simulator Room	Attach the motion-sensor cable to the BP measurement belt with tape.
76	Simulator Room	Ask subject to be seated in the driver's seat of the car.
77	Simulator Room	Move the waste basket and chair back to the rear of the room.
78	Simulator Room	Take amplifier boxes out of the closet and set them up.
79	Simulator Room	Connect the EEG and ECG cables and route them along the subject's shoulders.
80	Simulator Room	Connect the ECG and Respiration sensors.
81	Simulator Room	Attach the ground wire near the left trouser-leg.
82	Simulator Room	Attach pulseoximeter sensor to the left-hand ring finger and connect it to the cable.
83	Simulator Room	Start *Signal Preview* program on the Measurement Computer; press cursor button.
84	Simulator Room	Inject gel into the EEG electrodes and check the signal integrity.
85	Simulator Room	Check the ECG signals by observing the signal when the subject moves his eyes.
86	Simulator Room	Press *Auto* button; when letter A appears on the display, cover the clock.
87	Simulator Room	Close signal preview application on the Measurement Computer.
88	Simulator Room	Close the subject's lab coat, if subject desires.
89	Simulator Room	Start data recording software on the Measurement Computer.
90	Simulator Room	Fill out subject data in the software interface on the Measurement Computer.
91	Simulator Room	Press *Monitor Off* button on the Measurement Computer.
92	Simulator Room	Press right-cursor button on the Measurement Computer.
93	Simulator Room	Cycle the power on the Dynamics Box.
94	Simulator Room	Restart *STISIM* software on Simulation Computer 2.
95	Simulator Room	Load driving scenario on Simulation Computer 2.
96	Simulator Room	Have subject fill out Questionnaire 2.
97	Simulator Room	Clear off the top surface of the equipment cabinet.
98	Simulator Room	Push equipment cabinet back, but leave the door open.
99	Simulator Room	Turn off telephone bell.
100	Simulator Room	Close the window.
101	Simulator Room	Have subject read Instruction 2.
102	Simulator Room	Darken the room to the specified setting.
103	Simulator Room	Have subject center the steering wheel and apply brakes.
104	Simulator Room	Press *Start* button on the Simulator Computer 1 and the Measurement Computer.
105	Simulator Room	Check position of the IR camera.
106	Simulator Room	Remove lens caps on the video cameras.
107	Simulator Room	Switch computer control to the Observer Room.
108	Simulator Room	Turn off monitors on the Simulator Computers and the Measurement Computer.
109	Simulator Room	Remind subject to remain seated after the trial has ended.
110	Simulator Room	Turn out lights in the Simulator Room.
111	Observer Room	Test the mouse and keyboard.
112	Observer Room	Press *Record* on the VCR.
113	Simulator Room	Ask subject to relax and close his eyes for 20 seconds as if he were asleep.
114	Simulator Room	Set the count on the odometer to zero; go to Observer Room.
115	Observer Room	Using the intercom, ask subject to release the brakes and start driving.

Continued on next page

116	Observer Room	Enter the subject's questionnaire answers into the Observer Room workstation.
117	Observer Room	Observe trial and record observations in the subject folder.
118	Observer Room	At end of trial, press *Stop* button on the Observer Room workstation display.
119	Observer Room	Press *Stop* on the VCR.
120	Observer Room	Turn off the video monitor.
121	Simulator Room	Turn on the lights.
122	Simulator Room	Put lens caps on the video cameras.
123	Simulator Room	Open the window blinds.
124	Simulator Room	Open the window.
125	Simulator Room	Ask the subject to remain seated.
126	Simulator Room	Have subject fill out Questionnaire 3.
127	Simulator Room	Deactivate the automatic BP measuring equipment.
128	Simulator Room	Measure the subject's BP manually and record results.
129	Simulator Room	Pull the equipment cabinet forward.
130	Simulator Room	Disconnect the BP instrument and put it back in the closet.
131	Simulator Room	Turn on the video monitor.
132	Simulator Room	Switch computer control back to the Simulator Room.
133	Simulator Room	Turn off the video projectors and sound.
134	Simulator Room	Shut down the *Labview* software after data formatting is finished.
135	Simulator Room	Disconnect all cables at amplifier boxes.
136	Simulator Room	Put the amplifier boxes in the closet.
137	Simulator Room	Place the waste basket and chair next to the car and subject.
138	Simulator Room	Help the subject take off the lab coat.
139	Simulator Room	Ask the subject to sit down again on the driver's seat.
140	Simulator Room	Remove the motion sensor, respiration sensor, and skin-temperature sensor.
141	Simulator Room	Remove the pulseoximeter.
142	Simulator Room	Remove the BP cuff.
143	Simulator Room	Remove the electrodes on the subject's face.
144	Simulator Room	Open and remove the EEG cap.
145	Simulator Room	Put the cap with its sensors in the blue container.
146	Simulator Room	Remove the skin-resistance sensor.
147	Simulator Room	Disconnect the ECG.
148	Simulator Room	Remove the ECG electrodes.
149	Simulator Room	Remove the BP instrument belt.
150	Hallway	Help the subject to walk to the restroom, where he may wash up and get dressed. Wait for subject, and return with him to Simulator Room.
151	Simulator Room	Return the subject's jewelry, wristwatch, and cell phone to him.
152	Simulator Room	Pay the subject, and thank him for his participation.
153	Hallway	Accompany the subject to the building exit and sign his admission pass.
154	Simulator Room	Connect the BP instrument to the Measurement Computer.
155	Simulator Room	Start the BP software.
156	Simulator Room	Select the subject by name.
157	Simulator Room	Press *Edit*.
158	Simulator Room	Press *TM2430* button to start transmission.
159	Simulator Room	Clear memory if transmission is successful.
160	Simulator Room	Quit the software application.

Continued on next page

161	Simulator Room	Create a new dayfolder on Measurement Computer.
162	Simulator Room	Shut down all active software on the Simulation Computers.
163	Simulator Room	Transmit the simulator data to the Measurement Computer.
164	Simulator Room	Shut down the Simulation Computers.
165	Simulator Room	Copy .avi data file to the dayfolder.
166	Simulator Room	Copy *.dat, *.raw, *.sim files to the dayfolder.
167	Simulator Room	Copy BP data to the dayfolder.
168	Simulator Room	Check the dayfolder to make sure there are 6 files.
169	Simulator Room	Rename *.txt files to *.dat.
170	Simulator Room	Check to make sure all 6 files have the correct names preceding their extensions.
171	Simulator Room	Cover the carbon-dioxide sensor.
172	Simulator Room	Shut down the Measurement Computer.
173	Simulator Room	Turn off electrical power for the Measurement Computer.
174	Simulator Room	Turn off the breakers.
175	Simulator Room	Turn off the Dynamics Box.
176	Simulator Room	Push the equipment cabinet into the corner of the room.
177	Simulator Room	Turn on the telephone bell.
178	Simulator Room	Turn off the room lights.
179	Simulator Room	Remove the "No Entrance" sign from the door.
180	Simulator Room	Take all empty bottles from the Simulator Room to the Observer Room.
181	Observer Room	Remove the charged batteries.
182	Observer Room	Tidy up the BP equipment and the EEG cap.
183	Observer Room	Copy the payment receipt twice and leave both copies in the subject folder.
184	Observer Room	Tidy up the notes in the subject folder.
185	Observer Room	Label and tidy up the videotape.
186	Observer Room	Tidy up the laboratory and close the closet door.

APPENDIX G

An Algorithm for Discovery

This is an editorial written by David Paydarfar and William J. Schwartz for the 6 April 2001 issue of Science Magazine *(No. 5514). It is reprinted here with permission from the American Association for the Advancement of Science. The editorial reaffirms many of the principles set forth in this book.*

As academic physicians, we are experiencing the rush to restructure medical services and have participated in the development of algorithms for the evaluation and treatment of patients. It has been argued that such algorithms are a critical tool for evidence-based medicine, for improving patient management, and for raising the community standard of clinical care.

One day, during a particularly lengthy commute in our carpool, we began to wonder whether the process of creating new knowledge—asking the right question, pursuing the unknown, making discoveries—might also benefit from such an algorithmic approach. Surely a formula for boosting the rate and magnitude of discoveries would be most welcome. Of course, there are many great treatises on discovery in science, but we were thinking of something more compact for everyday use . . . that could be carried on a laminated card. Many carpools later . . . we unexpectedly discovered that its properties could be reduced to five simple principles.

1. **Slow down to explore.** Discovery is facilitated by an unhurried attitude. We favor a relaxed yet attentive and prepared state of mind that is free of the checklists, deadlines, and other exigencies of the workday schedule. Resist the temptation to settle for quick closure and instead

actively search for deviations, inconsistencies, and peculiarities that don't quite fit. Often hidden among these anomalies are the clues that might challenge prevailing thinking and conventional explanations.

2. **Read, but not too much.** It is important to master what others have already written. Published works are the forum for scientific discourse and embody the accumulated experience of the research community. But the influence of experts can be powerful and might quash a nascent idea before it can take root. Fledgling ideas need nurturing until their viability can be tested without bias. So think again before abandoning an investigation merely because someone else says it can't be done or is unimportant.

3. **Pursue quality for its own sake.** Time spent refining methods and design is almost always rewarded. Rigorous attention to such details helps to avert the premature rejection or acceptance of hypotheses. Sometimes, in the process of perfecting one's approach, unexpected discoveries can be made. An example of this is the background radiation attributed to the Big Bang, which was identified by Penzias and Wilson while they were pursuing the source of a noisy signal from a radio telescope. Meticulous testing is a key to generating the kind of reliable information that can lead to new breakthroughs.

4. **Look at the raw data.** There is no substitute for viewing the data at first hand. Take a seat at the bedside and interview the patient yourself; watch the oscilloscope trace; inspect the gel while still wet. Of course, there is no question that further processing of data is essential for their management, analysis, and presentation. The problem is that most of us don't really understand how automated packaging tools work. Looking at the raw data provides a check against the automated averaging of unusual, subtle, or contradictory phenomena.

5. **Cultivate smart friends.** Sharing with a buddy can sharpen critical thinking and spark new insights. Finding the right colleague is in itself a process of discovery and requires some luck. Sheer intelligence is not enough; seek a pal whose attributes are also complementary to your own, and you may be rewarded with a new perspective on your work. Being this kind of friend to another is the secret to winning this kind of friendship in return.

Although most of us already know these five precepts in one form or another, we have noticed some difficulty in putting them into practice. Many obligations appear to erode time for discovery. We hope that this essay can serve as an inspiration for reclaiming the process of discovery and making it a part of the daily routine. In 1936, in *Physics and Reality,* Einstein wrote, "The whole of science is nothing more than a refinement of everyday thinking." Practicing this art does not require elaborate instrumentation, generous funding, or prolonged sabbaticals. What it does require is a commitment to exercising one's creative spirit—for curiosity's sake.

INDEX